TODAY'S TECHNICIAN

Shop Manual for
Automotive Heating and Air Conditioning

Second Edition

Boyce H. Dwiggins

Jack Erjavec

Series Advisor
Professor Emeritus, Columbus State Community College
Columbus, Ohio

DELMAR

™

THOMSON LEARNING

Australia • Canada • Mexico • Singapore • Spain • United Kingdom • United States

NOTICE TO THE READER

Delmar Staff
Business Unit Director: Alar Elken
Executive Editor: Sandy Clark
Developmental Editor: Allyson Powell
Editorial Assistant: Matthew Seeley
Executive Marketing Manager: Maura Theriault

Channel Manager: Mona Caron
Marketing Coordinator: Brian McGrath
Executive Production Manager: Mary Ellen Black
Senior Production Coordinator: Karen Smith
Project Editor: Ruth Fisher
Art/Design Coordinator: Cheri Plasse

Library of Congress Cataloging-in-Publication Data

Dwiggins, Boyce H.
   Automotive heating and air conditioning / Boyce H. Dwiggins.— 2nd ed.
      v. cm. — (Today's technician)
   Includes index.
   Contents: [1] Classroom manual — [2] Shop manual.
   ISBN 0-7668-0934-X (alk. paper) — ISBN 0-7668-0935-8 (alk. paper)
   1. Automobiles—Heating and ventilation—Maintenance and repair.
   2. Automobiles—Air conditioning—Maintenance and repair.   I. Title. II. Series.
TL271 .D85 2000
629.2'772—dc21   00-065751

Asia (including India):
Thomson Learning
60 Albert Street, #15-01
Albert Complex
Singapore 189969
Tel 65 336-6411
Fax 65 336-7411

Australia/New Zealand:
Nelson
102 Dodds Street
South Melbourne, Victoria 3205
Australia
Tel 61 (0)3 9685-4111
Fax 61 (0)3 9685-4199

Latin America:
Thomson Learning
Seneca 53
Colonia Polanco
11560 Mexico D. F. Mexico
Tel (525) 281-2906
Fax (525) 281-2656

Canada:
Nelson
1120 Birchmount Road
Toronto, Ontario
Canada M1K 5G4
Tel (416) 752-9100
Fax (416) 752-8102

UK/Europe/Middle East:
Thomson Learning
Berkshire House
168-173 High Holborn
London WC1V 7AA
United Kingdom
Tel 44 (0)171 497-1422
Fax 44 (0)171 497-1426

Business Press
Berkshire House
168-173 High Holborn
London WC1V 7AA
United Kingdom
Tel 44 (0)171 497-1422
Fax 44 (0)171 497-1426

Spain:
Parainfo
Calle Magallanes 25
28015 Madrid
España
Tel 34 (0)91 446-3350
Fax 34 (0)91 445-6218

Distribution Services:
ITPS
Cheriton House
North Way
Andover,
Hampshire SP10 5BE
United Kingdom
Tel 44 (0)1264 34-2960
Fax 44 (0)1264 34-2759

International Headquarters
Thomson Learning
International Division
290 Harbor Drive, 2nd Floor
Stamford, CT 06902-7477
USA
Tel (203) 969-8700
Fax (203) 969-8751

CONTENTS

CHAPTER 1

CHAPTER 2

CHAPTER 3

CHAPTER 4

CHAPTER 5

CHAPTER 6

CHAPTER 7

Photo Sequences

JOB SHEETS

PREFACE

Thanks to the support that the *Today's Technician* series has received from those who teach automotive technology, Delmar Publishers is able to live up to its promise to provide new editions every three years. We have listened to our critics and our fans and present this new revised edition. By revising our series every three years, we can and will respond to changes in the industry, changes in the certification process, and to the ever-changing needs of those who teach automotive technology.

The *Today's Technician series,* by Delmar Publishers, features textbooks that cover all mechanical and electrical systems of automobiles and light trucks. Principal titles correspond with the eight major areas of ASE (National Institute for Automotive Service Excellence) certification. Additional titles include remedial skills and theories common to all of the certification areas and advanced or specialized subject areas that reflect the latest technological trends.

Each title is divided into two manuals: a Classroom Manual and a Shop Manual. Dividing the material into two manuals provides the reader with the information needed to begin a successful career as an automotive technician without interrupting the learning process by mixing cognitive and performance-based learning objectives.

Each Classroom Manual contains the principles of operation for each system and subsystem. It also discusses the design variations used by different manufacturers. The Classroom Manual is organized to build upon basic facts and theories. The primary objective of this manual is to allow the reader to gain an understanding of how each system and subsystem operates. This understanding is necessary to diagnose the complex automobile systems.

The understanding acquired by using the Classroom Manual is required for competence in the skill areas covered in the Shop Manual. All of the high priority skills, as identified by ASE, are explained in the Shop Manual. The Shop Manual also includes step-by-step instructions for diagnostic and repair procedures. Photo Sequences are used to illustrate many of the common service procedures. Other common procedures are listed and are accompanied with fine-line drawings and photographs that allow the reader to visualize and conceptualize the finest details of the procedure. The Shop Manual also contains the reasons for performing the procedures, as well as when that particular service is appropriate.

The two manuals are designed to be used together and are arranged in corresponding chapters. Not only are the chapters in the manuals linked together, the contents of the chapters are also linked. Both manuals contain clear and thoughtfully selected illustrations. Many of the illustrations are original drawings or photos prepared for inclusion in this series. This means that the art is a vital part of each manual.

Jack Erjavec, Series Advisor

Highlights of this Edition—Classroom Manual

The Classroom Manual content and organization has been based on ASE testing areas. Other content updates include information on refrigerant recovery, recycle and reclaim procedures, and the latest techniques for leak testing. All areas are expanded to include additional information relative to the latest service and troubleshooting techniques and technologies.

Highlights of this Edition—Shop Manual

The Shop Manual has been updated to correlate with the new content of the Classroom Manual. More information has been added on Environmental Protection Agency rules and regulations as well as current environmental issues.

Job sheets have been added to the end of each chapter. The Job Sheets provide a format for students to perform some of the tasks covered in the chapter. In addition to walking a student through a procedure, step-by-step, these Job Sheets challenge the student by asking why or how something should be done, thereby making the students thing about what they are doing.

Classroom Manual

To stress the importance of safe work habits, the Classroom Manual dedicates one full chapter to health and safety. Included in this chapter are common safety practices, safety equipment, and safe handling of hazardous materials and wastes. This includes information on MSDS sheets and OSHA regulations. Other features of this manual include:

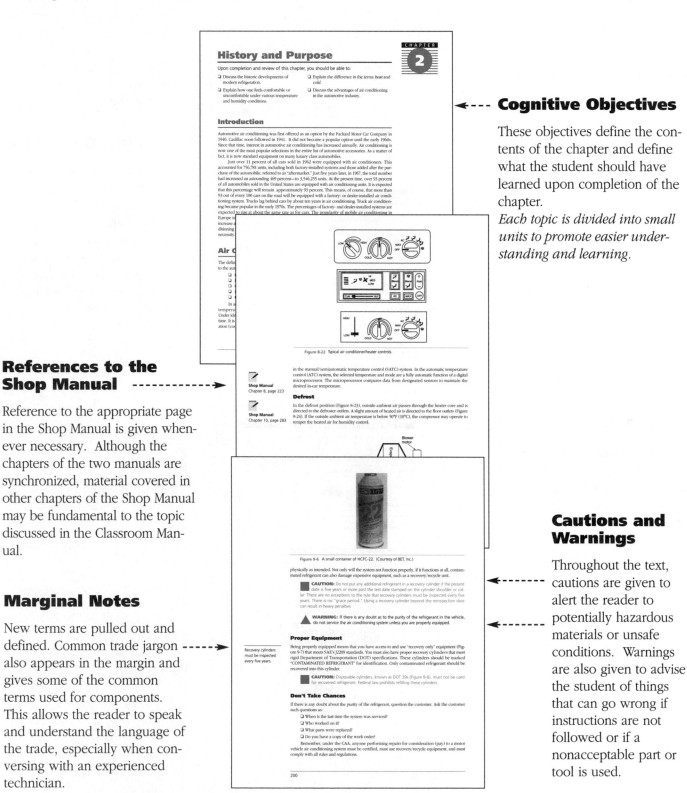

Cognitive Objectives

These objectives define the contents of the chapter and define what the student should have learned upon completion of the chapter.

Each topic is divided into small units to promote easier understanding and learning.

References to the Shop Manual

Reference to the appropriate page in the Shop Manual is given whenever necessary. Although the chapters of the two manuals are synchronized, material covered in other chapters of the Shop Manual may be fundamental to the topic discussed in the Classroom Manual.

Marginal Notes

New terms are pulled out and defined. Common trade jargon also appears in the margin and gives some of the common terms used for components. This allows the reader to speak and understand the language of the trade, especially when conversing with an experienced technician.

Cautions and Warnings

Throughout the text, cautions are given to alert the reader to potentially hazardous materials or unsafe conditions. Warnings are also given to advise the student of things that can go wrong if instructions are not followed or if a nonacceptable part or tool is used.

Terms to Know

A list of new terms appears next to the Summary. Definitions for these terms can be found in the Glossary at the end of the manual.

Summaries

Each chapter concludes with summary statements that contain the important topics of the chapter. These are designed to help the reader review the contents.

Review Questions

Short answer essay, fill-in-the-blank, and multiple-choice type questions follow each chapter. These questions are designed to accurately assess the student's competence in the stated objectives at the beginning of the chapter.

A Bit of History

This feature gives the student a sense of the evolution of the automobile. It not only contains nice-to-know information, but also should spark some interest in the subject matter.

For example, assume that the low-side pressure is 30 psig and the high-side pressure is 220 psig. When 15 is added to these values, they become 45 psia and 235 psia. Hence:

$$\frac{235}{45} = 5.2$$

Pressure ratios above 7.5:1 can cause early compressor failure because of the added load on bearings, pistons, and seals. Also, higher operating temperatures generated by the higher operating pressures can cause lubrication breakdown, which results in harmful deposits on the compressor's internal assembly.

Summary

Terms to Know
Auxiliary
Axial plate
Compression
Crankshaft
Discharge
Exhaust
High pressure
Intake
Low pressure
Rotary
Scroll
Serpentine
Suction
Swash plate
Wobble plate

☐ Reciprocating compressors have a piston or pistons that draw low-pressure heated refrigerant vapor into a chamber, increases its heat content and pressure, and "pump" it out as a high-pressure high-temperature vapor.

☐ A scroll compressor draws low-pressure heated refrigerant vapor through its suction port into a continuously rotating scroll where the vapor's pressure and temperature are increased, and are then forced out through its discharge port as a high-pressure high-temperature vapor.

☐ In a rotary compressor, a rotating vane draws in low-pressure heated refrigerant vapor through the suction port and increases its temperature and pressure before forcing it out through the discharge port. It is discharged as a high-temperature high-pressure vapor.

☐ An electromagnetic clutch is used to engage and disengage (turn on and off) a compressor, as desired, in present applications of automotive air conditioning systems.

Review Questions

Short Answer Essays

1. Describe the operating principals of a reciprocating compressor.
2. Why is a low pressure important?
3. What are other terms used in describing an axial plate?
4. How is a compressor clutch generally driven off an engine?
5. What are some of the design considerations for an automotive air conditioning compressor?
6. Describe the function of a variable displacement compressor.
7. Describe the operating principals of a rotary compressor.
8. What is the purpose of a magnetic clutch:
   a. in a fixed-displacement compressor system?

Terms to Know
Aftermarket
Ambient
ATC
Bilevel
Blend door
Bowden cable
Case
Defrost
Hi/LO
MIX
Mode
Mode door
Plenum
Recirculate
SATC
Vent

Review Questions

Short Answer Essays

1. What are one of the two purposes of the case/duct system?
2. How does the dealer-installed air conditioner differ from the manufacturer-installed system?
3. Where is the air taken from that is to be "conditioned" for MAX cooling?
4. What is the purpose of maintaining a slightly positive pressure in the vehicle's interior?
5. What are the main components of the plenum section?
6. How is air tempered to maintain a desired humidity?
7. Where is air directed in the distribution section?
8. Describe the air flow through the case/duct system when DEFROST is selected.
9. Describe one of the two types of control panels in use for system selection.
10. Define the term *bilevel*.

Fill-in-the-Blanks

1. Mode doors may be _____, cable, or _____ operated.
2. The MIX selection on some models is basically the same as the _____ selection on other models.
3. A fully automatic temperature control system relies on a _____ microprocessor.
4. A small amount of conditioned air may be directed to the _____ outlets to prevent windshield _____.
5. The compressor may operate in the heat mode to help maintain in-vehicle _____.
6. Cooled air is generally delivered into the passenger compartment through the _____ vents.
7. Heated air is generally delivered into the passenger compartment through the _____.
8. The air conditioning and heating system is provided to _____ the in-

---

A BIT OF HISTORY

Early automotive air conditioning systems did not have a convenient driver-operated means of engaging and disengaging the compressor. Most compressors were belt and pulley driven off the crankshaft or accessory device. If one wished to disable the compressor for the winter months, the belt was removed. It was then replaced for the warm months. In-car temperature was controlled by means of a hot gas bypass valve. The compressor operated any time the engine was running and the hot gas bypass valve simply routed the unwanted gas from the compressor **discharge** back to the **suction** side of the system. This inefficient method of temperature control has not been used for automotive service since the mid-1960s.

The hot gas is no longer used as a method of temperature control in automotive air conditioning systems.

Types of Compressors

According to a leading compressor rebuilder, there are currently over 160 makes and models of remanufactured compressors readily available for use in mobile air conditioning systems. The various types include reciprocating piston (Figure 6-15), scroll (Figure 6-16), rotary vane (Figure 6-17), and scotch yoke (Figure 6-18).

Reciprocating (Piston-Type) Compressors

Reciprocating, piston-type mobile air conditioning compressors, depending on their design, may have one, two, four, five, six, seven, or ten pistons (cylinders). Tecumseh manufactured a single-cylinder compressor for use with aftermarket conditioning systems in compact imports.

A two-cylinder V-type compressor was manufactured by Chrysler Air-Temp but was discontinued due to its heavy weight. Two-cylinder, in-line reciprocating compressors manufactured by Nippondenso, Tecumseh, and York may be found on some early model vehicles as well as some heavy duty and off-road equipment.

A four-cylinder, radial-design scotch yoke reciprocal compressor, manufactured by Harrison (Frigidaire) as their model R-4, is available in either standard or light-weight versions. A similar compressor, model HR-980 by Tecumseh, was produced through the late 1980s. A version of the

Chrysler Air-Temp manufactured the only V-type compressor for automotive use. It is now discontinued.

Figure 6-15 A typical reciprocating piston compressor details.

137

Shop Manual

To stress the importance of safe work habits, the Shop Manual also dedicates one full chapter to safety. Other important features of this manual include:

Performance Objectives

These objectives define the contents of the chapter and identify what the student should have learned upon completion of the chapter. These objectives also correspond to the list of required tasks for ASE certification. *Each ASE task is addressed.*

Although this textbook is not designed to simply prepare someone for the certification exams, it is organized around the ASE task list. These tasks are defined generically when the procedure is commonly followed and specifically when the procedure is unique for specific vehicle models. Imported and domestic model automobiles and light trucks are included in the procedures.

Photo Sequences

Many procedures are illustrated in detailed Photo Sequences. These detailed photographs show the students what to expect when they perform particular procedures. They also can provide a student a familiarity with a system or type of equipment, that the school may not have.

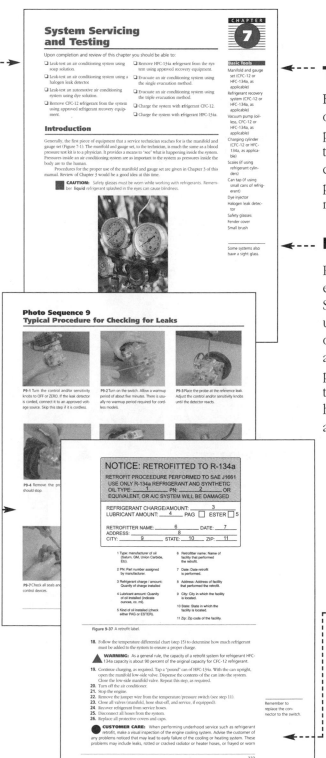

Tools Lists

Each chapter begins with a list of the Basic Tools needed to perform the tasks included in the chapter. Whenever a Special Tool is required to complete a task, it is listed in the margin next to the procedure.

Marginal Notes

Page numbers for cross-referencing appear in the margin. Some of the common terms used for components, and other bits of information, also appear in the margin. This provides an understanding of the language of the trade and helps when conversing with an experienced technician.

Customer Care

This feature highlights those little things a technician can do or say to enhance customer relations.

Service Tips

Whenever a shortcut or special procedure is appropriate, it is described in the text. These tips are generally those things commonly done by experienced technicians

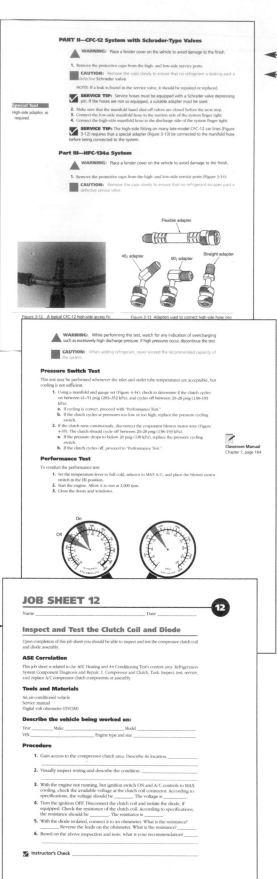

Cautions and Warnings

Throughout the text, cautions are given to alert the reader to potentially hazardous materials or unsafe conditions. Warnings are also given to advise the student of things that can go wrong if instructions are not followed or if a nonacceptable part or tool is used.

References to the Classroom Manual

Reference to the appropriate page in the Classroom Manual is given whenever necessary. Although the chapters of the two manuals are synchronized, material covered in other chapters of the Classroom Manual may be fundamental to the topic discussed in the Shop Manual.

Job Sheets

Located at the end of each chapter, the Job Sheets provide a format for students to perform procedures covered in the chapter. A reference to the ASE Task addressed by the procedure is referenced on the Job Sheet.

Case Studies

Case Studies concentrate on the ability to properly diagnose the systems. Each chapter ends with a case study in which a vehicle has a problem, and the logic used by a technician to solve the problem is explained.

ASE Style Review Questions

Each chapter contains ASE style review questions that reflect the performance objectives listed at the beginning of the chapter. These questions can be used to review the chapter as well as to prepare for the ASE certification exam.

Diagnostic Chart

Chapters include detailed diagnostic charts linked with the appropriate ASE task. These charts list common problems and most probable causes. They also list a page reference in the Classroom Manual for better understanding of the system's operation and a page reference in the Shop Manual for details on the procedure necessary for correcting the problem.

ASE Practice Examination

A 50-question ASE practice exam, located in the Appendix, is included to test students on the content of the complete Shop Manual.

Terms to Know

Terms in this list can be found in the Glossary at the end of the manual.

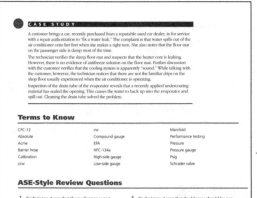

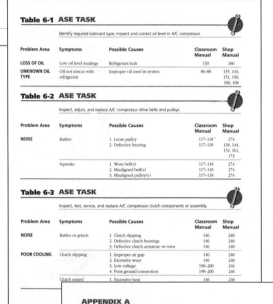

Reviewers

Special thanks to the following instructors for reviewing this material.

Richard J. Sweat, Statesboro High School

Glen Hammonds, Rancho Santiago College

Brian Coppola, Eastern Arizona Community College

Daniel Hall

Michael Francis, Owens Community College

Charles Ginthen

Ray Taylor, Copiah-Lincoln Community College

David Washington, Northwest Los Angeles Technical College

Contributing Companies

I would also like to thank these companies who provided technical information and art for this edition:

American Honda Motor Co., Inc.
AMMCO
Bendix Brakes
Brake Parts
Branick Industries, Inc.
Brodhead-Garrett
Central Tools, Inc.
Century/Lincoln Service Equipment
DaimlerChrysler
CRC Industries
Dalloz Safety
DuPont Automotive Finishes
EIS Brake Parts
Federal-Mogul Corp.

Fluke Corporation
The Ford Motor Company
General Motors*
Goodson Shop Supplies
Gutman Adversing Agency
Hunter Engineering Company
IDSC Holdings, Inc.
ITT Automotive
Lincoln Automotive
LucasVarity Automotive
Mettler-Toledo, Inc.
Mine Safety Appliances Co.
Mitchell Anti-Lock Brake Systems,
Mitchell Repair Information Co., LLC

Mitsubishi Motor Sales of America,
Inc.
NAPA
Nelson Australia Pty Ltd
Nissan North America, Inc.
Parker Hannifin Corp.
Pro-Tech Respirators, Inc.
Pullman/Holt
Raybestos/Brake Parts, Inc.
Rolero-Omega
Securall Safety Storage Equipment
Snap-on, Incorporated

Portions of materials contained herein have been reprinted with permission of General Motors Coporation, Service Operations.

A special thanks to Columbia-Greene Community College for letting us use their facilities to shoot some of the photos featured in this book.

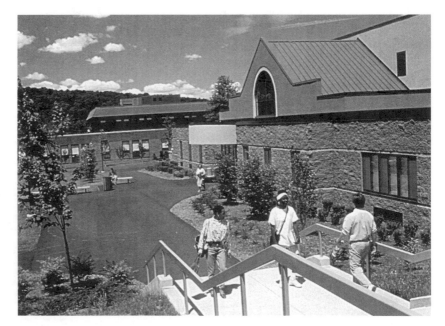

Health and Safety

Upon completion and review of this chapter you should be able to:

❏ Recognize the hazards associated with the automotive repair industry.

❏ Identify hazardous conditions that may be found in the automotive repair facility.

❏ Explain the need for a health and **safety** program.

❏ Discuss the philosophy regarding health and safety.

❏ Compare and identify unsafe and safe tools.

❏ Understand the limitations, by design, of hand tools.

Personal Safety

Technicians working in the automobile repair industry may be exposed to a wide variety of **hazards** in the form of gases, dusts, vapors, mists, fumes, and noise, as well as ionizing or nonionizing radiation. In the course of their work, automotive air conditioning technicians may not be directly exposed to all such hazards, but they must be aware of all the potential dangers that may exist in the facility. Some of the most common hazards include:

❏ Asbestos
❏ Carbon monoxide
❏ Caustics
❏ Solvents
❏ Paints
❏ Glues
❏ Heat and cold
❏ Oxygen deficiency
❏ Radiation
❏ Refrigerants

> There are many airborne hazards in the automotive shop.

Fibrous Material (Asbestos)

Perhaps one of the most serious of the hazards found in an automotive repair facility is the exposure to **asbestos** fibers. Asbestos has been used in brake linings and clutch friction plates for many years. Exposure to asbestos may result in asbestosis or lung cancer. The problem is so serious that work with asbestos must now be done in accordance with Standard 1910.93a of the Occupational Safety and Health Administration (**OSHA**). Materials that are less hazardous have now replaced most asbestos applications, but it must still be considered a hazard. Always follow all applicable procedures associated with good industrial hygiene any time there is a possibility of airborne fibers.

Technicians must not be exposed to unsafe levels of airborne asbestos. It is required that:

❏ Asbestos waste and debris must be collected in impermeable bags or containers.
❏ All asbestos and materials bearing asbestos must be appropriately labeled.
❏ Special clothing and approved **respirators** are to be worn when handling asbestos.
❏ Technicians handling asbestos must be given regular periodic physical checkups.
❏ Methods to limit technician exposure to asbestos include isolation and **ventilation** of dust-producing operations and wetting the material before handling.

> ■ **CAUTION:** The risk of cancer for people who smoke while working with asbestos is almost 90 times greater than for people who do not smoke.

> Smokers are at greater risk than nonsmokers by almost 10:1.

Carbon Monoxide

Alternate fuels such as **propane** and compressed natural gas (CNG), diesel and gasoline-powered vehicles, and some hot work operations such as welding all produce **carbon monoxide (CO)**. The technician's exposure to carbon monoxide may be excessive if such operations are conducted in low-ceilinged or confined areas. Corrective action must be taken if the levels exceed safe standards. The technician must always work in a well-ventilated area to avoid exposure to excessive vapors and fumes.

Caustics, Sovents, Paints, Glues, and Adhesives

Many caustics, acids, and solvents are used in the automotive industry for cleaning operations (Figure 1-1). Epoxy paints, resins, and adhesives are used regularly in body repair and refinishing shops. Always read the cautions printed on the container label before using the contents.

Some of the more common organic chemicals may cause dizziness, headaches, and sensations of drunkenness; they can also affect the eyes and respiratory tract. The use of many chemicals in this special trade industry can cause various types of skin irritations and, in extreme cases, dermatitis. Proper use and availability of appropriate protective equipment is essential. Such equipment includes:

❑ Gloves
❑ Goggles or face shields
❑ Aprons
❑ Respirators

Any hazardous materials that come in contact with skin should be washed off immediately. In addition, an eye wash fountain or safety shower (Figure 1-2) should be provided. For example, an exploding battery may saturate the entire body and clothing with sulfuric acid. The fastest way to reduce the effects of this type of contamination is to step into a shower while disrobing.

Adequate ventilation is necessary to avoid excessive exposure to fumes and vapors during operations in confined spaces. Spray booths with a personal respiratory system are required in many areas for production spray operations involving adhesives and paints.

Classroom Manual
Chapter 1, pages
17 and 19

Lack of oxygen is evidenced by dizziness and drunkenness.

Figure 1-1 Many hazardous solvents may be used in the automotive shop.

Figure 1-2 Showers should be available for personal safety.

Heat and Cold

Consideration should be given to ensure that the temperature of the work area assigned to the technician be maintained within acceptable narrow limits. When exposed to extremes of heat and cold, it has been found that the technician's work performance can suffer because of:

❑ Fatigue
❑ Sunburn
❑ Discomfort
❑ Collapse
❑ Other related health problems

Oxygen Deficiency

If proper precautions are not observed, many operations carried out in a confined space, such as the repair of an automotive air conditioning system, can be very dangerous. Not only may the technician be exposed to various toxic gases, but the atmosphere may also be deficient in oxygen, which would immediately pose a danger to life. Other vapors, which may not be harmful in themselves, displace the oxygen essential to life. Such potential hazards should be approached only when absolutely necessary and when there are adequate procedures outlining the proper precautions and safeguards, such as:

❑ Air line
❑ Respirator
❑ Lifeline
❑ Buddy system

Classroom Manual
Chapter 1, page 19

Radiation

Lasers, used for some alignment procedures, may produce intense nonionizing **radiation**. While it should be avoided, this minor radiation is not generally considered harmful. Welding, however, produces **ultraviolet (UV)** light that is hazardous to the eyes and skin. If proper safeguards are not observed, both ionizing and nonionizing radiation can be very hazardous. Only qualified and trained technicians should use such equipment. It is often a requirement that such technicians be licensed by a federal, state, or local authority. Welders, for example, may be certified by the American Welding Society (AWS).

Classroom Manual
Chapter 1, page 3

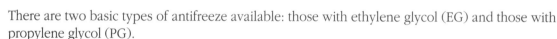

CAUTION: Refrigerants may be flammable.

Refrigerants

The primary problem that may occur during installation, modification, and repair of an automotive air conditioning system is the leakage of **refrigerant**. Refrigerants may be considered in the following classes:

❏ Nonflammable substances where the toxicity is slight, such as some fluorinated hydrocarbons—Refrigerant-12 (R-12), for example. Although considered fairly safe, this refrigerant may decompose into highly toxic gases, such as hydrochloric acid or chlorine, upon exposure to hot surfaces or open flames.

❏ Toxic and corrosive refrigerants such as ammonia, often used in recreational vehicle (RV) absorption refrigerators, which may be flammable in concentrations exceeding 3.5 percent by volume. Ammonia is the most common refrigerant in this category, and is very irritating to the eyes, skin, and respiratory system. In large releases of ammonia, the area must be evacuated. Reentry to evaluate the situation may only be made wearing appropriate respiratory protective devices and protective clothing. As ammonia is readily soluble in water (H_2O), it may be necessary to spray water in the room via a mist-type nozzle to lower concentrations of ammonia.

❏ Highly flammable or explosive substances, such as propane, must be used with strict controls and safety equipment. While propane is not used as a refrigerant in mobile refrigeration, it is a fuel often found in mobile applications.

If a refrigerant escapes, action should be taken to remove the contaminant from the premises. If ventilation is used, exhaust from the floor area must be provided for gases heavier than air and, similarly, from the ceiling for gases lighter than air. For an analytical analysis of the product, consult the **Material Safety Data Sheet (MSDS)** (Figure 1-3) provided by the manufacturer.

Antifreeze

There are two basic types of antifreeze available: those with ethylene glycol (EG) and those with propylene glycol (PG).

EG-Based Antifreeze. EG-based antifreeze is a danger to animal life. Properly handled and installed, however, EG antifreeze presents little problem. If it is carelessly installed, improperly disposed of, or leaks from a vehicle's cooling system, it can be very dangerous.

EG-based antifreeze causes thousands of accidental pet deaths in the United States each year. Animals are attracted to EG antifreeze because of its sweet taste. As little as two ounces can kill a dog and only one teaspoon is enough to poison a cat. This antifreeze can also be a hazard to small children in an undiluted quantity of as little as two tablespoons.

Toxicologists report that EG antifreeze inside the body is changed into a crystalline acid that attacks the kidneys. The effects are fast-acting and one must act immediately if it is suspected that an animal or child may have ingested EG antifreeze.

The signs of EG poisoning in pets include excessive thirst and urination, lack of coordination, weakness, nausea, tremors, vomiting, rapid breathing and heart rate, convulsions, crystals in urine, diarrhea, and paralysis.

Pets do not often survive EG poisoning because owners do not usually recognize the symptoms until it is too late for treatment.

PG-Based Antifreeze. A safer alternative is an antifreeze and coolant formulated with propylene glycol. Unlike EG antifreeze, PG-based antifreeze is essentially nontoxic and hence safer for animal

Ozone friendly HFC-134a is the refrigerant preferred by the automotive industry to replace CFC-12.

Classroom Manual
Chapter 1,
pages 14 and 19

Classroom Manual
Chapter 1, page 15

MATERIAL SAFETY DATA SHEET

CRC Industries, Inc. • 885 Louis Drive • Warminster, PA 18974 • (215) 674-4300

PRODUCT NAME CLEAN-R-CARB (AEROSOL)   #-MSDS05079
PRODUCT- 5079,5079T,5081,5081T
(Page 1 of 2)

1. INGREDIENTS

	CAS #	ACGIH TLV	OSHA PEL	OTHER LIMITS	%
Acetone	67-64-1	750 ppm	750 ppm		2-5
Xylene	1330-20-7	100 ppm	100 ppm		68-75
2-Butoxy Ethanol	111-76-2	25 ppm	25 ppm	(skin)	3-5
Methanol	67-56-1	200 ppm	200 ppm		3-5
Detergent	-	NA	NA		0-1
Propane	74-98-6	NA	1000 ppm		10-20
Isobutane	75-28-5	NA	NA	1000ppm	10-20

2. PHYSICAL DATA : (without propellent)

Specific Gravity : 0.865   Vapor Pressure : ND
                                  % Volatile   : > 99
Boiling Point   : 176 F initial   Evaporation Rate : Moderately fast
Freezing Point : ND   Vapor Density   : ND
Appearance and Odor: pH: NA
        A clear colorless liquid, aromatic odor

Solubility   :   Partially soluble in water.

3. FIRE AND EXPLOSION DATA

Flashpoint : -40 F Method : TCC
Flammable Limits : propellent   LEL:1.8   UEL:9.5
Extinguishing Media : CO2, dry chemical, foam
Unusual Hazards   : Aerosol cans may explode when heated above 120 F.

4. REACTIVITY AND STABILITY

Stability   : Stable
Hazardous decomposition products
                : CO2, carbon monoxide (thermal)

Materials to avoid : Strong oxidizing agents and sources of ignition.

5. PROTECTION INFORMATION

Ventilation :   Use mechanical means to insure vapor conc. is
                     below TLV.

Respiratory :   Use self-contained breathing apparatus above TLV.

        Gloves : Solvent resistant   Eye & Face : Safety glasses
Other Protective Equipment: Not normally required for aerosol product usage.

Figure 1-3 Material Safety Data Sheets (MSDS) are supplied by hazardous chemical manufacturers on request. (Courtesy CRC Industries)

life, children, and the environment. PG antifreeze is classified as "Generally Recognized as Safe" (GRAS) by the U.S. Food and Drug Administration (USFDA). Actually, propylene glycol is used in small quantities in the formula of many consumer products such as cosmetics, medications, snack food, and as a moisturizing agent in some pet food.

PG-based antifreeze coolant protects against freezing, overheating, and corrosion the same as conventional EG-based antifreeze coolants. PG-based antifreeze coolants should not be mixed with toxic EG-based antifreeze coolants (Figure 1-4) because the safety advantage will be lost.

Welding, Burning, and Soldering

Fumes from welding and other hot-work operations actually contain the metals being welded together, such as cadmium (Cd), zinc (Zn), lead (Pb), iron (Fe), or copper (Cu), as well as the filler material, flux, and the coating on the welding rods. Such operations may also generate other gases such as carbon monoxide (CO) and ozone (O_3) at concentrations which may be hazardous to health. When extensive hot-work operations such as welding are performed in confined areas, there can be an excessive fume exposure to these materials. Ventilation or respiratory protection may be needed for certain operations. Eye protection for the welder and for other technicians working in or near the vicinity of welding operations should be provided because of the UV light produced during such operations. Engineering controls such as local exhaust ventilation are required before use of personal protective equipment as a control measure. When effective

Flux is used to promote "wetting" during soldering.

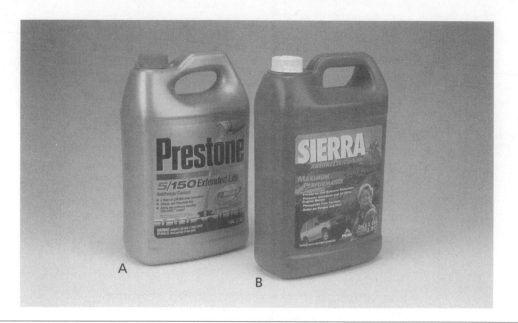

Figure 1-4 EG- (A), and PG-based (B), antifreeze.

engineering controls are not feasible or while they are being instituted, personal protective equipment is required.

Safety in the Shop

Classroom Manual
Chapter 1, page 9

In general, the automotive air conditioning technician may be involved in all phases of automotive service, including electrical and mechanical repairs, relating to air conditioning malfunctions.

Some of the common occupational safety and health problems found during walk-around surveys of typical service repair facilities include:

❏ Poor **housekeeping**: refuse and nonsalvageable materials not being removed at regular intervals; electrical cords and compressed-gas lines scattered on floors (Figure 1-5); and oily, greasy spots or water pools on floor areas.
❏ Ineffective and, in many cases, nonexistent guard rails and toe-boards around open pit areas.
❏ Use of unsafe equipment, such as damaged creepers with missing crosspieces and broken wheels.
❏ The unsafe stacking of stock and other material.
❏ Failure to identify "safety" zones.
❏ Pulleys, gears, and the "point of operation" of equipment without effective barrier guards or other guarding devices or methods.
❏ Inadequate ventilation or unacceptable respirator programs for operations in confined spaces.
❏ Handling of resins, cements, oils, and solvents without protection, causing skin problems or dermatitis
❏ Electrical hazards such as "U" **ground** prong missing from power tools (Figure 1-6); ungrounded extension cords and electrical equipment; and frayed, damaged, or misused power and extension cords.

All machine pulleys and belts should be guarded.

Figure 1-5 Keep electrical cords, gas lines, and air hoses off the floor of the work area.

❏ Unsecured and improper storage of compressed gas cylinders.
❏ Fire hazards caused by improper storage and use of flammable and combustible materials and the presence of various ignition sources.
❏ Improper lifting and material handling techniques (Figure 1-7).
❏ Unsafe work practices that could result in burns from hot-work operations such as welding, burning, and soldering.

Although the above deficiencies cover some of the most significant problem areas, the list should not be considered exhaustive.

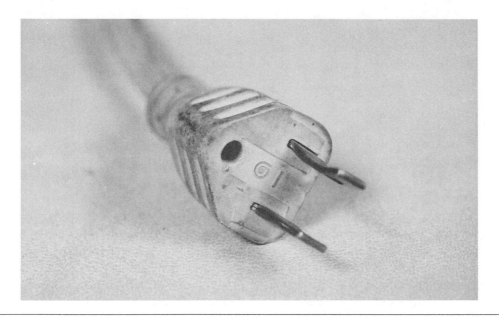

Figure 1-6 An unsafe male electrical plug.

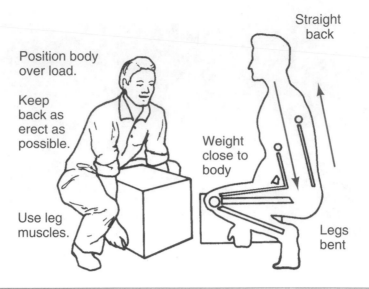

Position body over load.

Keep back as erect as possible.

Use leg muscles.

Straight back

Weight close to body

Legs bent

Figure 1-7 Use legs, not back, for lifting.

Health and Safety Program

Hazardous conditions or practices not covered in the standards promulgated by the Occupational Safety and Health Act (OSHA) of 1970 are covered under the general duty clause of the Act: "Each employer shall furnish to each employee a place of employment which is free from recognized hazards that are causing or are likely to cause death or serious physical harm."

A health and safety program is an effective method to help ensure a safe working environment. The purpose of such a program is to recognize, evaluate, and control hazards and potential hazards in the workplace. Hazards may be identified by:

❑ Performing self-inspections
❑ Soliciting employee input
❑ Interviews
❑ Suggestions
❑ Complaints
❑ Promptly investigating accidents
❑ Reviewing injury and illness records
❑ Other information sources

Typical examples of hazards are:

❑ Unsafe walking surfaces
❑ Unguarded machinery
❑ Electrical hazards
❑ Improper lifting
❑ Air contaminants

In the classroom, the instructor may assign students safety and health management responsibilities in the areas of both program development and implementation. Regular meetings and informal discussions should be held to discuss safety promotions and actual or potential hazards. To ensure program success the participation and cooperation of all class members is essential. Leadership is also necessary. The students assigned the responsibility for carrying out the program must be delegated the proper authority and have the instructor's support. All participants must be aware of the program activities through a systematic interchange of information. Students cannot take an interest in the program if they are unaware of what is occurring. Conversely, well-informed students will likely show interest and a desire to participate.

In the learning as well as the work environment, persons may be exposed to excessive levels of a variety of harmful materials. These include:

- ❏ Gases
- ❏ Dusts
- ❏ Mists
- ❏ Vapors
- ❏ Fumes
- ❏ Certain liquids and solids
- ❏ Noise
- ❏ Heat or cold

Of the illnesses reported, respiratory problems caused by dusts, fumes, and other toxic agents are the most prevalent. Often health hazards are not recognized because materials used are identified only by trade names. A further complication arises from the fact that materials tend to contain mixtures of substances, making identification still more difficult.

To begin identifying occupational health hazards, a materials analysis or product inventory should be made on which all hazardous substances are listed and evaluated. If the composition of a material cannot be determined, the information should be requested from the manufacturer or supplier who must provide a Material Safety Data Sheet (MSDS) for their product. These sheets contain safety information about materials, such as toxicity levels, physical characteristics, protective equipment requirements, emergency procedures, and incompatibilities with other substances.

A process analysis should be performed noting all chemicals used and all products and byproducts formed. When doing such an analysis, allied activities such as maintenance and service operations should be included:

- ❏ Welding performed around chlorinated materials, such as R-12, may cause the formation of toxic gases in addition to welding fumes.
- ❏ Exhaust gases from cars and trucks with internal combustion engines contain carbon monoxide.
- ❏ When certain cleaning agents are mixed, poisonous gases such as chlorine are sometimes formed.

It should be noted that skin conditions such as chemical burns, skin rashes, and dermatitis constitute over half of all occupational health problems. The use of protective creams or lotions, proper personal protective clothing (Figure 1-8) and other protective equipment, and good personal hygiene practices can often prevent these problems.

There are various control methods that can be used to prevent or reduce the technician's exposure to air contaminants. They are as follows:

- ❏ Substitution of less-toxic materials, for example, the use of non-CFC aerosols.
- ❏ Change of a process, for example, a change from gas (open flame) to an electronic leak detector.
- ❏ Isolation, or placing the potentially hazardous process in a separate room or in a corner of the building to reduce the number of technicians exposed.
- ❏ Ventilation, including local exhaust ventilation where contamination is removed at the point of generation and general mechanical ventilation.
- ❏ Administrative controls, such as limiting the total amount of time a technician may be exposed to a health hazard.
- ❏ Training and educating technicians as to the hazards to which they are to be exposed and how to reduce or limit that exposure.
- ❏ Practicing personal hygiene cannot be overemphasized. Technicians should wash their hands before eating. If chemicals such as caustic epoxies or resins get on the skin, they should be washed off immediately.
- ❏ Technicians should not eat around toxic chemicals or in contaminated areas.
- ❏ Clothing should be changed and washed daily if it becomes contaminated with toxic chemicals, dusts, fumes, or liquids.

The law requires suppliers to furnish MSDS sheets on request.

Classroom Manual
Chapter 1, page 19

Toxic gases may be found around welding operations.

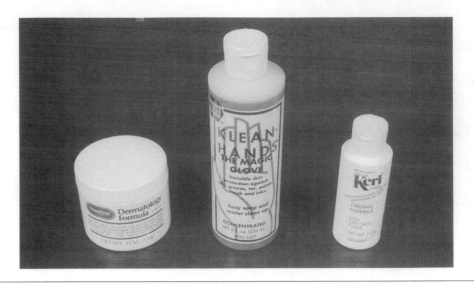

Figure 1-8 Protective creams and lotions may be used when necessary.

❏ Using personal protective equipment such as respirators, hearing protection devices (Figure 1-9), protective clothing, and latex gloves when appropriate.

General Philosophy. A health and safety program helps identify unsafe acts or conditions in the workplace. For many of these there may not be specific standards for rectifying the dangers they pose. Nevertheless, it is important to find a solution for these recognized problems.

During the analysis of the workplace for health and safety problems, it may also become apparent that "the letter of the law" is not being met. If it is apparent that the "intent" of the law is being met, instead of making changes, a variance may be requested from OSHA. The application

Figure 1-9 Use personal hearing protection. (Courtesy of Willson Safety Co..)

for a variance must show it is "as effective as" the OSHA standards in ensuring a safe work environment. The decision not to make changes must be made only with the concurrence of OSHA.

Even when a citation is issued, it is desirable that the employer have demonstrated the willingness to comply with the intent of the law by operating effective, on-going safety and health programs, by correcting imminent dangers in the workplace; and by maintaining records of purchases, installations, and other compliance-promoting activities. Therefore, after an OSHA compliance visit and a citation, the employer can substantiate the intent to provide a safe and healthy workplace for employees by producing records that document that purpose and may be given the benefit of having shown "good faith," which can serve to reduce penalties.

Technician Training

The following are suggestions that can help reduce unsafe acts and practices in the shop:

❏ Be constantly aware of all aspects of safety, particularly good housekeeping, and the elimination of slipping, tripping, and other such hazards.

❏ Be knowledgeable in the maintenance and operation of any special equipment. Do not attempt to use such equipment without first having been instructed in its proper use.

❏ Use appropriate personal protective and safety equipment, such as safety glasses (Figure 1-10) and, whenever necessary, respiratory apparatus.

❏ Develop and maintain check points to be observed as part of the standard and emergency procedures.

❏ Be knowledgeable in the proper use of portable fire extinguishes. Know where fire extinguishes are located.

❏ Know who is trained and responsible for emergency first aid treatment and the procedures for reporting an emergency. At least one technician should be trained in first aid on each shift at each site.

❏ Be aware of safe lifting practices.

Safety glasses must be worn at all times.

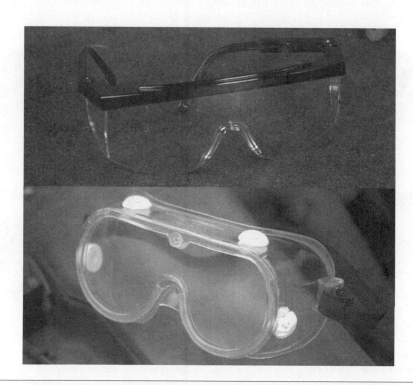

Figure 1-10 Use personal eye protection. (Courtesy of Goodson Shop Supplies)

Safety Rules for Operating Power Tools

The following rules apply to those who use and operate power tools. To ensure safe operation the technicians must:

- ❑ Know the application, limitations, and potential hazards of the tool used.
- ❑ Select the proper tool for the job.
- ❑ Remove chuck keys (Figure 1-11) and wrenches before turning on the power.
- ❑ Not use tools with frayed cords, or loose or broken switches.
- ❑ Keep guards in place and in working order.
- ❑ Have electrical ground prongs in place (See Figure 1-6).
- ❑ Maintain working areas free of clutter.
- ❑ Keep alert to potential hazards in the working environment, such as damp locations or the presence of highly combustible materials.
- ❑ Dress properly to prevent loose clothing from getting caught in moving parts.
- ❑ Tie back long hair or otherwise protect it from getting caught in moving parts.
- ❑ Use safety glasses, dust or face masks, or other protective clothing and equipment when necessary.
- ❑ Do not surprise or distract anyone using a power tool.

Machine Guarding

It is generally recognized that machine guarding (Figure 1-12) is of the utmost importance in protecting the technician. In fact, it could be said that the degree to which machines are guarded in an establishment is a reflection of management's interest in providing a safe workplace.

A technician cannot always be relied upon to act safely enough around machinery to avoid accidents. One's physical, mental, or emotional state can affect the attention paid to safety while working. It follows that even a well-coordinated and highly trained technician may at times perform unsafe acts that can lead to injury and death. Therefore, machine guarding is important.

Figure 1-11 Remove chuck key before turning on power to the drill press.

Figure 1-12 Machine guarding is essential for safety.

Good Housekeeping Helps Prevent Fires

Maintaining a clean and orderly workplace reduces the danger of fires. Rubbish should be disposed of regularly. If it is necessary to store combustible waste materials, a covered metal receptacle is required.

Cleaning materials can create hazards. Combustible sweeping compounds such as oil-treated sawdust can be a fire hazard. Floor coatings containing low-flash-point solvents can be dangerous, especially near sources of ignition. All oily mops and rags must be stored in closed metal containers (Figure 1-13). The contents should be removed and disposed of at the end of each day.

Some of the common causes of fires are:

❑ Electrical malfunctions
❑ Friction
❑ Open flames

Combustible material must be stored in special metal cabinets.

Figure 1-13 Oily rags are stored in metal containers approved for that purpose.

❑ Sparks
❑ Hot surfaces
❑ Smoking

The proper housekeeping and safety policies can help reduce these fire hazards. It can also help to eliminate many tripping hazards.

Walking and Working Surfaces

All areas, passageways, storerooms, and maintenance shops must be maintained in a clean and orderly fashion and be kept as dry as possible. Spills should be cleaned up promptly. Floor areas must be kept clear of parts, tools, and other debris. Areas that are constantly wet should have non-slip surfaces where personnel normally walk or work.

Every floor, working place, and passageway must be maintained free from protrusions, such as nails, splinters, and loose boards.

Where mechanical handling equipment is used, such as lift trucks, sufficient safe clearances must be provided for aisles at loading docks, through doorways, and wherever turns or passage must be made. Low obstructions that could create a hazard are not permitted in the aisle.

All permanent aisles must be easily recognizable. Usually aisles are identified by painting or taping lines on the floors (Figure 1-14).

Use of Tools

Use the proper tool. Do not use, for example, a metric tool on an English fastener.

The proper use of tools is an important consideration for today's technician. This notion may be divided into two major categories:

1. The use of safe tools.
2. The safe use of tools.

The two go hand-in-hand, for without one there can not be the other. The formula for tool safety comprises the following three general rules:

1. Use safe tools.
2. Maintain tools in a safe condition.
3. Use the right tool for the task.

Figure 1-14 Aisles should be identified for safety.

Rule #1: Use Safe Tools

The safe use of tools and equipment is fundamental to any automotive technician safety program. The first step in any tool safety program is to upgrade tools to maintain minimum safety standards. This requires inspecting all current tools on hand and replacing any of them that are defective or do not meet minimum quality standards.

Quality Standards for Tools. Tools that serve the automotive industry are put to rugged use. If they are to stand up, the tools must be designed and manufactured according to rigid quality standards.

Some of the more important points for consideration when selecting tools include:

❏ Tools should be made of alloy steel. Finer-grade alloys impart toughness to the metal used in the manufacture of tools. If a tool made of an alloy steel is inadvertently overloaded, it will deform before it will break, thus providing a warning to the user.

❏ Tools should be tempered by heat. Tool strength and lasting quality is enhanced by precision heat treatment of the metal.

❏ Tools should be machined accurately. If a tool is to fit the intended application accurately (Figure 1-15), without slip or binding, machining must be held to close tolerances.

❏ Tools should be designed for safety. Firm, safe tool control with a minimum of effort should be provided by a lightweight, balanced design. Design features should include those that prevent slipping or accidental separation of tool parts.

Unsafe Tools. Identifying and discarding unsafe tools is an important part of developing the habit of tool safety. In addition to those tools that are easily recognized as below standard, broken, or otherwise damaged, you should avoid using homemade and reworked tools. Few repair facilities are equipped to work steel into tools suitable for high-leverage automotive repair applications. Homemade tools are therefore often heavy and awkward to handle. Lighter and stronger tools, for virtually any purpose, are commercially available and should replace all homemade relics.

Grinding or otherwise reworking a tool to fit a particular application usually results in a tool that no longer measures up to safety requirements. Grinding a tool robs it of metal needed for strength. Heat created by grinding, bending, and brazing impairs the temper of the metal, which also weakens the tool. These tools, too, should be replaced.

Light-Duty Tools. There are many inexpensive tools (Figure 1-16) in the marketplace. Most are intended for amateur or do-it-yourself (DIY) mechanics or other light-duty applications. Examples

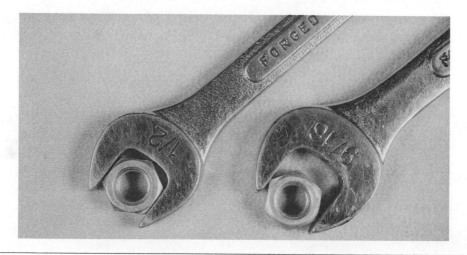

Figure 1-15 Tools should fit intended applications.

Figure 1-16 Light-duty tools are not used in the automotive shop.

include stamped or die-cast tools made of nonalloy carbon steel. These tools are not designed for professional use and are not suitable for general use in the automobile repair trade.

Rule #2: Maintain Tools in a Safe Condition

Keep tools in good repair.

When safe tools are used in the workplace, the next step is to keep them in safe condition. A routine inspection on a regular basis leads to repairing or replacing those tools that are worn or otherwise considered no longer safe. The following tips on specific tool care will help minimize the risk of personal injury due to tool failure.

Ratchets. Ratchets are mechanical devices, and as such are subject to mechanical failure. Frequent causes of failure are worn parts and dirt. The results of failure are slippage, which in turn can lead to possible injury.

To reduce the risk of ratchet failure, a program of preventive maintenance—cleaning and lubricating the ratchet mechanism—should be performed at least once every six months.

Screwdrivers. Screwdrivers with worn, chipped, or broken tips (Figure 1-17) are a potential menace to the technician who works with them. Such tools have little grip on the screw head and frequently jump the slot, leaving the technician open to injury.

Use the proper screwdriver for the screw.

Regular inspection and replacement of screwdrivers is a must, because of the performance expected of these tools. The following procedure can be used to repair screwdrivers with slightly worn or nicked tips:

1. Grind down the flat surfaces using the grinder very lightly to avoid overheating and destroying the temper of the metal (Figure 1-18). The amount of material removed should be minor. "Quench" the tip in water (H_2O) every few seconds, to keep it cool.

Figure 1-17 Damaged screwdrivers should be repaired or replaced.

Figure 1-18 Grind down the flat surfaces.

2. Square off the edges and the tip (Figure 1-19).
3. Test the tip for correct fit (Figure 1-20).
4. Use a file or an oil stone to remove burrs.

▲ **WARNING:** If excessive grinding is required it is best to replace the screwdriver. Only the tip of the screwdriver is tempered and excess grinding will remove the tempered area.

Figure 1-19 Square off the edges and the tip.

Figure 1-20 Test the tip for correct fit.

Wrenches and Sockets. Tools that show signs of "old age" are prime candidates for replacement. Because worn-out wrenches and sockets get only a partial "bite" on the corners of a nut, they are often likely to slip on a heavy pull.

A regular inspection of the tool box for worn tools will prevent a lot of mishaps. Look for:

- ❑ Open-end wrenches with battered, spread-out jaw openings.
- ❑ Sockets or "box-sockets" whose walls have been battered and rounded by use.
- ❑ Tools that have been abused, such as standard thin-wall (hand-use) sockets with lapped-over metal around square drive opening (revealing their use on impact wrenches) and wrenches or handles bearing hammer marks. These tools should also be scrapped because hammer or impact shock leads to metal fatigue, which substantially weakens tools of this type.

Keep tools clean.

It is important, too, that dirt and grit are not caked inside sockets. Such debris may prevent the socket from seating fully on the nut or bolt head. This would concentrate the twisting force at the very end of the socket, possibly causing the socket to break even with a moderate pull.

Other Tools. Routine inspection of tool boxes will also uncover other unsafe tool conditions that could result in accidents:

- ❑ Hammers with cracked heads or handles
- ❑ Pliers with smoothly worn gripping sections
- ❑ Pliers with rivets or nut-and-bolt assemblies that have become sloppy

Many hand tool accidents can be traced to poor housekeeping. Safety, as well as good workmanship, dictates that tools be properly stored and cleaned.

A misplaced tool frequently is the cause of a technician's tripping or being hit by a falling object. Tools should always be kept in tote trays, boxes, or chests when not being used.

Tools with oily handles can be slippery and dangerous. Technicians should get in a habit of wiping off tools with a dry shop rag before starting each job or, better yet, before putting them away after use. With tools, as with everything else, good housekeeping makes good safety sense.

Rule #3: Use the Right Tool for the Task

Safe tools, in safe condition, are only half of the tool safety story. The other half rests with the technicians who use the tools. Every technician in the shop who works with tools should understand the type, size, and capacity of the tools they use.

Hand tools are available in an endless assortment of types, styles, shapes, and sizes—perhaps over 5,000 in all. Each tool is designed to do a certain job quickly, safely, and easily. Here are three important considerations in selecting tools for a task:

Type of Tool. The safest tool is the tool that is specifically designed for the particular task. The use of makeshift tools just to "get by" is one of the major causes of hand tool accidents.

Size and Shape of Tool. The safest tool is one that fits the job squarely and snugly. Misuse can lead to the tool's slipping or breaking. This can lead to injury.

Capacity of Tool. Every tool has a design safety limit. Exceeding the design limit can result in tool failure.

The following sections look more carefully at specific tools commonly used in the workshop.

Wrenches and Socket Wrenches

A combination wrench has a box on one end and is open on the other.

These are the safest tools for turning bolts. Some bolt-turning tools have definite safety advantages over others. Here is a list of tools suitable for bolt turning in order of preference:

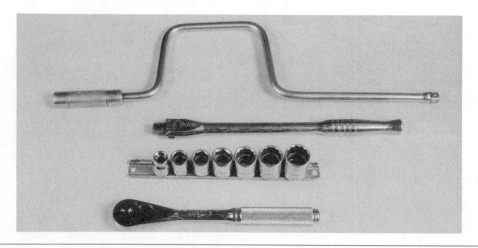

Figure 1-21 An assortment of sockets and drivers.

1. Box-sockets and socket wrenches (Figure 1-21). These tools are preferred for bolt-turning jobs where a heavy pull is required and safety is a critical consideration. A socket or box-socket completely encircles the hex nut or bolt and grips it securely at all six corners. It cannot slip off laterally, and there is no danger of springing jaws.

2. Open-end and flare-nut wrenches (Figure 1-22). Firm, strong jaws make open-end wrenches a very satisfactory tool for medium-duty bolt-turning work. Many technicians work with combination wrenches, using the open end to speed the nut on or off and the box end for breaking loose or final tightening. Flare-nut wrenches are recommended for those jobs where sockets or box-sockets can not be used. Flare-nut wrenches are a "must" for use in servicing automotive air conditioning hoses.

3. Adjustable wrenches. The adjustable wrench, commonly referred to by its trade name, *Crescent,* is recommended only for light-duty applications where time is an important factor and the proper tool is not readily available.

Adjustable wrenches (Figure 1-23) are prone to slip because of the difficulty encountered in setting the correct wrench size. They also have a tendency for the jaws to "work" as the wrench is being used. For these reasons, an adjustable wrench should not be considered an all-purpose tool.

NOTE: Though often used for that purpose, pliers are not on the list of tools recommended for bolt turning.

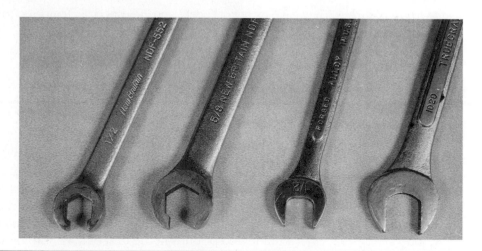

Figure 1-22 Flare-end (A) and open-end (B) wrenches.

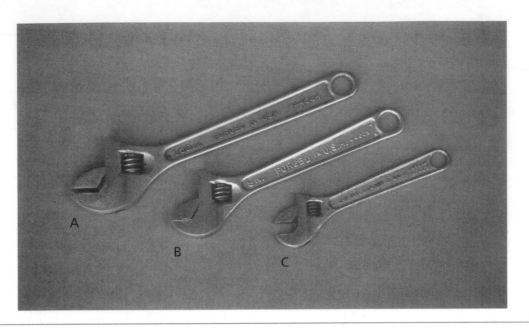

Figure 1-23 Adjustable wrenches: 10 in. (A), 8 in. (B), and 6 in. (C).

Overloading Wrenches and Socket Wrenches. The "safety limit" of a wrench or socket wrench is determined by the length of its handle. Use of a pipe extension or other "cheater" to move a tightly rusted nut can overload the tool past its safety limit.

When the tool being used cannot turn the nut, a heavier-duty tool is required. Both open-end and box-socket wrenches are available in a heavy-duty series that can be safely used with tubular handles from 15 in. to 36 in. in length and can be substituted for a wrench that is too light for the job.

Table 1-1 presents a breakdown of government minimum proof loads for ratchets and other socket wrench handles. These figures can be used as a guide for the safe use of socket wrenches.

Use of Hammer with Wrenches. Wrenches and socket wrench handles should never be used as hammers, nor should a hammer be used on these tools to break loose a tightly seized nut or bolt. Hammer abuse weakens the metal and can cause the tool to fail under a heavy load. Sometimes, however, hammer shock is the only cure for an extrastubborn nut or bolt. In such cases, here are the suggested tools to use:

❏ *Sledge-type box-sockets.* They have plenty of "beef," and are especially tempered for use with a sledge hammer.

TABLE 1-1 MAXIMUM TORQUE LOAD FOR SOCKET WRENCHES

DRIVE	MINIMUM TORQUE (in.-lb.)	PROOF LOAD (ft.-lb.)	LOAD APPROXIMATE EQUIVALENT
1/4 in.	450	37	Pulling 100 pounds at the end of a 4.5-in. handle
3/8 in.	1,500	125	Pulling 200 pounds at the end of a 7.5-in. handle
1/2 in.	4,500	375	Pulling 200 pounds at the end of a 22.5-in. handle
3/4 in.	9,500	792	Pulling 200 pounds at the end of a 47.5-in. handle
1 in.	17,000	1,417	Pulling 200 pounds at the end of an 85-in. handle

Figure 1-24 Thin-wall chrome standard sockets (A) are not intended for impact use. A thick-wall socket (B) is used with an impact wrench.

Use thick-wall sockets with an impact driver.

❏ *Cupped-anvil box-socket.* These air-driven tools were developed to meet the requirements of heavy processing industries.

❏ *Impact driver.* This tool transmits a hammer blow into rotary shock. It is especially useful for smaller nuts and bolts and can also be used on screws.

Sockets for Impact Use. Sockets used with impact wrenches or impact drivers should be the thick-wall power type (Figure 1-24). Thin-wall chrome standard sockets designed for hand use will weaken under impact shock and are likely to fail on a high-leverage application. Impact abuse is one of the most frequent causes of socket failure.

Selecting the Correct Size. Selection of the correct wrench size is a necessary part of safe bolt-turning work. A wrench or socket one size too large will not grip the corners of the nut securely. The result can be a bad slip during a heavy pull.

There is a correct wrench size available for virtually every nut or bolt made in the United States, Canada, Great Britain, and Europe. Wrench size is determined by measuring the nut or bolt head across the flats.

Danger of "Cocking." Sockets and box-sockets should also fit squarely. When these tools are "cocked," they are likely to break even under a moderate load. This is due to "binding" that concentrates the entire strain at one point, rather than spreading it evenly over the tool. The point at which the strain is concentrated becomes vulnerable to failure.

Cocking is a frequent cause of tool failure. It can usually be avoided by using different arrangements of sockets, flex-sockets, and extensions, or by substituting box-sockets of different lengths and offsets.

Use a six-point hex socket on worn nuts and bolts.

Selecting Open-End Wrenches. For the most secure grip, open-end wrench jaws should contact the entire length of two flat surfaces of the nut or bolt head (Figure 1-25). When it is necessary to reach the fastening at extreme angles, there is a danger that the wrench will slide off. This can usually be avoided by the use of crowfoot, offset-head, or taper-head open-end wrenches.

Correct Style of Socket or Box-Socket. Here are some rules that govern the selection of the safest tool for the job:

❏ When turning a fitting where corners are rounded by wear or corrosion, single-hex sockets or box-sockets offer more protection because they grip a larger amount of the surface of the fitting.

❏ On square fittings, use a single- or double-square, not a double-hex.

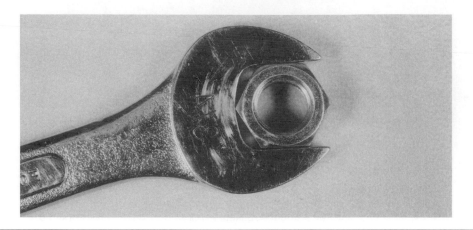

Figure 1-25 An open-end wrench should contact the entire flat surface.

❏ Where bolt clearance is a problem, avoid tool breakage by using deep, extra-length sockets.

Special Purpose Tools. Many special purpose tools are designed for jobs where a critical clearance problem exists and tool jaws or walls are extra thin. Examples are wrenches and sockets for removing and replacing air conditioning compressor clutch retaining nuts and bolts.

These tools are plenty safe for the job intended but should not be used for general bolt-turning work. The usual result is overload and failure.

Pliers

Pliers (Figure 1-26) are often misused as general purpose tools. Their use should be limited to gripping and cutting operations for which they were designed. Pliers are not recommended for bolt-turning work for two reasons: (1) because their jaws are flexible, they slip frequently when used for this purpose; and (2) they leave tool marks on the nut or bolt head, often rounding the corners so badly that it becomes extremely difficult to service the fittings in the future.

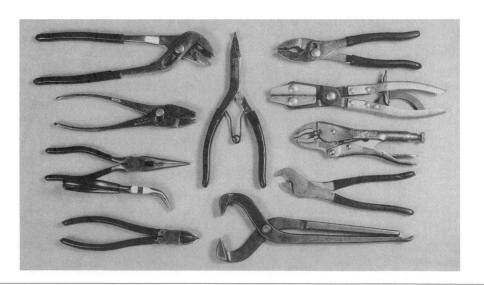

Figure 1-26 Typical pliers.

Keep Plier Jaws Parallel. For a firm, safe grip with a minimum of effort, plier jaws should be as nearly parallel as possible. Use of the right size pliers and proper positioning make this possible.

Avoid Overloading Cutting Pliers. To avoid overloading the tool, the user should select the plier that will cut a wire using the strength of only one hand. *Another tip:* The inside of the cutting jaws should point away from the user's face to prevent injury from flying cuttings.

Screwdrivers

The screwdriver is not an all-purpose tool, although some attempt to use them as such in place of lining-up punches, chisels, and pry bars. The usual result is a damaged tool and a possible injury. The use of screwdrivers should be limited to screw turning only.

Correct Tool for Phillips Screws. It is very common for those not familiar with tools to try to turn a Phillips screw with a standard tip screwdriver designed for use on slotted screw heads. They usually end up with the tool slipping off the job, a nicked screwdriver tip, and a hopelessly chewed-up fastener. Only a Phillips tip screwdriver (Figure 1-27) should be used on a Phillips screw.

Phillips vs. Reed and Prince. The tools to turn these "look-alike" screws are not interchangeable (Figure 1-28). A screwdriver of one type will not seat properly in the other screw head.

Selecting the Right Size. Selection of screwdriver tip size is an important factor in the safe use of these tools. An oversized tip will tend to jump from the slot. A screwdriver with an undersized tip is also likely to twist out. In either case, a slip of the tip can result in a trip to the first aid kit.

Figure 1-27 A Phillips screwdriver is used with a Phillips screw.

| Keystone | Cabinet | Phillips | Torx® | Clutch Head | Hex Head | Reed & Prince (Frearson) | Square Recess |

Figure 1-28 Typical screw head types.

TABLE 1-2 PHILLIPS SCREWDRIVER SELECTION GUIDE

PHILLIPS	MACHINE SCREW DIAMETER	SHEET-METAL SCREW DIAMETER
1	#4 and smaller	#4
2	#5 to #10	#5 to #10
3	#12 to 5/16 in.	#12 to #14
4	3/8 in. and larger	

Here is an easy rule to remember: Use the largest screwdriver that will fit snugly in the slot. The length of the tip should be the same as that of the slot. Table 1-2 may be used as a guide in determining which size Phillips should be used for a particular fastener.

Lining-up with the Screw. Screwdrivers should line up with the screw on which they are being used to provide sufficient contact between the tip and the screw head. To avoid an out-of-line application, substitute a different length or an offset-type screwdriver.

Punches

Use the correct size punch and chisel.

There are several types of punches (Figure 1-29), each designed to do certain jobs properly and safely. Misuse often ends up in tool breakage. The jobs for which each punch tool was designed are:

❏ *Starter* or *drift punch*: For starting tightly jammed pins and bolts and for driving pins clear through a hole after they have been started.
❏ *Pin punch*: A speedy combination punch for starting and driving pins through hole, to be used only on light-duty jobs.

Correct Punch or Chisel Size. The greatest tool life and safety will result from the selection of the chisel whose cutting edge is the same width or wider than the area to be cut. This avoids unnecessary strain on a small chisel trying to do a big job.

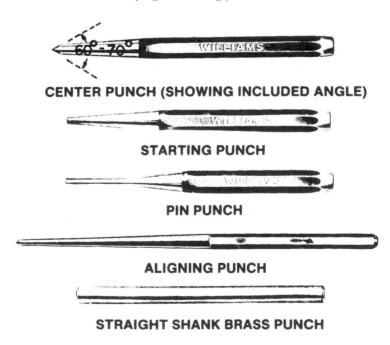

CENTER PUNCH (SHOWING INCLUDED ANGLE)

STARTING PUNCH

PIN PUNCH

ALIGNING PUNCH

STRAIGHT SHANK BRASS PUNCH

Figure 1-29 Typical punches. (From Erjavec, *Automotive Technology*, © 1996 Delmar Publishers Inc).

Figure 1-30 A typical puller.

When punches are used, the largest punch that will fit the job without binding should be used. Use of an undersized punch is apt to result in wedging the part being driven as well as tool failure.

Pullers

The puller is the only quick, easy, and safe tool for forcing a gear, wheel, pulley, or bearing off a shaft (Figure 1-30). Use of pry bars or chisels often causes the part to cock on the shaft, making it even more difficult to remove. Also, with the wrong tools for the job, the operator must exert a great deal of force that is difficult to control, thereby creating an unnecessary hazard.

When a puller is used, the technician enjoys a mechanical advantage that reduces the amount of force required. Furthermore, the puller is so designed that the force that is used is always under control.

Selecting the Correct Size. Selection of the correct size puller can prevent serious accidents. Some important considerations are:

- ❏ The jaw capacity of the puller should be such that when the tool is applied to the job, the jaws press tightly against the part being pulled.
- ❏ In pulling gears, the jaws should be wide enough to cover as many gear teeth as possible to minimize the danger of breakage.
- ❏ Use a puller with as large a pressure screw as possible, but avoid using one that is larger than the hole in the part that is to be pulled.
- ❏ Power capacity of pullers is stated in tons. To avoid the danger of overloading, it is best to use the largest capacity puller that will fit the job.

The Value and Techniques of Safety Sense

Safety sense with tools pays off. The technician should think safety whenever applying a tool to the task. Some of the tips presented in this chapter may seem to be nothing more than common safety sense; they are included because technicians who overlook them are apt to be injured.

Safety sense reminds the technician to protect against the possibility of something going wrong. Whenever tools are used, there is a risk of tools breaking or slipping. And there is also a risk that the part on which tools are used may break loose too.

Bracing against a Backward Fall

Always pull on a wrench handle, never push on it. It is far easier to brace against a backward fall than against a sudden lunge forward should the tool slip or break. To brace against a backward fall when pulling on a wrench, place one foot well behind the other.

 CAUTION:: Never push a wrench.

Have you ever pulled an open-end wrench right off the nut or bolt? This danger can be minimized by using a wrench of the proper size and by making sure it is positioned so that the jaw opening faces in the direction of pull.

Haste Makes Waste

A technician attempted to make a speedy adjustment to the shift linkage while the engine was running. When the technician lost control of his wrench, he was lucky to get off with nothing more serious than a few bruises.

Engines should be turned off when making adjustments whenever possible. There are, of course, adjustments that must be made with the engine running. The secret, then, is safety. . . . Be safe. . . . Take care. . . . Take time. . . . Haste makes waste.

Safety Accessories

Safety glasses are essential eye protection when metal strikes metal, such as in using punches and chisels, or when grinding metal tools or parts on a power grinder. Safety glasses are recommended for everyone working in a shop. In many situations safety glasses and hard hats are required for anyone entering the premises.

Putting Safety Sense into Action

- ❏ Make a thorough check of the tool box. Discard all tools that do not meet minimum safety standards. Where necessary replace them with quality tools.
- ❏ Investigate procurement standards to make certain that only professional-quality tools are purchased by your company.
- ❏ Instruct one technician or the tool room attendant in the repair of ratchets, screwdrivers, and other tools.
- ❏ Instruct technicians in the care of hand tools at a regular departmental safety meeting. Use bad-example tools picked from tool boxes to illustrate the hazards. Advise those using ratchets to bring them into tool room at regular periods for service.
- ❏ Set up a small stock of repair parts for ratchets, screwdrivers, pliers, and other small tools with replaceable parts.
- ❏ Make spot checks for correct tool application.
- ❏ Review shop tooling to ensure that an adequate selection of tools is readily available for all jobs. This step is important to the elimination of makeshift tool procedures.
- ❏ Investigate tool applications involving moving machinery and correct or minimize any hazards noted.
- ❏ Incorporate safety sense tips into tool safety education programs in departmental meetings.

Start a health and safety program in your shop.

Classroom Manual
Chapter 1, page 21

Summary

- ❏ The many hazards associated with the automotive repair industry include exposure to asbestos, carbon monoxide, radiation, caustics, solvents, glues, and paints.

- ❏ Hazardous conditions—such as cluttered floors, missing guard rails, lack of safety zones, inadequate ventilation, and improper storage of combustible materials—are often found in the repair facility.
- ❏ A health and safety program is an effective method to assist in providing for safe working conditions.
- ❏ Through the use of a health and safety program unsafe acts and conditions become apparent and may be corrected.
- ❏ Substandard tools, those not made of alloy steel, should be replaced with tools made of industry-standard high-alloy steel.
- ❏ Tools must never be "worked" beyond their design capabilities.

CASE STUDY

A diesel technician using a heavy-duty wrench avoided serious injury when, without warning, a bolt sheared. He would have fallen backward to the floor had he not braced himself. This injury was avoided by one simple precaution. When pulling on a wrench the technician should always brace against a backward fall. This is easily accomplished by placing one foot well behind the other.

Pull—don't push.

Terms to Know

Asbestos	Housekeeping	Radiation	Technician
Carbon monoxide	MSDS	Refrigerant	Ultraviolet
Grounded	OSHA	Respirator	Ventilation
Hazard	Propane	Safety	

ASE-Style Review Questions

1. *Technician A* says that asbestos fibers are a hazard in the automotive shop.
   Technician B says that proper procedures eliminate the risk of asbestos contamination in the automotive repair facility.
   Who is right?
   A. A only
   B. B only
   C. Both A and B
   D. Neither A nor B

2. *Technician A* says that asbestos hazards are reduced by the use of proper isolation and ventilation procedures.
   Technician B says that hazards are reduced by wetting the material before handling.
   Who is right?
   A. A only
   B. B only
   C. Both A and B
   D. Neither A nor B

3. *Technician A* says that a highly toxic refrigerant, ammonia, is used in RV refrigerators.
   Technician B says that a highly flammable gas, propane, is used in RV refrigerators.
   Who is right?
   A. A only
   B. B only
   C. Both A and B
   D. Neither A nor B

4. *Technician A* says that exhaust gases from internal combustion engines contain carbon dioxide gas.
   Technician B says that exhaust gases from internal combustion engines contain phosgene gas.
   Who is right?
   A. A only
   B. B only
   C. Both A and B
   D. Neither A nor B

5. The shop's safety program is being discussed.
Technician A says that the program will be ineffective if the technician does not work with care.
Technician B says that the program will be ineffective if the rules are not observed.
Who is right?
A. A only   C. Both A and B
B. B only   D. Neither A nor B

6. Tool quality is being discussed.
Technician A says that mechanics' tools should be made of alloy steel.
Technician B says that mechanics' tools should be tempered.
Who is right?
A. A only   C. Both A and B
B. B only   D. Neither A nor B

7. Technician A says that "homemade" tools are often the handiest tool in the tool box.
Technician B says that "homemade" tools are often the least safe tool in the tool box.
Who is right?
A. A only   C. Both A and B
B. B only   D. Neither A nor B

8. *Technician A* says that his "handiest" tool is an adjustable wrench.
Technician B says that the slip joint plier is her "handiest" tool.
Who is right?
A. A only   C. Both A and B
B. B only   D. Neither A nor B

9. *Technician A* says that a combination wrench may be used on a flare nut.
Technician B says that an open-end wrench may be used on a flare nut.
Who is right?
A. A only   C. Both A and B
B. B only   D. Neither A nor B

10. *Technician A* says that placing one foot behind the other braces oneself against a fall if the wrench slips while pulling.
Technician B says that placing one foot in front of the other braces oneself if the wrench slips while pushing.
Who is right?
A. A only   C. Both A and B
B. B only   D. Neither A nor B

JOB SHEET 1

Name _____ Date _____

Compare and Identify Safe and Unsafe Tools

Upon completion of this job sheet you should be able to identify unsafe tools.

Tools and Materials

Miscellaneous and assorted hand tools.

Procedure

Lay out the tools and separate them into two groups—safe and unsafe. Briefly describe and inventory the tools as follows:

TOOL	SAFE	UNSAFE	WHY (UNSAFE)
1. Screwdrivers	_____	_____	_____
2. Pliers	_____	_____	_____
3. Open-end wrench	_____	_____	_____
4. Box-end wrench	_____	_____	_____
5. Punch/chisel	_____	_____	_____
6. Hammer	_____	_____	_____
7. Socket wrench	_____	_____	_____
8. Snap-ring tools	_____	_____	_____
9. *_____	_____	_____	_____
10. *_____	_____	_____	_____

* Other (describe) _____

_____

What can be the results of using an unsafe tool, such as:

11. Hammer? _____

12. Screwdriver? _____

13. Plier?_____

14. Wrench:

Open-end? _____

Box-end? _____

Socket? _____

15. Punch or chisel? _____

☑ **Instructor's Check** _____

JOB SHEET 2

Name _____ Date _____

The Need for Health and Safety

Upon completion of this worksheet you should be capable of participating in a health and safety program.

Tools and Materials

None required

Procedure

Briefly describe your plan to eliminate or avoid the following health and safety hazards:

1. Engine exhaust fumes. _____

_____

2. Caustic chemicals. _____

_____

3. Liquid refrigerant. _____

_____

4. Hot engine parts. _____

_____

5. Cooling fan start without notice. _____

_____

6. Coolant boil over. _____

_____

7. Oil spill on floor. _____

_____

8. A discharged fire extinguisher. _____

_____

9. Spontaneous combustion. _____

_____

10. Electrical shock. _____

_____

✓ **Instructor's Check** _____

JOB SHEET 3

Name _____ Date _____

Identify and Correct Hazardous Conditions

Upon completion of this worksheet you should be able to identify hazardous conditions and to make recommendations for correction.

Tools and Materials

None required

Procedure

Inspect your work area and identify five hazardous or potentially hazardous conditions and briefly describe your plan to prevent or eliminate them.

1. _____
   _____

2. _____
   _____

3. _____
   _____

4. _____
   _____

5. _____
   _____

Inspect adjoining areas and identify five hazardous or potentially hazardous conditions and briefly describe your plan to prevent or eliminate them.

6. _____
   _____

7. _____
   _____

8. _____
   _____

9. _____
   _____

10. _____
    _____

☑ **Instructor's Check** _____

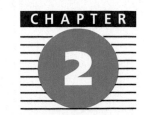

Introduction to Basic Shop Practices

Upon completion and review of this chapter you should be able to:

❏ Identify the responsibilities of the employer and employee.

❏ Identify required and alternative services and the special tools required.

❏ Discuss how to use and interpret service manual service procedures and specifications.

❏ Compare the English and metric system of measurement as related to the automotive technologies.

Shop Rules and Regulations

There are many rules and regulations that are imposed in an automotive repair facility. Some have to do with the Occupational and Safety Health Administration (OSHA), safety of the **customer** or technician, and fire and local ordinances; others are at the discretion of management. In any event, it is expected that everyone associated with the facility will help ensure that all rules and regulations are followed. For example, it is generally posted that customers are not permitted in the service area (Figure 2-1). Reasons given include insurance requirements, fire codes, or local ordinances. The reason, some feel, is that customers simply get in the way, ask foolish questions, and slow down production. Understandably, the average customer does not have a knowledge of the operation of an automobile and, much less, the routine procedures associated with an automotive repair facility. The real reason, then, is for customer safety.

Employee-Employer Relationship

It is very important that a good rapport exists between the technician and those for whom he or she works. This is known as employer-employee (facility owner or representatives-technician) relationship. Note that the word *representatives* is plural. That means that the beginning technician may have to be accountable to and answer to several supervisors. Many look at this as a great advantage, for after finishing formal training, the real learning experiences begin—on the job.

CAUTION

AUTHORIZED PERSONNEL ONLY

Service Bays Are A Safety Area.
Eye Protection Is Required At All Times.

Insurance regulations prohibit customers
in the service bay area during work hours.
We suggest you check out our everyday low prices until
your technician has completed the work on your car.
Thank you for your cooperation.

Figure 2-1 Rules and regulations are posted to provide for safety.

Figure 2-2 The service bay area should be well organized.

Employer-employee relationships are a two-way street. There are certain assumed obligations of both parties. Having a good understanding of these obligations eliminates any problems that may arise relating to what you may expect or what your employer may expect of you.

Work Area

First, and perhaps foremost, you are entitled to a clean, safe place to work (Figure 2-2). There should be **facilities** for your personal **hygiene** and accommodations for any handicap that you may have.

Opportunity

OSHA-required "Right to Know" law should be posted.

Available positions should be posted.

The work environment should provide you an opportunity to successfully advance. This could be in the form of in-house training or an incentive to further your studies in vocational education programs such as Ford's Automotive Service Student Educational Training (ASSET) or General Motors' Automotive Service Education Program (ASEP), which leads to a two-year Associate of Science degree in automotive technology. Others, such as Chrysler, Honda, and Toyota, have similar programs. Opportunity includes fair treatment. This means that all technicians must be considered and treated equally without prejudice or favoritism.

Supervision

There should be a competent and qualified supervising technician who can lead you in the right direction if you have problems, suggest alternative methods, and tell you when you are correct.

Wages and Benefits

There are other aspects of your employment, other than how much you will be paid, that should be known. For example, how often and on what day are you paid? Is it company policy to hold back pay? If you are being paid a commission, do you have a guarantee? What are the **fringe benefits**? If there is a health plan, what is the employee's contribution, if any? Is there a waiting period before being eligible? Does the company have a tool purchase plan? If there is a paid vacation plan, get the details; generally, one week is given after one year and two weeks after three years or more. And finally, while it does not seem important now, is there a retirement plan? If so, get the details; if not, it is time to consider a personal plan for the future.

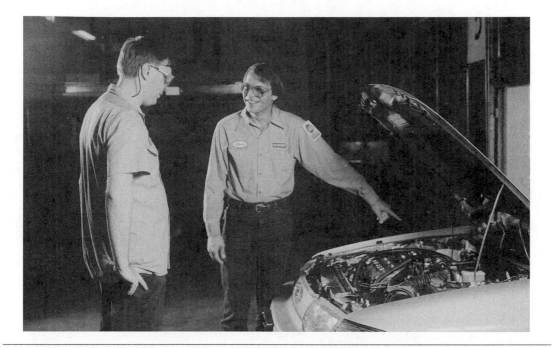

Figure 2-3 Listen to and follow the directions of your supervising master technician.

Take pride in everything you do.

Be dependable.

Employee Obligations

There are also employee obligations. They are really more simple than employer obligations, for all you have to do is be a caring, loyal employee. This probably begins with your ability to follow directions (Figure 2-3). Remember, you are being paid to follow instructions. Doing it your way may not be in the best interest of the employer. If you have any questions, or any doubts, *ask*. "I didn't know what you meant" is not a good response after something is botched.

Attitude. Proceed with a positive attitude. *Your* attitude can have an effect on your fellow technicians as well as your supervisors. Saying, "That's not the way we did it in school" does not portray a positive attitude.

Responsibility. Be responsible and take pride in your work. Regardless of the task assignment, remember, someone has to sweep the floor—it may be you. Never forget that your primary responsibility is to make your employer a profit. Always be busy and productive. Be willing to learn and take advice from the senior technicians. You may be surprised how little time it takes to become one of their peers.

Dependability. Be dependable. Often repair orders for the following day are scheduled in advance. Habitual lateness or absenteeism cannot be tolerated.

Pride. Remember, the most important technician in that facility is you; and you work for the most important and best company in town—perhaps anywhere.

OSHA

The Occupational Safety and Health Admistration (OSHA) was established in 1970 to ensure safe and healthful conditions for every American worker. The agency's enforcement, educational, and partnership efforts are intended to reduce the number of occupational injuries, illnesses, and deaths in America's workplaces. Since its inception, the workplace death rate has been cut in half. Still, about 17 Americans die on the job every day.

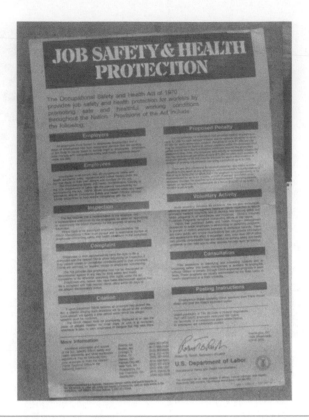

Figure 2-4 OSHA poster in employee's "common" area.

OSHA is committed to a common-sense strategy of forming partnerships with employers and their employees. They conduct firm but fair inspections, develop easy-to-understand regulations, and eliminate unnecessary rules to assist employers in developing quality health and safety programs for their employees.

State consultants, authorized and funded largely by OSHA, conduct consultation visits with employers who request assistance in establishing safety and health programs or in identifying and dealing with specific hazards at their workplaces. OSHA will also conduct unannounced inspections at work sites under its jurisdiction. An inspection is made when three or more workers are hospitalized because of injury or if a job-related death occurs. An inspection will also occur based on an employee complaint. Only half of their staff of about 2,200 are safety and health officers, so many inspections are handled by telephone and fax, often without the requirement for an onsite inspection.

In 1996 and 1997 OSHA simplified the written text outlining its regulations by eliminating almost 1,000 pages and by putting over 600 other pages into plain English. Employers must post a full-sized 10 x 16 in. (254 x 406 mm) OSHA or state-approved poster, such as that shown in Figure 2-4, where required. This is generally in a "common" area where it will be seen by all employees.

Service Tools

Special tools are needed to perform service, testing, and many repair procedures on most automotive air conditioning systems. In addition to common mechanic's hand tools, such as pliers, screwdrivers, wrenches, and **socket** sets, you will need a **manifold and gauge set** with hoses, a refrigerant **can tap**, a thermometer, and safety glasses or other suitable eye protection.

In addition to the technician's tools, the shop should have the required refrigerant recovery, recycle, and recharge systems; antifreeze recovery, recycle, and recharge systems; electronic scales; electronic leak detector; and electronic thermometer. Other specialized tools may also be required for the services to be performed.

Manifold and Gauge Set

The manifold and gauge set (Figure 2-5) generally consists of a manifold with two hand valves and two gauges, a compound gauge and a pressure gauge, and three hoses. There are several types of manifold and gauge sets. Regardless of the type, they all serve the same purpose.

Note that the manifold has fittings for the connection of three hoses. The hose on the left, below the compound gauge, is the low-side hose. On the right, below the pressure gauge, is the high-side hose. The center hose is used for system service, such as for evacuating and charging (Chapter 5). Some manifolds may be equipped with two center hoses: a small hose, generally 1/4 in. or 6 mm, for refrigerant recovery and charging; and a large hose, generally 5/16 in. or 8 mm, for evacuation.

Regardless of the type selected, a separate and complete manifold and gauge set with appropriate service hoses, having unique fittings, are required for each type of refrigerant that is to be handled in the service facility. This means that a minimum of three sets are generally required: one set for CFC-12 refrigerant, one set for HFC-134a refrigerant, and a set for contaminated refrigerant.

Low-Side Gauge. The low-side gauge, also referred to as the compound gauge, will indicate either a vacuum or pressure (Figure 2-6). Generally, this gauge will be calibrated from 30 in. Hg (0 kPa absolute) vacuum to 250 psig (1,724 kPa) pressure. Actually, the pressure calibration is to 150 psig (1,035 kPa) with respect to 250 psig (1,724 kPa) maximum. That means that pressures to 150 psig (1,035 kPa) may be read with reasonable accuracy, while pressures to 250 psig (1,724 kPa) may be applied without damage to the gauge movement. The low-side gauge is found at the left of the manifold.

> Dedicated manifold and gauge sets with hoses must be provided for each of the two types of refrigerant currently used for automotive air conditioning service.

Figure 2-5 The manifold and gauge set.

Figure 2-6 The low-side (compound) gauge.

Pressure Gauge. The high-side gauge (Figure 2-7) is usually calibrated from 0 psig (0 kPa) to 500 psig (3,448 kPa). Insomuch as the high side of the system will never go into a vacuum, pressures below 0 psig (0 kPa) are not indicated on the high-side gauge. The high-side gauge is also often referred to as the pressure gauge. The high-side gauge is found at the right of the manifold.

The Manifold. Note the "circuits" in the manifold. When both hand valves are closed (Figure 2-8), both the low- and high-side hose ports are connected only to the low- and high-side gauges. If, in this case, the hoses are connected from the manifold to an air conditioning system, the low-side gauge will indicate the pressure on the low side of the system. The high-side gauge will indicate the pres-

Figure 2-7 The high-side (pressure) gauge.

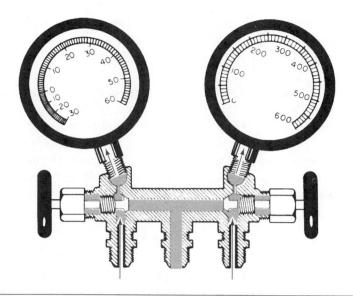

Figure 2-8 Manifold circuit with both hand valves closed.

sure on the high-side of the system. The gauges will always show respective system pressures when the manifold hand valves are closed.

If either the low- or high-side manifold hand valve is cracked open and the center hose is not connected to anything, refrigerant will escape from the system. This is a procedure known as purging.

If only the low-side or high-side manifold hand valve is cracked, both gauges will still indicate system pressure. If, however, both manifold hand valves are cracked, the pressures will equalize in the manifold and neither of the gauges will accurately indicate system pressure.

The center hose is used to evacuate, purge, or charge the air conditioning system. Procedures for this service and system problems relating to gauge pressure indications are given in Chapter 7 of this manual as well as the classroom text.

Hoses

Specific requirements for service hoses used in automotive air conditioning service are given in the Society of Automotive Engineers (SAE) standards J2196 and J2197. The following is a brief overview of those standards as they apply to refrigerants CFC-12 and HFC-134a.

It must be noted that there are no less then eight other alternate refrigerants that have been approved for automotive use by the Environmental Protection Agency (EPA). Each of these alternate refrigerants require its own unique fittings. The refrigerant manufacturer will supply any information about this requirement on request.

High-Side Service Hose. The high-side service hose is connected between the system high side and the manifold or service equipment. For CFC-12 service, this hose is to include a 1/4-in. female refrigeration flare (FFL) nut on both ends. For HFC-134a refrigerant, it is to include a high-side coupling (defined in SAE J639) and a 1/2-in. ACME female nut on the other end. A shut-off valve must be placed within 1 ft. or 30 cm of the end connected to the system of both types of hose. CFC-12 hoses shall also include a valve depressor. The high-side hose for CFC-12 is generally solid red, though it may be black with a red stripe. The HFC-134a high-side hose must be solid red with a black stripe.

Low-Side Service Hose. The low-side service hose is connected between the system low side and the manifold or service equipment. For CFC-12 service, this hose is to include a 1/4-in. FFL nut on both ends. For HFC-134a refrigerant, it is to include a low-side coupling (as defined in SAE J639) and a 1/2-in. ACME female nut on the other end. A shut-off valve must be placed within 1 ft. or 30 cm of the end connected to the system of both types of hose. CFC-12 hoses shall also include a valve depressor. The low-side hose for CFC-12 is generally solid blue, though it may be black with a blue stripe. The low-side hose must be solid blue with a black stripe for HFC-134a refrigerant.

Utility Hose. The utility hose, often called the service hose, is connected between the manifold gauge set and the service equipment, such as vacuum pump, recovery equipment, or charging system. For CFC-12 service, this hose is to include a 1/4-in. FFL nut on both ends. For HFC-134a refrigerant, it is to include a 1/2-in. ACME female nut on both ends. A shut-off valve must be placed within 1 ft. or 30 cm of the end connected to the system of both types of hose. The utility hose for CFC-12 is generally solid yellow or white, though black with yellow or white stripe is preferred. For HFC-134a service, the utility hose must be solid yellow with a black stripe.

Can Tap

The can tap is used to dispense refrigerant from a "pound" can. "Pound" disposable cans actually contain 12 or 14 oz. (340 or 397 g) of refrigerant and are available in either screw top or flat top (Figure 2-9). Some can taps are designed to fit either type. To install the can tap on either flat-top or screw-top cans, using a universal-type can tap, proceed as follows:

1. Wear suitable eye protection.
2. Hold can at arm's length, in an upright position.
3. Affix clamp-type fixture on can top.
4. Turn can tap handle fully counterclockwise (ccw).
5. Screw handle assembly into clamp fixture.

When ready to dispense refrigerant, as outlined in Chapter 7, turn the can tap handle fully clockwise (cw). This pierces the can. The can tap should not be removed until the contents of the can have been dispensed.

It should be noted that small "pound" cans of refrigerant are not legally sold in some states. Other state requirements limit sales of small containers of refrigerant to certified and properly licensed shops and technicians only.

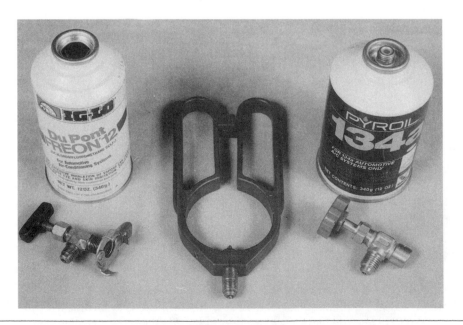

Figure 2-9 Typical can taps with "pound" cans of refrigerant.

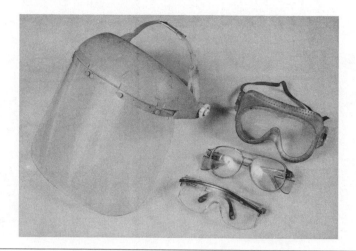

Figure 2-10 Typical safety glasses and goggles: face shield (A), goggle (B), safety glasses with side shield (C), and safety glasses (D).

Safety Glasses

There are several types of safety glasses (Figure 2-10) available. A safety shield-type goggle may be used with or without eyeglasses. It is important to note that glasses or goggles selected should be a type that is approved for working with liquids or gases, meeting the ANSI Z87.1-1989 standard.

Not long ago, while removing an air conditioning system compressor, a technician accidentally allowed an open-end wrench to come into contact with the battery terminals. The resulting short caused a spark which, in turn, caused the battery to explode. The technician was wearing safety goggles to keep his prescription glasses in place; he never thought they might serve an even more important purpose. Although he suffered facial burns, his eyes were protected by the safety goggles. Wearing prescription glasses alone would have offered very little protection since they generally have no peripheral shielding.

Manufacturers' specifications must be used. A service procedure for a 1998 Sable, for example, may be different than that for a 1999 Sable.

Hand Tools

Common hand tools, such as wrenches, pliers, screwdrivers, punches, and hammers, are necessary. Other tools, such as a 3/8-inch drive socket set, are helpful but not a necessity. Many of the hand tools referred to throughout this text may be found around the house. If not, it is suggested that they only be purchased as needed. It is not necessary, for example, to purchase a complete set if only a few sizes are needed.

Special Tools

Basically, three special tools will expand your service and repair capabilities considerably. It is helpful to have a thermometer, a leak detector, and a **vacuum pump**.

Thermometer

A glass-type or a dial-type thermometer may be used. The glass type is usually less expensive, but it is more easily broken. Regardless of the type (Figure 2-11), it is suggested that the temperature range be from 0°F to 220°F (–17.8°C to 104.4°C). Inexpensive thermometers purchased in housewares or automotive departments in large department stores may not be as accurate as a refrigeration thermometer. For this reason, they are not recommended. For more accurate and reliable service, an electronic digital thermometer is recommended.

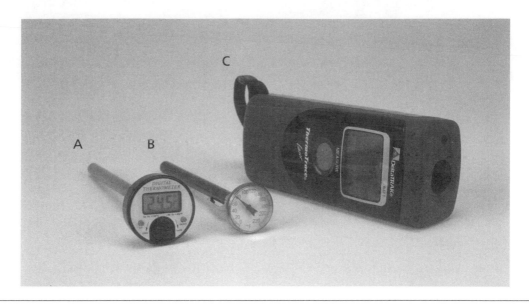

Figure 2-11 Typical thermometers: digital pocket thermometer (A), dial pocket thermometer (B), and infrared electronic (C).

Leak Detector

Leaks can, in most cases, be detected by the use of a soap solution. A good dishwashing liquid mixed with an equal amount of clean water and applied with a small brush will indicate a leak by bubbling. A commercially available product, such as "Leak Finder" (Figure 2-12), can also be used.

Halide. A halide leak detector for use in detecting CFC and HCFC leaks, once a popular "tool of the trade" is no longer used in automotive work because it will not detect an HFC-134a refrigerant leak. This flame-type leak detector (Figure 2-13) includes a propane gas cylinder, shut-off valve, chimney, and reactor plate. Ambient air for the flame is admitted through the search hose. Whenever the hose passes a CFC or HCFC refrigerant leak, vapor is mixed with the air causing the flame to change color. If passed over an HFC-134a leak, there is little to no effect on the flame. If passed over a leak

Figure 2-12 A typical liquid dye leak detector. (Courtesy of BET, Inc.)

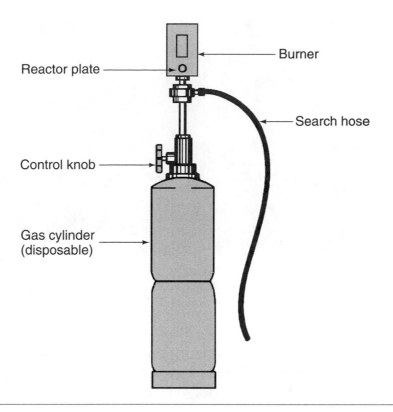

Figure 2-13 A halide leak detector.

ing refrigerant that contains butane or isobutane, there may be a slight flare-up of the flame. Because some approved (as well as most unapproved) refrigerants contain flammable material, this type leak detector is considered dangerous by many and its use is prohibited in many areas.

Electronic. Electronic leak detectors, called halogen leak detectors, though considerably more expensive, are desirable because they offer great sensitivity and can pinpoint a leak as slight as 0.5 oz. (14 g) per year. It should be noted that halogen leak detectors are available that can be used to test either CFC-12 or HFC-134a refrigerants (Figure 2-14). When refrigerant vapor enters a halogen leak detector's search probe, the device emits an audible or visual signal.

Figure 2-14 A typical electronic leak detector. (Robinair, SPX Corporation)

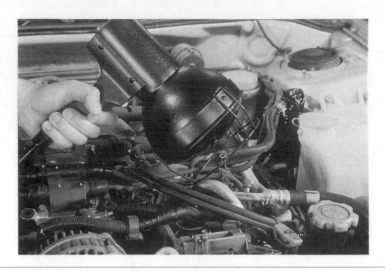

Figure 2-15 A typical fluorescent leak detector. (Tracer Products)

Fluorescent. Fluorescent leak detectors (Figure 2-15) are becoming more and more popular. A fluorescent dye is injected into the system, where it remains without affecting cooling performance. When a leak is suspected, an ultraviolet lamp will quickly and efficiently pinpoint the problem area. Many manufacturers now use refrigerant containing a fluorescent dye for the initial charge of the air conditioning system.

Nitrogen. If time permits, many technicians prefer to hold a standing pressure test using nitrogen to determine the integrity of an air conditioning system after extensive leak repairs have been made. This test is especially helpful if the leak was difficult to locate. It provides added assurance that the leak was located and repaired.

CAUTION: Nitrogen is under very high pressure. Make no attempt to disperse nitrogen without having proper pressure regulators in place.

Take care of the special tools and equipment provided by the service facility. Treat them like they were your own.

To perform this test, the air conditioning system is evacuated and then pressurized to 100 psig (689.5 kPa) and allowed to "rest" overnight. Since the nitrogen is dry and stable, the pressure should be within a few pounds (kiloPascals) of 100 psig (689.5 kPa) the following morning. Nitrogen poses no threat to the environment and may be purged from the air conditioning system to the atmosphere. The air conditioning system is then evacuated to remove any residual nitrogen and air from the system.

The proper use of the more popular leak detectors is given in Chapter 5.

CAUTION: A halide or halogen leak detector must not be used in a space where explosives, such as gases, dust, or vapor, are present. Use halide and halogen leak detectors in a well ventilated area only. Byproducts of decomposing CFC and HCFC refrigerants, hydrochloric and hydrofluoric acid, are a health hazard.

CAUTION: Take care not to inhale these fumes. To minimize the danger, work in a well-ventilated area when leak checking an air conditioning system.

Vacuum Pump

It is necessary to remove as much moisture and air from the system as possible before charging it with refrigerant. This is best accomplished with the use of a vacuum pump (Figure 2-16). The

Figure 2-16 A typical high-vacuum pump. (Courtesy of Robinair, SPX Corporation)

vacuum pump is one of the most expensive pieces of service equipment required. It is usually provided by the service facility.

Some refrigerant recovery equipment has a vacuum pump incorporated into the equipment. High-volume service facilities, however, cannot generally tie up an expensive piece of equipment for the time required to adequately evacuate an automotive air conditioning system.

Refrigerant Identifier

To determine what type refrigerant is in a system, a refrigerant identifier (Figure 2-17), should be used prior to servicing the refrigeration system of any vehicle. The refrigerant identifier is used to identify the purity and quality of a gas sample taken directly from a refrigeration system or a refrigerant storage container. The identifier, such as Rotunda's Refrigerant Analyzer, will display:

- ❏ R-12: If the refrigerant is CFC-12 and its purity is better than 98 percent by weight.
- ❏ R-134a: If the refrigerant is HFC-134a and its purity is 98 percent or better by weight.
- ❏ FAIL: If neither CFC-12 or HFC-134a have been identified or if it is not at least 98 percent pure.
- ❏ HC: If the gas sample contains hydrocarbon, a flammable material. A horn will also sound.

After the analysis is completed the identifier will automatically purge the sampled gas and be ready for the next sample for analysis. If neither CFC-12 nor HFC-134a has been identified or either is not at least 98 percent pure, consider the refrigeration system to be contaminated and perform all service accordingly.

It is important to always refer to the equipment manufacturer's instructions for specific and proper tool usage, and to specific local regulations relating to refrigerant handling.

CAUTION: If the sample reveals that the refrigeration system contains a flammable hydrocarbon, do not service the system unless extreme care is taken to avoid personal injury.

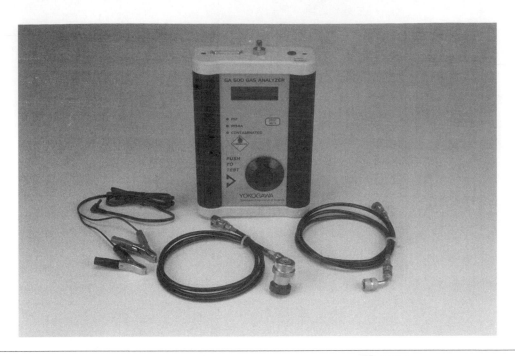

Figure 2-17 A typical refrigerant indentifier.

The high cost of the equipment required for removal and storage as well as the proper disposal of contaminated refrigerant often discourages the customer from having repairs made. There are not many sites in the United States that dispose of contaminated refrigerants. Disposal is usually accomplished by burning at a very high temperature and requires expensive equipment.

For more information about contaminated refrigerant disposal on a local level, check with the local automotive refrigerant supplier. If they can offer no assistance, and often they cannot, consult a major commercial refrigeration supply house.

Other Special Tools

Other special tools available to the service technician are generally supplied by the service facility. These tools include a refrigerant recovery and/or recycling system, an antifreeze recovery and recycle system, an electronic thermometer, and an electronic scale. Also, special testers are available for automatic temperature control (ATC) testing, and special tools are available for servicing and repairing compressors.

The following is a brief description of these tools. They are covered in more detail in the appropriate chapters of both the classroom text and shop manual.

Refrigerant Recovery and Recycle System. The service center must have a recovery, recycle, and recharge machine for each type refrigerant to be serviced. Some refrigerant recover, recycle, and recharge machines, however, may be used for both CFC-12 and HFC-134a. A system, similar to the one shown in Figure 2-18, is a single-pass system with an onboard microprocessor that controls the evacuation time as well as the amount of refrigerant charged into the system. The mixing of CFC-12 and HFC-134a refrigerants is prevented by the use of a sliding lock-out panel allowing only one set of manifold hoses to be connected at any time. Also, the fittings on the hoses prevent them from being connected to the wrong port. Each type of refrigerant has a separate dedicated set of hoses and recovery tank. A self-clearing loop removes residual refrigerant from the machine before connecting the other set of hoses and recovery tank.

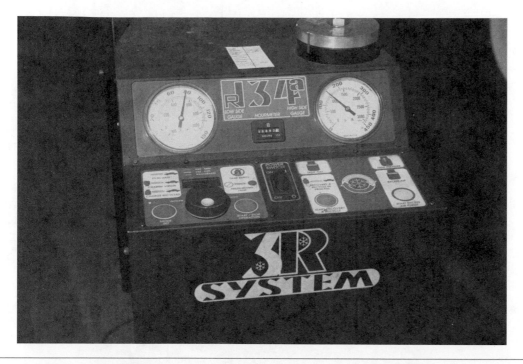

Figure 2-18 A typical refrigerant recovery/recycle machine. (Courtesy of Robinair SPX Corp.)

The recovered refrigerant in the tank is always clean and ready to reuse, having made an initial pass through the filter-drier on its way to the recovery tank. The refrigerant is then recirculated through the filter-drier during evacuation to provide the cleanest possible refrigerant with no extra time or procedures involved. Other desirable features of a recover, recycle, recharge system include an automatic air purge, a high-performance vacuum pump, and an automatic shut-off when the tank is full.

Antifreeze Recovery and Recycle System. An antifreeze recovery and recycle machine, such as Prestone's ProClean Plus™ Recycler (Figure 2-19), is a self-contained system that drains, fills, flushes, and pressure tests the cooling system. It can also be used to recycle coolant. A typical cooling system drain, recycle, and refill takes about 20 minutes.

This particular system adds additives during the recycle phase to bond heavy metals, such as lead (Pb) and other contaminants. This renders them into nonleachable solids that are not hazardous as defined by the EPA. Additives separate the contaminants so they can be easily removed. Also, inhibitors are added to protect against corrosion and acid formation. The recycled coolant exceeds ASTM and SAE performance standards for new antifreeze. This eliminates the problems of waste disposal such as costs and the necessity of storing and hauling used coolant.

Other onboard functions of the illustrated machine include standard coolant exchange, flushing procedures, pressure testing for leaks, and vacuum fill for adding coolant to an empty system. Its tank-within-a-tank design holds 40 gallons of used coolant for recycling in the inner section, and up to 60 gallons of recycled coolant in the outer section.

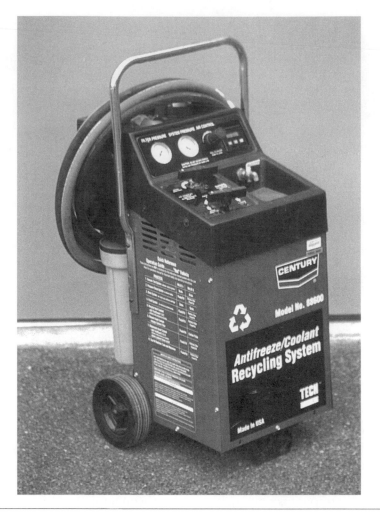

Figure 2-19 A typical antifreeze recovery/recycle machine. (Courtesy of Robinair SPX Corp.)

Electronic Thermometer. A single- or two-probe, hand-held electronic thermometer, such as shown in Figure 2-20, is commonly used in automotive air conditioning system diagnosis and service. The two-probe model is used to quickly and accurately measure superheat as required for critically charging some HFC-134a air conditioning systems.

The range for most battery-powered digital electronic thermometers are generally on the order of –50°F to 2,000°F (–46°C to 1,093°C) and have an accuracy greater than ±0.3 percent with a switchable resolution of 1.0 to 0.1 degrees in either °C or °F scale.

The desirable effective operating range for a thermometer for automotive air conditioning system use is 32°F (0°C) to 120°F (48.9°C). The digital readout for a hand-held electronic thermometer should be no less than 1/2 in. (12.7 mm) for easy reading.

Electronic Scale. An electronic scale (Figure 2-21) may be used for CFC-12 or HFC-134a refrigerants to deliver an accurate charge by weight, manually or automatically. Automatic charging is generally accomplished by programming the amount of refrigerant to be charged into the onboard solid-state microprocessor. The charge is stopped and an audible tone signals that the programmed weight has been dispersed. A liquid crystal display is used to keep track of refrigerant dispersed. Some models have a switchable pounds/kilograms readout with a resolution of 0.05 lb. (0.02 kg). The 10-in. (2.5-cm) scale platform will handle up to a 50-lb. (23-kg) bulk tank of refrigerant and is equipped with control panel fittings and two hoses to accommodate both CFC-12 and HFC-134a refrigerants.

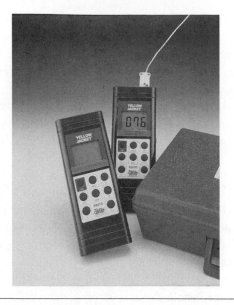

Figure 2-20 A typical electronic thermometer. (Courtesy Ritchie Engineering.)

Figure 2-21 A typical electronic scale.

Automatic Temperature Control (ATC) Testers (Scan Tool). There are many different types of testers available. The scan tool (Figure 2-22) is very popular and is used to enhance troubleshooting efforts to quickly get to the root of a problem. They are available in a wide variety of brands, prices, and capabilities. One good feature is that not only can a scan tool be used to retrieve trouble codes, some allow the technician to monitor and view sensor and computer information. This feature, known as *serial data* or *the data stream,* helps to pinpoint a heating, ventilation, or air conditioning (HVAC) problem. A scan tool can sometimes even take the role of a manifold and gauge set by obtaining system pressure readings through transducers in refrigerant lines.

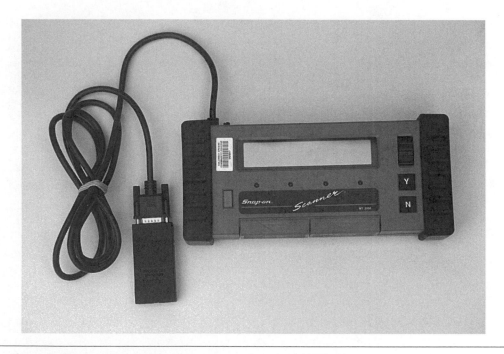

Figure 2-22 A typical automatic temperature control (ATC) tester.

Depending on the vehicle, the scan tool, and the software, the serial data that is obtained from an ATC system can include information such as blend door position and blower motor voltage. Some scan tools have a feature known as *bidirectional function* that enables the technician to activate various air conditioning system components, such as cooling fan and compressor clutch relay. Some scan tools have a *recorder* mode that is very useful in diagnosing intermittent problems. To use the recorder mode the technician hooks up the scan tool and drives the vehicle. When the intermittent problem is experienced, the technician pushes the appropriate button and the malfunction will generally be captured and stored in short-term memory. Back at the shop the stored information can be retrieved for evaluation.

Compressor Tools. There are several special tools (Figure 2-23) required for compressor clutch and shaft seal service. Clutch plate tools are used to remove the clutch plate to gain access to the shaft seal. They are also used for reinstalling the clutch plate after service. These tools should be compact in design for working in close quarters so that it may often be possible to service the compressor clutch and shaft seal without having to remove the compressor from the vehicle.

Basically, a shaft seal service kit includes an adjustable spanner wrench, clutch plate remover/installer, snap-ring pliers, ceramic seal remover/installer, seal seat remover/installer, shaft seal protector, seal assembly remover/installer, thin wall socket, O-ring remover, and O-ring installer. For clutch pulley and bearing service, service tools will include a pulley puller, pulley installer, bearing remover/installer, and rotor and bearing installer.

Special compressor service tools are designed to fit a particular application. Though some are interchangeable, most are not. For example, the seal seat remover/installer used on GM's models R4 and A6 compressors may also be used on Diesel Kiki models DKS-12 and DKS-15 compressors as well as on Sanden/Sankyo model 507, 508, and 510 compressors. If there is any doubt about the application of any particular tool do not use force. If it does not fit freely it may not be the correct tool for the task.

Often, good service tools are found at a very low cost at flea markets and garage sales. Nevertheless, the cost of having the required tool is often more than offset by the savings in time required to accomplish a task.

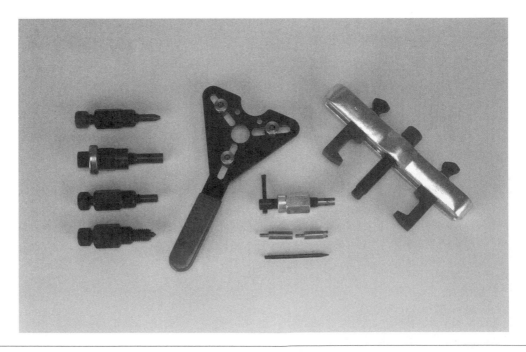

Figure 2-23 Special tools required for compressor service.

There are many other sources for information available to the automotive technician. These include, but are not limited to, Mitchell and Chilton Service Manuals, model-specific repair guides by Hayes available at local automotive parts stores, compact disks such as Mitchell-on-Demand, and aftermarket manufacturer's inhouse publications. There are many Internet connections such as ALDATA and International Automotive Technician's Network (iATN). Also, the local library generally has an automotive book section. One should not forget the valuable information that appears in the monthly publications by the International Mobile Air Conditioning Association (IMACA) and the Mobile Air Conditioning Society (MACS) made available to their members.

Finally, if you are taking any secondary or postsecondary automotive classes be sure to take notes and to save all of the handouts provided by your instructor, which often contain valuable information not available from other sources. Be sure to index your class notes and the handouts in a notebook for future reference. Although it may not seem so at first, this information can become more relevant and important for your ongoing study and practice, and can serve as valuable reference material.

Service Manual Procedures and Specifications

It is not possible to provide detailed information on all of the various **service procedures** that may be performed on the many different makes and models of automobile in service today. It is, therefore, generally necessary to consult manufacturers' service manuals for any particular application. This would be the case particularly when the procedures are not familiar to the technician. Service manuals are generally pretty much straightforward and relatively simple to follow. They generally provide information in a step-by-step basis and provide alternative methods whenever feasible.

Service Procedures

To find a particular service procedure, it is first necessary to locate the general group in which the procedure is to be found. Since we are concerned with automotive air conditioning, let us locate the procedure for replacing the heater blower motor in a 1994 Ford Taurus.

First, tab to the "Group Index" (Figure 2-24) of the 1994 Taurus/Sable Service Manual. Note that the "Climate Control System" is listed in "Group 12." Climate control includes heating and cooling of the in-car air.

Next, note the numeric or alphanumeric reference at the top of each page, such as 01-10-14 or 01-14B-23, and so on. The first two digits are the "group" numbers. Remember, we are looking for group 12, so tab through the manual until you find 12-00-01 (Figure 2-25). The second two numbers, 00, in this case, refer to the section within the group. The last number(s) refers to the page number within the section. Look at the "Section Titles" and note, "Air Conditioning System-Manual A/C Heater," to be found on page 12-03A-l.

Tab to page 12-03A-1 (Figure 2-26) and note another index, identified as "Subject." Under the heading of "Removal and Installation," find "Heater Blower Motor and Blower Motor Wheel Assembly." This information, according to the subject index, will be found on page 12-03A-37.

The proper removal and installation procedures are outlined on page 12-03A-37 and continue on page 12-03A-38 (Figure 2-27). If these step-by-step procedures are followed, there should be no problem with the removal and installation of the blower motor assembly.

Specifications

Specifications are found in the same manner as service procedures. Suppose, for example, that the cooling system capacity must be known in order to properly add 50 percent antifreeze solution for maximum winter protection. The "Group Index" tells us that the Service Information is in Group 00. Group 00 index (Figure 2-28) tells us that "Maintenance and Lubrication" will be found on page 00-03-1. The "Subject Index" of page 00-03-1 (Figure 2-29) refers us to "Specifications" to be found

Specifications provide information on system capacities. This information is also generally given in the Owner's Manual.

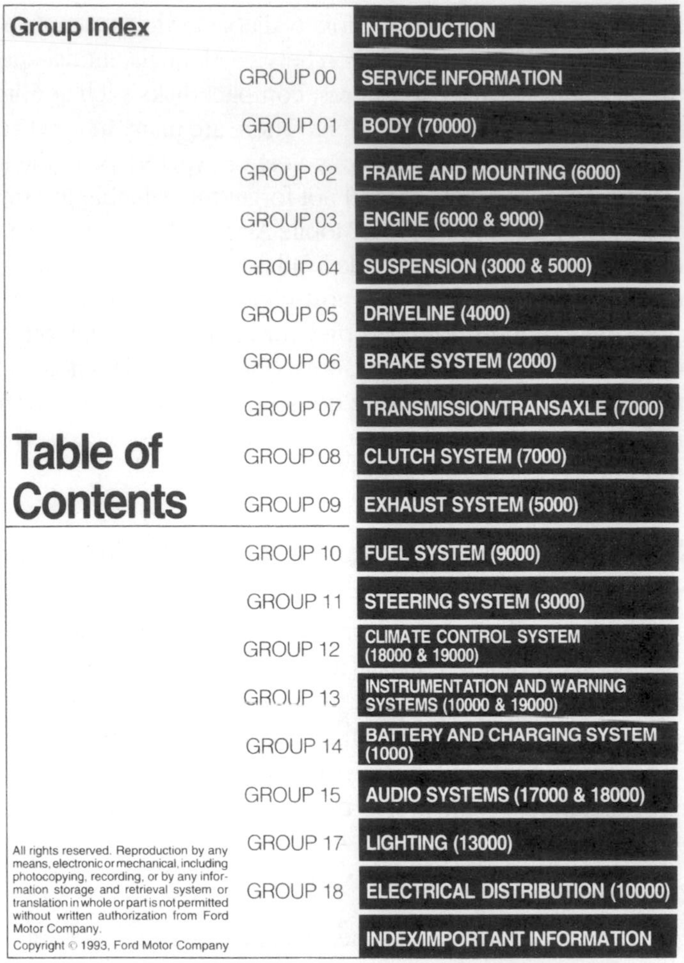

Group Index		INTRODUCTION
	GROUP 00	SERVICE INFORMATION
	GROUP 01	BODY (70000)
	GROUP 02	FRAME AND MOUNTING (6000)
	GROUP 03	ENGINE (6000 & 9000)
	GROUP 04	SUSPENSION (3000 & 5000)
	GROUP 05	DRIVELINE (4000)
	GROUP 06	BRAKE SYSTEM (2000)
	GROUP 07	TRANSMISSION/TRANSAXLE (7000)
Table of Contents	GROUP 08	CLUTCH SYSTEM (7000)
	GROUP 09	EXHAUST SYSTEM (5000)
	GROUP 10	FUEL SYSTEM (9000)
	GROUP 11	STEERING SYSTEM (3000)
	GROUP 12	CLIMATE CONTROL SYSTEM (18000 & 19000)
	GROUP 13	INSTRUMENTATION AND WARNING SYSTEMS (10000 & 19000)
	GROUP 14	BATTERY AND CHARGING SYSTEM (1000)
	GROUP 15	AUDIO SYSTEMS (17000 & 18000)
	GROUP 17	LIGHTING (13000)
	GROUP 18	ELECTRICAL DISTRIBUTION (10000)
		INDEX/IMPORTANT INFORMATION

Figure 2-24 Group index. (Reprinted with the permission of Ford Motor Company)

beginning on page 00-03-12. Ultimately, "Refill Capacities" are easily referenced on the following page (Figure 2-30).

If the engine size, which is necessary to determine cooling system capacity, is not known, this information may also be obtained. As outlined, the eighth position in the **VIN** (Vehicle Identification Number) identifies the engine size by comparing it to the chart found on page 00-01-6 (Figure 2-31).

> ◼ **CAUTION:** Neither the manufacturer's service manuals nor any school text, such as this one, can anticipate all conceivable ways or conditions under which a particular service procedure may be performed. It is therefore impossible to provide precautions for every possible hazard that may exist. The technician must always exercise extreme caution and pay heed to every established safety practice when performing automotive air conditioning service procedures.

The Metric System

The United States is slowly but surely joining the rest of the world in a uniform system of physical measurement known as the metric system.

In the metric system, speed is measured in kilometers per hour (km/hr), pressure is measured in **kiloPascals** absolute (kPa absolute) or kiloPascals (kPa), liquid is measured in liters (L), tempera-

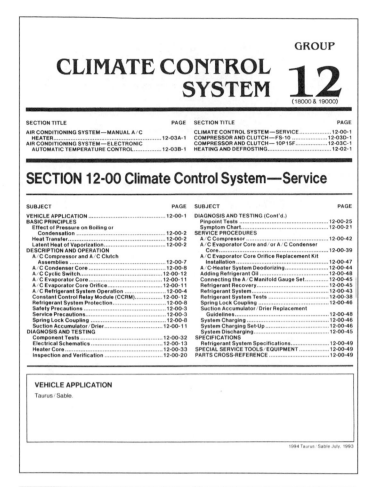

Figure 2-25 Section index. (Reprinted with the permission of Ford Motor Company)

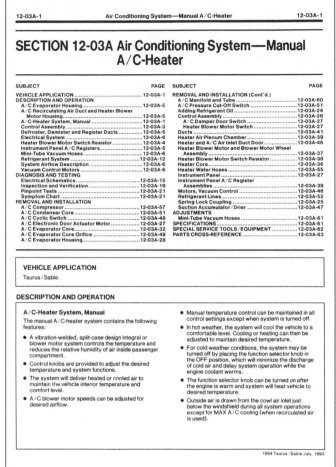

Figure 2-26 Subject index. (Reprinted with the permission of Ford Motor Company)

ture in degrees **Celsius** (°C), length in millimeters (mm) or meters (m), and weight in grams (g) or kilograms (kg). Whenever practical, measures in this text are given in both English and metric units. The metric equivalent is given in parentheses following the English measure. For example: The freezing point of water, at atmospheric pressure is 32°F (0°C) and its boiling point is 212°F (100°C).

There are several terms commonly used in the English system of measure that are the same as those used in the metric system of measure. Those most familiar to the automotive technician include ohm, volt, and ampere.

Some English standard measures cannot be converted to the standard metric measure. For example, there is no standard metric measure for 3/8 in. An English 3/8-in. measure is actually equal to 9.53 mm in the metric measure. Therefore, if you needed to remove 3/8-in. cap-screws from a plate, a 9-mm wrench would be too small, while a 10-mm wrench would be too large.

For the purpose of conversion, the metric equivalents in this text are held to one or two decimal places. For example, a temperature of 69°F converts to 20.5555555555°C. Little is gained by carrying the conversion to three or even two decimal places. The difference between one and ten decimal places in this example amounts to only 8/100°F (0.08°F) or 44/1,000°C (0.044°C), hardly worth consideration for the purpose of practical application.

As in the example given, the conversion of 69°F to °C will then be given as 20.6°C, rounded off to the nearest decimal place.

The metric system is known as "Systeme International d'Unites," a French term that literally translates to "International System of Units."

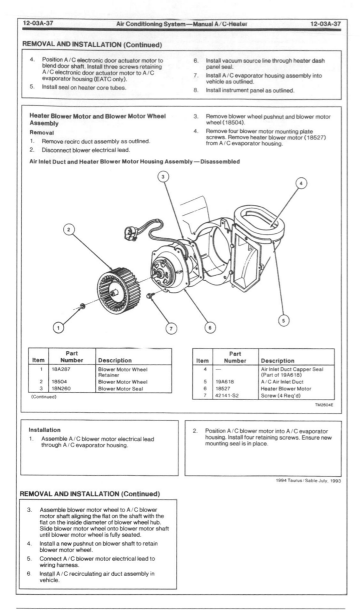

REMOVAL AND INSTALLATION (Continued)

4. Position A/C electronic door actuator motor to blend door shaft. Install three screws retaining A/C electronic door actuator motor to A/C evaporator housing (EATC only).

5. Install seal on heater core tubes.

6. Install vacuum source line through heater dash panel seal.

7. Install A/C evaporator housing assembly into vehicle as outlined.

8. Install instrument panel as outlined.

Heater Blower Motor and Blower Motor Wheel Assembly

Removal

1. Remove recirc duct assembly as outlined.

2. Disconnect blower electrical lead.

3. Remove blower wheel pushnut and blower motor wheel (18504).

4. Remove four blower motor mounting plate screws. Remove heater blower motor (18527) from A/C evaporator housing.

Air Inlet Duct and Heater Blower Motor Housing Assembly — Disassembled

Item	Part Number	Description
1	18A287	Blower Motor Wheel Retainer
2	18504	Blower Motor Wheel
3	18N260	Blower Motor Seal

(Continued)

Item	Part Number	Description
4	—	Air Inlet Duct Capper Seal (Part of 19A618)
5	19A618	A/C Air Inlet Duct
6	18527	Heater Blower Motor
7	42141-S2	Screw (4 Req'd)

TM2604E

Installation

1. Assemble A/C blower motor electrical lead through A/C evaporator housing.

2. Position A/C blower motor into A/C evaporator housing. Install four retaining screws. Ensure new mounting seal is in place.

1994 Taurus/Sable July, 1993

REMOVAL AND INSTALLATION (Continued)

3. Assemble blower motor wheel to A/C blower motor shaft aligning the flat on the shaft with the flat on the inside diameter of blower wheel hub. Slide blower motor wheel onto blower motor shaft until blower motor wheel is fully seated.

4. Install a new pushnut on blower shaft to retain blower motor wheel.

5. Connect A/C blower motor electrical lead to wiring harness.

6. Install A/C recirculating air duct assembly in vehicle.

Figure 2-27 Service procedures. (Reprinted with the permission of Ford Motor Company)

GENERAL SERVICE INFORMATION   00

SECTION 00-01 Identification Codes

VEHICLE APPLICATION

Taurus/Sable.

DESCRIPTION AND OPERATION

Vehicle Identification Number

The official Vehicle Identification Number (VIN) for title and registration purposes is stamped on a metal tab that is fastened to the instrument panel close to the windshield on the LH side of the vehicle and is visible from outside.

1994 Taurus/Sable July, 1993

Figure 2-28 Group 00 section index. (Reprinted with the permission of Ford Motor Company)

Pressure in the metric system is measured in terms of the Pascal (Pa). One pound per square inch (1 psi) is equal to 6.895×10^3 Pascals. For a more practical application, the kiloPascal (kPa) is used. One pound per square inch (1 psi) is equal to 6.895 kiloPascals (6.895 kPa). This equation applies to both the absolute (psia) and atmospheric (psig) pressure conversions.

A conversion chart for English to metric and metric to English values is given in Figure 2-32.

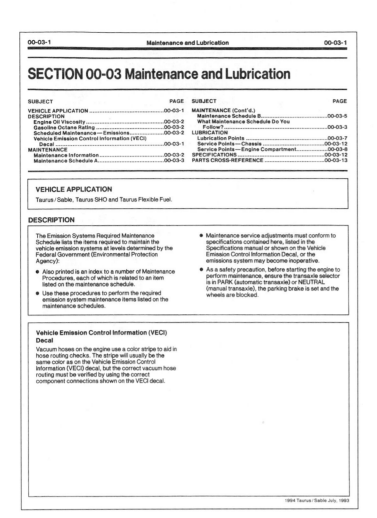

SECTION 00-03 Maintenance and Lubrication

VEHICLE APPLICATION

Taurus / Sable, Taurus SHO and Taurus Flexible Fuel.

DESCRIPTION

The Emission Systems Required Maintenance Schedule lists the items required to maintain the vehicle emission systems at levels determined by the Federal Government (Environmental Protection Agency):

- Also printed is an index to a number of Maintenance Procedures, each of which is related to an item listed on the maintenance schedule.
- Use these procedures to perform the required emission system maintenance items listed on the maintenance schedules.

- Maintenance service adjustments must conform to specifications contained here, listed in the Specifications manual or shown on the Vehicle Emission Control Information Decal, or the emissions system may become inoperative.
- As a safety precaution, before starting the engine to perform maintenance, ensure the transaxle selector is in PARK (automatic transaxle) or NEUTRAL (manual transaxle), the parking brake is set and the wheels are blocked.

Vehicle Emission Control Information (VECI) Decal

Vacuum hoses on the engine use a color stripe to aid in hose routing checks. The stripe will usually be the same color as on the Vehicle Emission Control Information (VECI) decal, but the correct vacuum hose routing must be verified by using the correct component connections shown on the VECI decal.

1994 Taurus / Sable July, 1993

Figure 2-29 Group 00 subject index. (Reprinted with the permission of Ford Motor Company)

SPECIFICATIONS (Continued)

LUBRICANT AND MAINTENANCE MATERIALS SPECIFICATIONS (Cont'd)

Description	Part Number	Ford Part Number	Ford Specification
Clutch Cable Connection Transaxle End Clutch Release Lever—At Fingers (Both Sides and Fulcrum) Clutch Release Bearing Retainer	Premium Long-Life Grease	XG-1-C or XG-1-K	ESA-M1C75-B
Outboard CV Joints / Inboard CV Joints	CV Joint Bearing Grease	E43Z-19590-A	ESP-M1C207-A

REFILL CAPACITIES

Component	U.S.	Imp.	Metric
Cooling System (including heater)			
3.0L engine	11.0 qts.	9.1 qts.	10.4 liters
3.0L SHO engine	11.6 qts.	9.6 qts.	11.0 liters
3.2L SHO engine	11.4 qts.	9.4 qts.	10.8 liters
3.8L engine	12.1 qts.	10.1 qts.	11.5 liters
Engine Oil (with filter change)			
3.0L/3.2L SHO engine	5.0 qts.	4.2 qts.	4.7 liters
3.0L engine	4.5 qts.	3.8 qts.	4.25 liters
3.8L engine	4.5 qts.	3.8 qts.	4.25 liters
Transaxle Automatic Overdrive (3.0L / 3.8L)	12.25 qts	10.2 qts.	11.6 liters
Manual (SHO)	6.1 pts.	5.1 pts.	2.9 liters
Power Steering	.85 qt.	.71 qt.	.80 liter
SHO engines	1.00 qt.	.60 qt.	.95 liter
Fuel tank Standard	16.0 gals.	13.3 gals.	60.6 liters
3.0/3.2L SHO models (optional fuel tank)	18.4 gals.	15.3 gals.	69.6 liters
Washer Reservoir Windshield	80 ozs.	83 ozs.	2400 cc

(Continued)

REFILL CAPACITIES (Cont'd)

Component	U.S.	Imp.	Metric
Wagon rear window	128 ozs.	133 ozs.	3840 cc
SHO model w/shield	64 ozs.	66 ozs.	1920 cc
Radiator Cap	16 psi	16 psi	110 kPa

TORQUE SPECIFICATIONS

Description	N-m	Lb-Ft
Spark Plugs (3.0L and 3.8L)	9-20	7-15
Spark Plugs (3.0L / 3.2L SHO)	20-30	15-22
Oil Pan Drain Plug Torque (3.0L)	11-16	9-12
Oil Pan Drain Plug Torque (3.0L / 3.2L SHO)	20-33	15-24
Oil Pan Drain Plug Torque (3.8L)	20-34	15-25

PARTS CROSS-REFERENCE

Base Part #	Part Name	Old Part Name
1125	Front Disc Brake Rotor	
1126	Brake Drum	
12405	Spark Plug	
2001	Brake Shoe and Lining	
2200	Rear Brake Shoe and Lining	
2C026	Rear Disc Brake Rotor	
6731	Oil Filter	
6A666	Positive Crankcase Ventilation Valve	
8620	Drive Belt	
9601	Air Cleaner Element	

Figure 2-30 Specifications. (Reprinted with the permission of Ford Motor Company)

Figure 2-31 Vehicle identification number (VIN) is observed through the windshield.

METRIC TO ENGLISH

Multiply	By	To Get
Celsius (°C)	1.8 (+32)	Fahrenheit (°F)
gram (g)	0.035 3	ounce (oz)
kilogram (kg)	2.205	pound (lb)
kilometer (km)	0.621 4	mile (mi)
kilopascal (kPa)	0.145	lb/in² (psi)
liter (L)	0.264 2	gallon (gal)
meter (m)	3.281	foot (ft)
milliliter (mL)	0.033 8	ounce (oz)
millimeter (mm)	0.039 4	inch (in)

ENGLISH TO METRIC

Fahrenheit (°F)	(−32) 0.556	Celsius (°C)
foot (ft)	0.304 8	meter (m)
fluidounce (fl oz)	29.57	milliliter (mL)
gallon (gal)	3.785	liter (L)
inch (in)	25.4	millimeter (mm)
mile (mi)	1.609	kilometer (km)
ounce (oz)	28.349 5	gram (g)
pound (lb)	0.453 6	kilogram (kg)
lb/in² (psi)	6.895	kilopascal (kPa)

Figure 2-32 English/metric conversion chart.

Terms to Know

Can tap	Fringe benefits	Socket
Caution	Hygiene	Specifications
Celsius	KiloPascal	Vacuum pump
Customer	Manifold and gauge set	VIN
Dependability	Responsibility	
Facilities	Service procedure	

ASE-Style Review Questions

1. *Technician A* says that rules are made to protect the customer.
   Technician B says that rules are made to protect the technician.
   Who is right?
   A. A only
   B. B only
   C. Both A and B
   D. Neither A nor B

2. *Technician A* says that Ford Motor Company sponsors the ASSET program.
   Technician B says that General Motors Corporation sponsors the ASEP as well as the ASSET program.
   Who is right?
   A. A only
   B. B only
   C. Both A and B
   D. Neither A nor B

3. *Technician A* says that one manifold and gauge set may be used for all automotive air conditioning service. *Technician B* says that one hose set may be used for all automotive air conditioning service. Who is right?
   - **A.** A only
   - **B.** B only
   - **C.** Both A and B
   - **D.** Neither A nor B

4. *Technician A* says if the low-side manifold hand valve is cracked, system pressure may be read on the low-side gauge. *Technician B* says that the low-side manifold hand valve does not have to be cracked to read low-side pressure. Who is right?
   - **A.** A only
   - **B.** B only
   - **C.** Both A and B
   - **D.** Neither A nor B

5. *Technician A* says that an OSHA poster must be displayed in a common place, such as the customer's waiting area. *Technician B* says that an OSHA poster must be displayed in the employee's common area, such as the break room. Who is right?
   - **A.** A only
   - **B.** B only
   - **C.** Both A and B
   - **D.** Neither A nor B

6. *Technician A* says that safety glasses should be approved for gases. *Technician B* says that safety glasses should be approved for liquids. Who is right?
   - **A.** A only
   - **B.** B only
   - **C.** Both A and B
   - **D.** Neither A nor B

7. *Technician A* says that refrigerant-12 in contact with an open flame produces a toxic vapor. *Technician B* says that one should work in a well-ventilated area. Who is right?
   - **A.** A only
   - **B.** B only
   - **C.** Both A and B
   - **D.** Neither A nor B

8. *Technician A* says that the English/metric conversion for pressure is psig/kP. *Technician B* says that the English/metric conversion for temperature is Fahrenheit/Centigrade. Who is right?
   - **A.** A only
   - **B.** B only
   - **C.** Both A and B
   - **D.** Neither A nor B

9. *Technician A* says that a vacuum pump is used to remove moisture from a system. *Technician B* says that an air pump is used to remove air from a system. Who is right?
   - **A.** A only
   - **B.** B only
   - **C.** Both A and B
   - **D.** Neither A nor B

10. *Technician A* says that customers are not allowed in the shop area because they get in the way. *Technician B* says that customers may be injured if allowed in the shop area. Who is right?
    - **A.** A only
    - **B.** B only
    - **C.** Both A and B
    - **D.** Neither A nor B

JOB SHEET 4

4

Name _____ Date _____

Identify the Responsibilities of the Employee

Upon completion of this job sheet you should understand the obligations of an employee.

Tools and Materials

Pad and pencil

Procedure

1. There are four primary employee obligations: attitude, responsibility, dependability, and pride. Can you think of any other obligations that may be an asset to the employee? If so, list them in the space provided.

   _____

2. Give several examples of employee attitude in the workplace.

   _____   _____

   _____   _____

3. Give several examples of employee responsibility in the workplace.

   _____   _____

   _____   _____

4. Give several examples of employee dependability in the workplace.

   _____   _____

   _____   _____

5. Give several examples of employee pride in the workplace.

   _____   _____

   _____   _____

6. Give examples of the other obligations you thought of in step 1.

   _____   _____

   _____   _____

✓ **Instructor's Check** _____

JOB SHEET 5

Name _____ Date _____

Use a Manufacturer's Service Manual

Upon completion of this job sheet you should understand how to use a service manual.

Tools and Materials

A late-model vehicle
A service manual for the vehicle

Procedure

Assume that you are to replace a circuit breaker in the vehicle you selected.

1. What vehicle did you select?

Make _____Model _____Year _____

2. Which service manual do you have?

Title _____   Year _____

3. Does the service manual cover the vehicle that you selected?_____

If not, explain:_____

4. Locate the Group Index and determine which group includes the circuit breaker.

What group did you select? _____

5. Is the circuit breaker in the group that you selected?_____ If not, what group did you find it in ?_____

6. Using information found in the manual, were you able to find the circuit breaker? _____

If not, what problems were encountered? _____

7. Were you able to determine the difference between the circuit breaker and the hazard signal flasher unit?_____ Were they similar? _____

8. How were they different? _____

☑ **Instructor's Check** _____

JOB SHEET 6

Name _____ Date _____

Compare the English and Metric System of Measure

Upon completion of this job sheet you should understand the English and metric system of measure

Tools and Materials

Miscellaneous and assorted nuts and bolts
Set of English (fractional inch) open-end wrenches
Set of metric (millimeter) open-end wrenches

Procedure

Ask your instructor to identify for you a 1/4-28 bolt and a 5/16-24 bolt.

1. What size wrench fits the head of the 1/4-28 bolt? _____

2. What size wrench fits the head of the 5/16-24 bolt? _____

3. Using the formula given in the shop manual, convert 1/4 in. to metric millimeters.

    a. First, convert 1/4 in. to a decimal value: What is the decimal value? _____

    b. Next, multiply the decimal value by the formula. _____

4. Is there a metric wrench the size determined in step 3b? _____ Explain: _____
_____

5. Is there a metric fastener the size as determined in step 3b? _____ Explain: _____
_____

6. Is there a metric fastener close to the size determined in step 3b? _____ Explain:
_____

7. Will the English wrench fit the fastener selected in step 6? _____ Explain: _____
_____

8. Are the two fasteners, determined in steps 1 and 6, interchangeable? Explain:
_____

9. Is the metric fastener identified in step 6 interchangeable with the 5/16-24 English fastener? _____ Explain: _____
_____

10. Is the metric fastener identified in step 6 closer in size to the 1/4-28 fastener or the 5/16-24 fastener? _____

☑ **Instructor's Check** _____

The Manifold and Gauge Set

Upon completion and review of this chapter you should be able to:

❏ Describe the nomenclature and function of the manifold and gauge set.

❏ Identify the scaling of the low- and high-side gauges in English and metric units.

❏ Calibrate a gauge.

❏ Connect a manifold and gauge set into an automotive air conditioning system.

❏ Hold a performance test on an automotive air conditioning system.

CHAPTER 3

The Manifold and Gauge Set

The **manifold and gauge set** is to the automotive air conditioning service technician what the stethoscope and ameroid manomometer is to the physician. Both are tools that are essential for the diagnosis of internal conditions that cannot otherwise be observed.

The manifold and gauge set is used to diagnose and troubleshoot various system malfunctions based on low- and high-side system **pressures**.

A basic tool for the air-conditioning service technician is the manifold and gauge set (Figure 3-1). The pressure measurement of an air conditioning system is a means of determining system performance. The manifold and gauge set is an essential tool for making these measurements. The servicing of automotive air conditioning systems requires the use of a two-gauge manifold set.

One gauge is used to observe pressure on the low (suction) side of the system. The second gauge is used to observe pressure on the high (discharge) side of the system. The service technician should have two gauge sets—one for **HFC-134a** and one for **CFC-12**. There is no significant difference between gauges designated for HFC-134a refrigerant and those designated for CFC-12 refrigerant. However, the two refrigerants are not compatible and two sets should be used. The general description, however, is the same for both types.

Basic Tools

Manifold and gauge set with hoses

Safety glasses or goggles

Fender cover

Basic tool set

Manifold and gauge sets must be dedicated; one for CFC-12 and one for HFC-134a.

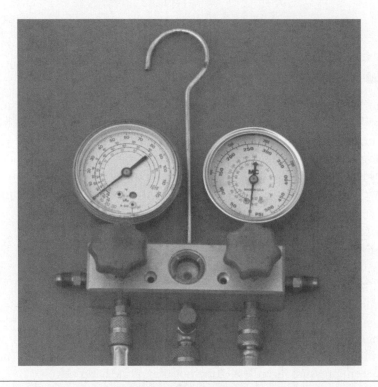

Figure 3-1 A typical CFC-12 manifold and gauge set.

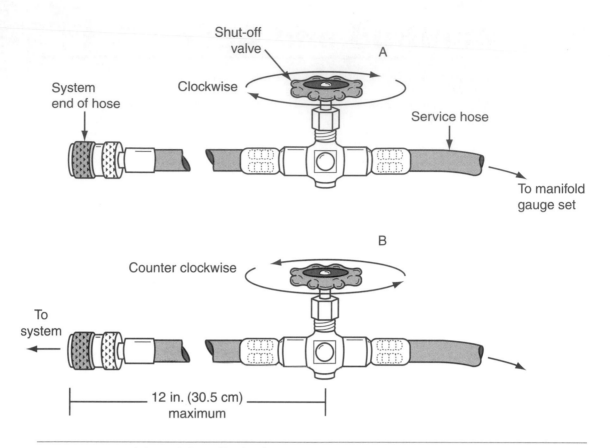

Figure 3-2 (A) Hand shut-off valve illustrating clockwise to close. (B) Hand shut-off valve illustrating counter clockwise to open.

Manifold

The low- and **high-side gauges** are connected into the air conditioning system through a manifold and two high-pressure hoses. The manifold has unique fittings to which hoses can be connected; 3/8 in. SAE for CFC-12 and 1/2 in. **Acme** for HFC-134a. Two hand valves are provided for controlling the fluid passages through the manifold: one for the low-side service hose and the other for the high-side service hose. Some manifolds have a third valve for connecting a vacuum service or refrigerant service hose through the manifold. Yet other manifolds may have four valves:

1. Low-side service hose
2. High-side service hose
3. Vacuum pump service hose
4. Refrigerant service hose

Turning the hand valve fully in a clockwise (**cw**) direction (Figure 3-2a) closes the fluid circuit in the manifold. The gauge will still indicate system pressure. The fluid passage is opened by turning the hand valve in the counterclockwise (**ccw**) direction (Figure 3-2b). When a hand valve is open, the fluid circuit is open to the center hose port of the manifold set. The respective gauge will record system pressure regardless of the position of the hand valve.

Figure 3-3 A typical low-side gauge.

Figure 3-4 A typical high-side gauge.

Low-Side Gauge

The **low-side gauge** (Figure 3-3), used to monitor low-side system pressure, is called a **compound gauge**. A compound gauge is designed to give both vacuum and pressure indications. This gauge is connected to the low side of the air conditioning system through the manifold and low-side hose.

Calibration of the vacuum scale of a compound gauge is from 30 in. Hg to 0 **psig**. The pressure scale is calibrated to indicate pressures from 0 psig to 120 psig. The compound gauge is constructed in such a manner so as to prevent any damage to the gauge if the pressure should reach a value as high as 250–350 psig.

Low-side pressures above 80 psig are rarely experienced in an operating system. Such pressures may be noted, however, if the low-side manifold service hose is accidentally connected to the high-side fitting. (Even experienced service technicians are known to make this type of error.)

The metric gauge used on the low side of the system is scaled in **absolute** kiloPascal (kPa) units. For conversion, 1 psi is equivalent to 6.895 kPa.

Low-side system operating pressures are generally 15–35 psig (103–241 kPa).

High-Side Gauge

The high-side gauge (Figure 3-4) indicates pressure in the side of the system. Under normal conditions, pressures in the high side seldom exceed 300 psig. As a safety factor, however, it is recommended that the maximum indication of the high-side gauge be 500 psig. The high-side gauge, though not calibrated below 0 psig, is not damaged when pulled into a vacuum.

High-side metric **pressure gauges** are scaled in kPa gauge or absolute.

The correct conversion is to kPa, whereby 1 psi equals 6.895 kPa. Atmospheric pressure at sea level (14.696 psia) is 101.32892 kPa, rounded off to 101 kPa.

$$14.696 \text{ psi} \times 6.895 \text{ kPa} = 101.32892$$

High-side system operating pressures are generally 160–220 psig (1,103–1,517 kPa).

Gauge Calibration

Most quality gauges have a provision for a calibration adjustment. Generally, a gauge is accurate to about 2 percent of its total scale when calibrated so the needle rests on zero with atmospheric pressure applied.

To calibrate a gauge:

1. Remove the hose after recovering refrigerant (if applicable).
2. Remove the retaining ring (bezel) and/or plastic lens cover.
3. Locate the adjusting screw.
4. Use a small screwdriver (Figure 3-5) to turn the adjusting screw in either direction until the pointer is lined up with the zero mark.

> ⚠️ **WARNING:** Do not force the adjusting screw; to do so may damage the gauge or alter its accuracy.

5. Replace the plastic lens and/or bezel.

Hoses

A refrigeration service hose (Figure 3-6) is constructed to withstand maximum working pressures of 500 psi (3,448 kPa). Some service hoses have a minimum burst pressure rating of up to 2,500 psi (17,238 kPa).

Hoses that are designated for CFC-12 service are available in several colors such as white, yellow, red, blue, and black. Blue is used on the low side, red on the high side, and white or yellow for the service (center) manifold port. Service hose fittings match manifold fittings that have 1/4-in. SAE threads.

Color coding for hoses designated for HFC-134a service is similar: blue for the low side, red for the high side, and yellow for the service hose. Service hose fittings match manifold fittings that have 1/2-in. Acme, designated 0.555-16, threads.

There must be a positive shut-off provision located within 12 in. (30.5 cm) of the service end (Figure 3-7). The shut-off valve is to be closed at all times when the hose is not in use. This valve is intended to help reduce the amount of refrigerant that may be vented to the atmosphere during a service procedure.

Figure 3-5 Use a small screwdriver to calibrate a gauge. (Courtesy of BET, Inc.)

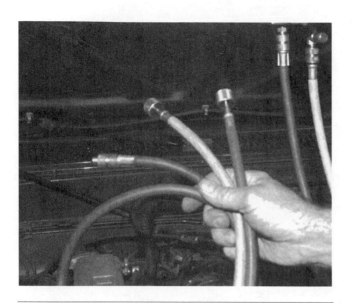

Figure 3-6 A typical service hose.

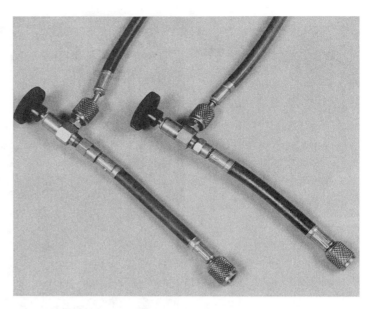

Figure 3-7 There must be a shut-off valve within 12 in. (30.5 cm) of the service hose end.

The fittings at the ends of standard CFC-12 service hoses are designed to fit the SAE flare fittings of the manifold set and compressor access ports. These hoses have a pin on one end for use on Schrader-type access ports. Hoses that are designated for HFC-134a service are not interchangeable with CFC-12 service hoses. Service valves require a special adaptor and manifold fittings are ACME threads. The differences in the two systems are intended to reduce the danger of cross-contamination. Only the proper hose may be connected to its respective fitting. This is an aid to prevent the intermixing of the two refrigerants.

Prior to recent environmental concerns, hoses were slightly porous and allowed a minimum amount of refrigerant to leak through. Current regulations, known as the SAE J2196 Standard, require that all hoses now be equipped with an impervious barrier to reduce the possibility of refrigerant leakage. These hoses are known as **barrier hoses**.

Connecting the Manifold and Gauge Set

This procedure is used when connecting the manifold and gauge set on the automotive air conditioning system to perform any one of the many operational tests and service procedures.

The transition from an ozone-depleting refrigerant (CFC-12) to an ozone-friendly refrigerant (HFC-134a) took place in some car lines during the 1993 model year. In other car lines, the transition took place during the 1994 model year.

By the 1995 model year, nearly all vehicles were equipped with the new refrigerant. Before proceeding with the installation of the manifold and gauge set, it is important to identify which type of refrigerant is in the air conditioning system.

Manifold and gauge sets for CFC-12 and HFC-134a are not interchangeable. The two refrigerants and their oils are not compatible. Mixing refrigerants, oils, or components could result in serious damage to the air conditioning system.

 CAUTION: It is important to use extreme care and observe all service and safety precautions when working with refrigerants and refrigeration oil.

Figure 3-8 Connecting the manifold and gauge set to an R-134a system.

Procedure

This procedure is given in three parts. Part I outlines the procedure to be followed when installing a manifold and gauge set into a CFC-12 system equipped with hand shut-off service valves. Part II procedures are to be used when installing a manifold and gauge set into a CFC-12 system equipped with Schrader-type service valves. Part III procedure (Photo Sequence 1) is used when connecting a manifold and gauge set into an HFC-134a system (Figure 3-8).

> **CAUTION:** Safety glasses must be worn at all times while working with refrigerants. Remember, liquid refrigerant sprayed in the eyes can cause blindness.

> **WARNING:** The Environmental Protection Agency (**EPA**) requires positive shut-off provisions within 12 in. (30.5 cm) of the service end of each hose. There are various methods of accomplishing this; some are manual and some are semiautomatic.

Part—CFC-12 System with Hand Valves

The high-side and low-side compressor-mounted hand-operated manual service valves were used on some early air conditioning systems—both factory and aftermarket installed—and had a cast iron or cast aluminum compressor manufactured by Tecumseh or York.

Though not used for years for automotive service, they are found in many over-the-road applications, such as Diamond Reo, Kenworth, Mack, and Peterbilt. They are also found in offroad applications such as Allis Chalmers, Caterpillar, International Harvester, and John Deere. The Tecumseh and York compressors, equipped with hand shut-off service valves are available new or rebuilt for CFC-12 or HFC-134a refrigerants having a lubricant charge of mineral oil, ester or Poly-Alkaline Glycol (PAG), as required.

> **WARNING:** Place a fender cover on the vehicle to avoid damage to the finish.

1. Remove the protective caps from the service valve stems (Figure 3-9), if equipped.

Figure 3-9 Remove the protective caps from the service valves.

2. Remove the protective caps from the service ports (Figure 3-10).

CAUTION: Remove the caps slowly to ensure that no refrigerant is leaking past the service valve.

NOTE: If a leak is found in the service valve stem it should be repaired or replaced. If there is a leak at the service port, the service valve should be replaced.

3. Connect the low-side manifold hose, finger tight, to the suction side of the system.
4. Connect the high-side manifold hose, finger tight, to the high side of the system.

SERVICE TIP: Make certain that the hand shut-off valves are closed on the manifold set before the next step.

5. Using a service valve wrench (Figure 3-11), rotate the suction-side service valve stem two or three turns clockwise.
6. Repeat the procedure of step 5 with the discharge service valve stem.

Special Tool

Service valve wrench

Figure 3-10 Remove the protective caps from the service ports.

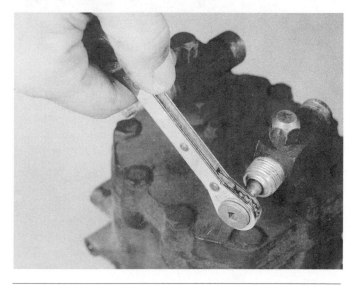

Figure 3-11 Use a service valve wrench to turn the service valve stem.

Photo Sequence 1
Typical Procedure for Connecting a Manifold and Gauge Set to an R-134a Air Conditioning System

P1-1 Typical location of low-side service valve on an R-134a air conditioning system.

P1-2 Typical location of high-side access fittings on an R-134a air conditioning system.

P1-3 Ensure that manifold and gauge set with hoses comply with SAEJ2211.

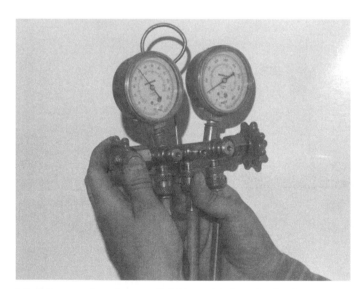

P1-4 Ensure that the manifold low-side hand valve is closed—turned fully clockwise.

Photo Sequence 1
Typical Procedure for Connecting a Manifold and Gauge Set to an R-134a Air Conditioning System (continued)

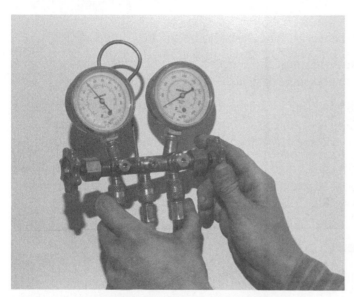

P1-5 Ensure that the manifold high-side hand valve is closed—turned fully clockwise.

P1-6 Remove protective cap from the low-side service valve fitting. Repeat this procedure with the high-side service valve fitting.

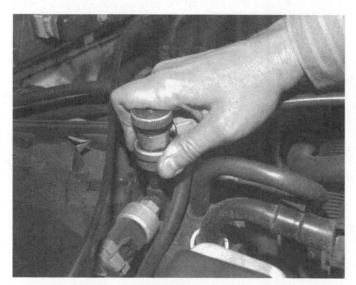

P1-7 Connect the low-side service hose quick connect adaptor to the low-side service valve fitting. Repeat this procedure with the high-side service valve fitting.

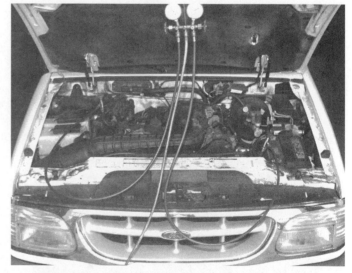

P1-8 The manifold and gauge set is ready to be used for servicing the air conditioning system.

PART II—CFC-12 System with Schrader-Type Valves

 WARNING: Place a fender cover on the vehicle to avoid damage to the finish.

1. Remove the protective caps from the high- and low-side service ports.

 CAUTION: Remove the caps slowly to ensure that no refrigerant is leaking past a defective **Schrader valve**.

NOTE: If a leak is found in the service valve, it should be repaired or replaced.

✔ **SERVICE TIP:** Service hoses must be equipped with a Schrader valve depressing pin. If the hoses are not so equipped, a suitable adapter must be used.

Special Tool

High-side adaptor, as required

2. Make sure that the manifold hand shut-off valves are closed before the next step.
3. Connect the low-side manifold hose to the suction side of the system finger tight.
4. Connect the high-side manifold hose to the discharge side of the system finger tight.

✔ **SERVICE TIP:** The high-side fitting on many late-model CFC-12 car lines (Figure 3-12) requires that a special adapter (Figure 3-13) be connected to the manifold hose before being connected to the system.

Part III—HFC-134a System

 WARNING: Place a fender cover on the vehicle to avoid damage to the finish.

1. Remove the protective caps from the high- and low-side service ports (Figure 3-14).

 CAUTION: Remove the caps slowly to ensure that no refrigerant escapes past a defective service valve.

Figure 3-12 A typical CFC-12 high-side access fitting. (Courtesy of BET, Inc.)

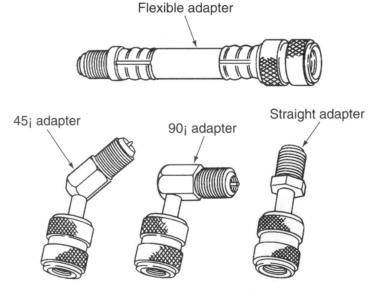

Figure 3-13 Adapters used to connect high-side hose into a CFC-12 system..

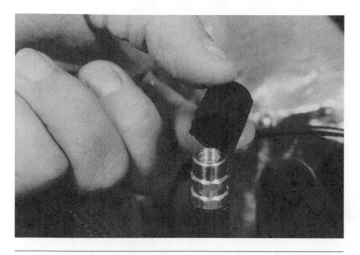

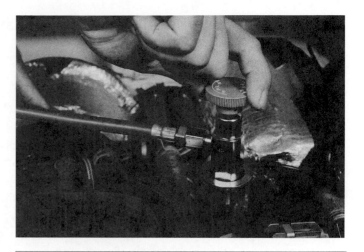

Figure 3-14 Remove the protective caps slowly to ensure no refrigerant loss.

Figure 3-15 Connecting the hose fitting to an \HFC-134a system.

NOTE: If a leak is found in the service valve, it should be repaired or replaced.

2. Turn the low-side hose hand valve fully counterclockwise to retract the Schrader depressor.
3. Connect the low-side hose (Figure 3-15).
   a. Press a quick-joint-type hose fitting onto a matching system fitting.
   b. Push firmly until a clicking sound is heard, ensuring that it is locked in place.
4. Repeat steps 2 and 3 with the high-side hose.

Performance Testing the Air Conditioning System

Humidity is an important factor in the quality and temperature of the air delivered to the interior of the car. The service technician must understand the effect that relative humidity (RH) has on the performance of the system. Relative humidity is the term that is used to denote the amount of moisture in the air. For example, a relative humidity of 80 percent means that the air contains 80 percent of the moisture that it can contain at a given temperature.

When the relative humidity is high, the evaporator has a double function. It must lower the air temperature as well as the temperature of the moisture carried in the air. The process of condensing the moisture in the air transfers a great amount of heat energy in the evaporator. Consequently, the amount of heat that can be absorbed from the air in the evaporator is greatly reduced.

The evaporator capacity required to reduce the amount of moisture in the air is not wasted, however. Lowering the moisture content of the air in the vehicle adds to the comfort of the passengers. The average person is comfortable at a temperature of 78–80°F at a relative humidity (RH) of 45–50 percent.

The following procedure is to serve as a guide for the service procedures required for **performance testing** the air conditioning system. The service technician must refer to the manufacturer's service manuals for specific data for any particular car model.

Moisture in the air is known as relative humidity (RH).

Preparing and Stabilizing the System

1. Ensure that both manifold hand valves are in the closed position to prevent refrigerant venting.

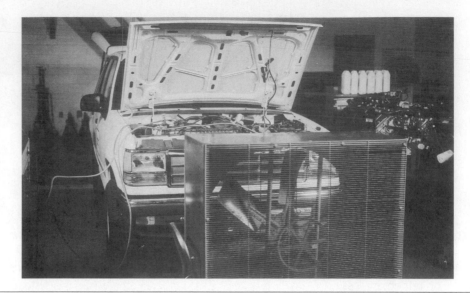

Figure 3-16 A fan is placed in front of the vehicle to provide additional air while performance testing.

2. Connect the manifold and gauge set into the system. Access the system: Set the compressor high-side service valve in the cracked position. Set the low-side service valve in the cracked position.

   ✔ **SERVICE TIP:** If the system is equipped with Schrader valves, they are automatically cracked when the hoses are connected.

3. Start the engine; set the speed to about 1,500–1,700 rpm.
4. Place a fan in front of the radiator to assist the ram air flow (Figure 3-16).
5. Turn on the air conditioner; set all controls to maximum cooling; set the blower speed on HI.
6. Insert a thermometer in the air conditioning duct as close as possible to the evaporator core (Figure 3-17).

   ✔ **SERVICE TIP:** Make certain of blower location to avoid damage to the thermometer and/or blower.

Visual Check of the Air Conditioner

1. The average low-pressure gauge reading should be in the range of 20–30 psig (239–310 kPa absolute). **NOTE:** The term *average* must be considered: for example, 15–25 psig (103–172 kPa) "averages" 20 psig (239 kPa); 25–35 psig (172–241 kPa) "averages" 30 psig.
2. The high-side gauge should be within the specified range of 160–220 psig (1,103–1,517 kPa) depending on the ambient temperature.
3. The discharge air temperature should be within the specified range of 40–50°F (4.4–10°C).

Inspect the High- and Low-Sides for Even Temperatures

1. Feel the hoses and components in the high side of the system to determine if the components are evenly heated.

   ■ **CAUTION:** Certain system malfunctions cause the high-side components to become superheated to the point that a serious burn can result if care is not taken when handling these components.

Schrader-type valves cannot be front seated.

An evaporating pressure of 30 psig (207 kPa) corresponds to a temperature of 32°F (0°C) for CFC-12 and 34.5°F (1.4°C) for HFC-134a.

A condensing pressure of 190 psig corresponds to a temperature of 93°F (33.9°C) for CFC-12 and 127°F (52.8°C) for HFC-134a.

Figure 3-17 A dial thermometer is used to check evaporator temperature.

2. Note the inlet and outlet temperatures of the receiver/drier assembly. A change in the temperature is an indication of a clogged or defective receiver/drier.

✔ **SERVICE TIP:** A change in temperature of the inlet and outlet of the accumulator is to be expected and is acceptable.

3. All lines and components on the high side should be warm to the touch (see CAUTION following step 1).
4. All lines and components on the low side of the system should be cool to the touch.
5. Note the condition of the thermostatic expansion valve (TXV) or fixed orifice tube (FOT).

✔ **SERVICE TIP:** Cold or frosted at the outlet side is to be expected and is acceptable. Cold or frosted at the inlet side, however, is an indication of a defective metering device or that there is excess moisture in the system.

Test the Thermostats and Control Devices

1. Refer to the service manual for the performance testing of the particular type of control device used.
2. Determine that the thermostat or low-pressure switch engages and disengages the clutch. There should be about a 12°F (6.7°C) temperature rise between the cut-out (off) and cut-in (on) point. Guides for determining the proper gauge readings and temperatures are shown (Figure 3-18). The relative humidity (RH) at any particular temperature is a factor in the quality of the air (Figures 3-19A and 3-19B). These figures should be regarded as guides only.

Air conditioning temperature control is by thermostat or low-pressure control.

Ambient Air Temperature, °F	70	80	90	100	110
Average Compressor Head Pressure, psig	150–190	170–220	190–250	220–300	270–370
Average Evaporator Temperature, °F	38–45	39–47	40–50	42–55	45–60

(A) English

Ambient Air Temperature, °C	21	27	32	38	43
Average Compressor Head Pressure, kPa	1 034–1 310	1 172–1 517	1 310–1 724	1 517–2 069	1 862–2 551
Average Evaporator Temperature, °C	3.3–7.2	3.9–8.3	4.4–10	5.5–12.8	7.2–15.6

(B) Metric

Figure 3-18 Hard pressure performance charts: (A) English; (B) metric.

Ambient Temperature, °F	70			80			90			100		
Relative Humidity, %	50	60	90	50	60	90	40	50	60	20	40	50
Discharge Air Temperature, °F	40	41	42	42	43	47	41	44	49	43	49	55

(A) English

Ambient Temperature, °C	21			27			32			38		
Relative Humidity, %	50	60	90	50	60	90	40	50	60	20	40	50
Discharge Air Temperature, °C	4.4	5	5.5	5.5	6.1	8.3	5	6.6	9.4	6.1	8.3	12.7

(B) Metric

Figure 3-19 Relative humidity performance charts: (A) English; (B) metric.

3. Complete performance testing. Refer to the manufacturer's service manual for specific requirements.

Return the System to Service

1. Return the engine speed to normal idle.
2. Back seat the high- and low-side compressor service valves, if equipped.
3. Close the service hose valves.

CAUTION: If service hoses are equipped with manual shut-off, be sure to close them before disconnecting from the system.

4. Remove the service hoses.
5. Replace the protective caps to prevent a loss of refrigerant in case the service valve leaks.
6. Turn off the air conditioning controls.
7. Stop the engine.

A customer brings a car, recently purchased from a reputable used car dealer, in for service with a repair authorization to "fix a water leak." The complaint is that water spills out of the air conditioner onto her feet when she makes a right turn. She also notes that the floor mat on the passenger side is damp most of the time.

The technician verifies the damp floor mat and suspects that the heater core is leaking. However, there is no evidence of antifreeze solution on the floor mat. Further discussion with the customer verifies that the cooling system is apparently "sound." While talking with the customer, however, the technician notices that there are not the familiar drips on the shop floor usually experienced when the air conditioner is operating.

Inspection of the drain tube of the evaporator reveals that a recently applied undercoating material has sealed the opening. This causes the water to back up into the evaporator and spill out. Cleaning the drain tube solved the problem.

Terms to Know

CFC-12	cw	Manifold
Absolute	Compound gauge	Performance testing
Acme	EPA	Pressure
Barrier hose	HFC-134a	Pressure gauge
Calibration	High-side gauge	Psig
ccw	Low-side gauge	Schrader valve

ASE-Style Review Questions

1. *Technician A* says that the performance test determines if system pressures are proper.
   Technician B says that the test determines if the system temperatures are proper.
   Who is right?
   A. A only **C.** Both A and B
   B. B only **D.** Neither A nor B

2. *Technician A* says that humidity has an effect on system performance.
   Technician B says that poor airflow has an effect on system performance.
   Who is right?
   A. A only **C.** Both A and B
   B. B only **D.** Neither A nor B

3. *Technician A* says that the blower should be run on high speed for the performance test.
   Technician B says it does not matter at what speed the fan is run.
   Who is right?
   A. A only **C.** Both A and B
   B. B only **D.** Neither A nor B

4. *Technician A* says that a temperature change at the inlet and outlet of the receiver/drier indicates a restriction.
   Technician B says that a temperature change at the inlet and outlet of an accumulator indicates a restriction.
   Who is right?
   A. A only **C.** Both A and B
   B. B only **D.** Neither A nor B

5. *Technician A* says that a change in temperature from the inlet to the outlet of the thermostatic expansion valve (TXV) is not acceptable.
Technician B says that a change in temperature from the inlet to the outlet of a fixed orifice tube (FOT) is acceptable.
Who is right?

   A. A only   **C.** Both A and B
   B. B only   **D.** Neither A nor B

6. *Technician A* says that a CFC-12 manifold and gauge set may be used on an HFC-134a system if it has not been used on a CFC-12 system.
Technician B says that an HFC-134a manifold and gauge set may be used on a CFC-12 system if it has not been used on an HFC-134a system.
Who is right?

   A. A only   **C.** Both A and B
   B. B only   **D.** Neither A nor B

7. *Technician A* says that the Hg/kPa scale is used to denote pressure on the metric gauges.
Technician B says that the Hg/psig scale is used to denote pressure on the English scale.
Who is right?

   A. A only   **C.** Both A and B
   B. B only   **D.** Neither A nor B

8. *Technician A* says that the maximum working pressure of service hoses should be 500 psig (3,448 kPa).
Technician B says that 500 psig (3,448 kPa) is also the burst pressure.
Who is right?

   A. A only   **C.** Both A and B
   B. B only   **D.** Neither A nor B

9. *Technician A* says that the low-side gauge may be used on the high side, if necessary.
Technician B says that the high-side gauge may be used on the low side, if necessary.
Who is right?

   A. A only   **C.** Both A and B
   B. B only   **D.** Neither A nor B

10. *Technician A* says that a pin is provided in the end of a service hose to access a Schrader-type service valve.
Technician B says that a special adaptor is required to access some Schrader-type service valves.
Who is right?

   A. A only   **C.** Both A and B
   B. B only   **D.** Neither A nor B

ASE Challenge

1. All of the following statements about a Schrader-type service valve are correct, EXCEPT:
   A. The valve may be front seated
   B. The valve may be back seated
   C. The valve may be midpositioned
   D. The valve may be cracked

2. Which of the following water/antifreeze mixtures are recommended for maximum overall cooling system protection?
   A. 45/55 percent
   B. 50/50 percent
   C. 55/45 percent
   D. 60/40 percent

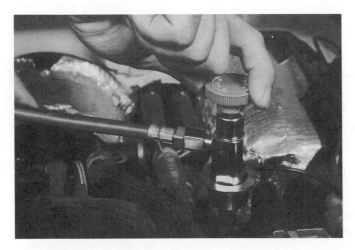

3. The service valve adaptor shown above is turned in the _____ direction to allow access to the air conditioning system.
   A. Clockwise (cw)
   B. Counterclockwise (ccw)
   C. Either A or B
   D. Neither A nor B

4. The theoretical dividing line between the low and high sides of an air conditioning system are the:
   A. Condenser and evaporator
   B. Accumulator and condenser
   C. Compressor and metering device
   D. Receiver-drier and compressor

5. Which of the following statements best describes latent heat?
   A. Latent heat cannot be felt but may be measured on a thermometer
   B. Latent heat can be felt but cannot be measured on a thermometer
   C. Latent heat can be felt and can be measured on a thermometer
   D. Latent heat cannot be felt nor can it be measured on a thermometer

Table 3-1 ASE TASK

Identify system type and conduct performance tests on the A/C system; determine needed repairs.

Problem Area	Symptoms	Possible Causes	Classroom Manual	Shop Manual
REFRIGERANT TYPE IS NOT KNOWN	System pressure is higher than expected	1. Incorrect refrigerant in the system		
		2. Air in the system	89	111
		3. Contaminated (mixed) refrigerant in system	81–89	114
	System pressure is lower than expected	1. Incorrect refrigerant in the system	81–89	114
		2. Undercharge of refrigerant in the system	86	110
		3. No refrigerant in the system	86	110

Table 3-2 ASE TASK

Diagnose A/C system problems indicated by refrigerant flow past the sight glass (for systems using a sight glass); determine needed repairs.

Problem Area	Symptoms	Possible Causes	Classroom Manual	Shop Manual
	Sight glass is clear	1. System is fully charged with refrigerant	166, 168	114–123
		2. System is empty	166, 168	110
	Sight glass is cloudy	1. System is low on refrigerant	86, 168	110
		2. Flow of refrigerant is partially restricted	86, 168	111

Table 3-3 ASE TASK

Diagnose A/C system problems indicated by pressure gauge readings; determine needed repairs.

Problem Area	Symptoms	Possible Causes	Classroom Manual	Shop Manual
ABNORMAL GAUGE PRESSURE	High-side gauge reading is excessive (high)	1. Air in the system	89	111
		2. Refrigerant overcharge	89	111
		3. High-side restriction of refrigerant	87	77–78
		4. Restriction of air across the condenser	89	111
	High-side gauge reading is too low	1. Refrigerant undercharge	86	110
		2. Defective compressor	87	107
	Low-side gauge reading is excessive (high)	1. Defective compressor	87	107
		2. Defective metering device (FOT or TXV)	86	108–109
		3. Refrigerant overcharge	89	111
	Low-side gauge reading is too low	1. Defective metering device (FOT or TXV)	86	108–109
		2. Refrigerant undercharge	86	110
		3. Refrigerant restriction in low side of system	87	110
		4. Airflow across evaporator is restricted	84	107

Table 3-4 ASE TASK

Diagnose A/C system problems indicated by sight, sound, smell, and touch procedures; determine needed repairs.

Problem Area	Symptoms	Possible Causes	Classroom Manual	Shop Manual
VISUAL INSPECTION	Oily hoses and/or fittings	Refrigerant leak	86	110
	Worn or abraded hose	1. Hose improperly routed	123, 264	110
		2. Loose or defective retainer clamps	123, 264	110
	Icing condition of component, such as drier	Partial restriction at point of icing	84	110
UNUSUAL ODOR	Musty smell from evaporator assembly	1. Partially clogged drain tube		
		2. Mold or mildew on evaporator core		
		3. Heater core leaking	267	107
TOUCH TEST	Temperature change	Partial restriction at point of temperature change	84, 87	110

JOB SHEET 7

Name _____ Date _____

Interpreting Gauge Pressure

Upon completion of this job sheet you should be able to troubleshoot an air conditioning system based on gauge pressure readings.

ASE Correlation

This job sheet is related to the ASE Heating and Air Conditioning Test's content area: *A/C System Diagnosis and Repair*, Task: *Diagnose A/C system problems indicated by pressure gauge readings. Determine needed repairs.*

Tools and Materials

None required

Procedure

The gauge pressure reading is given. Identify the problem and suggest the remedy.

High-side gauge reading too high:

PROBLEM	REMEDY
1. _____	_____
2. _____	_____
3. _____	_____

High-side gauge reading is too low:

PROBLEM	REMEDY
4. _____	_____
5. _____	_____

Low-side gauge reading is too high:

PROBLEM	REMEDY
6. _____	_____
7. _____	_____
8. _____	_____

Low-side gauge reading is too low:

PROBLEM	REMEDY
9. _____	_____
10. _____	_____

☑ **Instructor's Check** _____

JOB SHEET 8

Name _____ Date _____

Interpreting System Conditions

Upon completion of this job sheet you should be able to troubleshoot an air conditioning system based on visual, smell, and touch.

ASE Correlation

This job sheet is related to the ASE Heating and Air Conditioning Test's content area: *A/C System Diagnosis and Repair*, Tasks: *Diagnose A/C system problems indicated by refrigerant flow past the sight glass (for systems using a sight glass); determine needed repairs. Diagnose A/C system problems indicated by sight, sound, smell, and touch procedures; determine needed repairs.*

Tools and Materials

None required

Procedure

The condition is given. Identify the problem and suggest the remedy.

Sight glass is clear:

	PROBLEM	REMEDY
1.	_____	_____
2.	_____	_____

Sight glass is cloudy:

	PROBLEM	REMEDY
3.	_____	_____
4.	_____	_____

Oily hose and/or fittings; worn or abraded hose:

	PROBLEM	REMEDY
5.	_____	_____
6.	_____	_____

Icing condition or temperature change at a component:

	PROBLEM	REMEDY
7.	_____	_____
8.	_____	_____

Musty smell from evaporator:

	PROBLEM	REMEDY
9.	_____	_____
10.	_____	_____

☑ **Instructor's Check** _____

_____

JOB SHEET 9

Name _____ Date _____

Identifying Refrigeration System Type

Upon completion of this job sheet you should be able to determine the type of refrigerant in the system.

ASE Correlation

This job sheet is related to the ASE Heating and Air Conditioning Test's content area: *A/C System Diagnosis and Repair.* Task: *Identify system type and conduct performance test on the A/C system; determine needed repairs.*

Tools and Materials

None required

Procedure

The refrigerant type is not known. The system condition is given. Identify the probable cause and suggest the remedy.

System pressure is higher than expected:

PROBLEM	REMEDY
1. _____	_____
2. _____	_____
3. _____	_____

System pressure is lower than expected:

PROBLEM	REMEDY
4. _____	_____
5. _____	_____
6. _____	_____

Expected low-side pressure range for a properly operating system:

	CFC-12	HFC-134a
7.	_____ to _____ psig	_____ to _____ psig
8.	_____ to _____ kPa	_____ to _____ kPa

Expected high-side pressure range for a properly operating system:

	CFC-12	HFC-134a
9.	_____ to _____ psig	_____ to _____ psig
10.	_____ to _____ kPa	_____ to _____ kPa

☑ **Instructor's Check** _____

Servicing the System

Upon completion and review of this chapter you should be able to:

❏ Determine if the air conditioning malfunction is due to an electrical or mechanical failure.

❏ Determine the "state-of-charge" of refrigerant in the air conditioning system.

❏ Determine if the air conditioner is a cycling clutch or noncycling clutch system.

❏ Perform functional testing of the electrical and mechanical systems.

❏ Understand general troubleshooting procedures and practices.

Air Conditioning Diagnosis

Servicing the automotive air conditioning system requires a good working knowledge of the purpose and function of the individual components that make up the total system. This includes the action and reaction of both the mechanical and electrical systems and subsystems.

Air conditioning systems vary from vehicle to vehicle by year and model; therefore, no standard diagnostic procedures are possible. All automotive air conditioning system diagnostics, however, share a few common prerequisites. These prerequisites include:

❏ Determine if the problem is electrical, mechanical, or both.

❏ Determine that the system is properly charged with refrigerant, that it is not undercharged or overcharged

 SERVICE TIP: This determination may be made by performing a procedure known as an "insufficient cooling quick check."

❏ Determine the type system: cycling clutch or noncycling clutch type.

❏ Perform a **functional test** of the air distribution system to determine proper operation. This should be accomplished before proceeding with the **diagnosis** procedures. The components of the air distribution system include the blower motor, switches, vacuum lines, air **ducts**, and mode doors.

❏ Hold a performance test of the system to verify correct low- and high-side pressures based on the temperature-pressure chart.
NOTE: Consider the ambient temperature and relative humidity conditions.

System Inspection

If the air conditioning system malfunction is due to abnormally high or low system pressures, inspect the following (Photo Sequence 2):

1. Visually check the condenser. Make certain that the airflow is not blocked by dirt, debris, or other foreign matter.
2. Inspect between the condenser and radiator. Check for any foreign matter. Clean, if necessary.
3. Visually check for bends or kinks in the tubes and lines that may cause a restriction. Check the condenser, refrigerant hoses, and joining tubes.
4. Using a proper leak detector (Figure 4-1), check for refrigerant leaks. It is good practice to leak check the air conditioning system any time it has a low charge or a leak is suspected. It should also be checked for leaks whenever service operations have been performed that require "opening" the system.

Basic Tools

Basic mechanic's tool set

Manifold and gauge set

Flare nut wrench set

Springlock coupling tool set

Vacuum pump

Refrigerant recovery system

Test lamp

Jumper wire

Thermometers (2)

Large fan (if required)

When in doubt, refer to the manufacturer's specifications.

Photo Sequence 2
Typical Procedures for Inspecting an Air Conditioning System

P2-1 Visually check for signs of a refrigerant leak as may be noted by an oil stain.

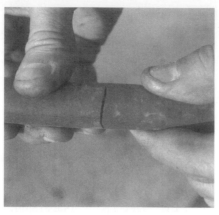

P2-2 Inspect the hoses for cuts or other obvious damage.

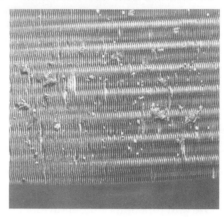

P2-3 Inspect the condenser for debris or other foreign matter.

P2-4 Carefully feel the hoses to determine if there is a temperature differential.
CAUTION: Some hoses may get extremely hot.

P2-5 A restriction will often result in frosting at the point of restriction.

P2-6 Ensure that the fan(s) are operating properly.

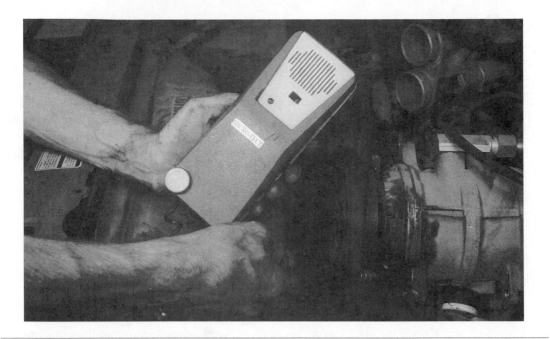

Figure 4-1 Check for refrigerant leaks.

 CAUTION: Follow specific instruction when using a leak detector.

5. Check the air distribution duct system for leaks, restrictions, or binding mode doors. Insufficient airflow may also indicate a clogged or restricted **evaporator core**.
6. Visually check for proper clutch operation. A slipping clutch may be caused by low voltage due to a loose wire or defective control device.
7. Use a belt **tension gauge** to check for proper drive belt tension. Consult the manufacturer's specifications for proper belt tension (Figure 4-2).

A visual inspection often reveals a problem.

Figure 4-2 Using a belt tension gauge to check for proper belt tension.

ELECTRICAL DIAGNOSIS AND TESTING

Remove rings, watches, and jewelry when working on electrical components.

The following is a typical systematic approach for applying the diagnostic procedures used to determine air conditioning system electrical problems. Those who lack a general knowledge of automotive electricity and electronics may review this information in Chapter 2 of the Classroom Manual and Chapter 3 of the Shop Manual in the *Today's Technician* series, *Automotive Electricity and Electronics*.

There are eight basic conditions that one may encounter relating to the electrical system of the automotive air conditioner. These conditions are:

1. Everything works electrically, but there is poor or no cooling
2. Nothing works
3. Only the clutch works
4. Only the evaporator blower works
5. Only the cooling fan works
6. The clutch does not work
7. The evaporator blower does not work
8. The cooling fan does not work

Because of the complex electrical wiring of most factory-installed air conditioning systems (Figure 4-3 and Figure 4-4), it is generally necessary to consult electrical schematics of appropriate manufacturer's shop manuals for proper troubleshooting and testing procedures. The following, however, may be considered typical for basic testing of most systems.

If everything works but there is poor or no cooling, the problem is likely to be in the vacuum, refrigeration, or duct system. One might also suspect an engine cooling system problem. Refer to Chapter 11 of this manual as well as of the Classroom Manual for suggested troubleshooting techniques and procedures.

If nothing works, the complaint is obvious. The system is not cooling and there is no air moving over the evaporator. A visual inspection will reveal that neither the compressor nor the evaporator fan motor is working, although the cooling fan may or may not be working. It will also be obvious that the blower motor is not running. The logical conclusion, then, is that the main fuse, circuit breaker, or fusible link is defective. The problem could also be in the master control or wiring.

Troubleshooting the Blower Motor Circuit

Before troubleshooting, check the battery charge (Figure 4-5). The voltage should be 12 volts or more. If not, recharge or replace the battery before proceeding. If the engine starts and runs, however, this test is not necessary. Proceed as follows for systematic testing of the blower motor electrical circuit:

Testing the Fuses and Blower Motor Relay (Figure 4-6)

Wires separated inside a connector are not easily detected by visual inspection.

1. Connect one test lead of a digital volt ohmmeter (DVOM) to **ground** and the other test lead to the hot side of the fuse or circuit breaker, point "A." Turn the ignition switch to the ON position, the air conditioning switch to A/C, and the blower motor switch to high speed. Note the DVOM reading. Next, disconnect the test lead from point "A" and connect it to point "B." Note the reading and disconnect the DVOM.
   a. If there are 0 volts at either point "A" or "B," a broken, disconnected, or defective wire or a defective fusible link is indicated.
   b. If there are less than 12 volts at either point, look for a loose or corroded (high-resistance) connection before the fuse that is causing the low voltage.
   c. If there are 12 volts or more at both points, proceed with step 2.
2. Connect the DVOM from ground to the other side of the fuse, point "C," and note the reading.

96

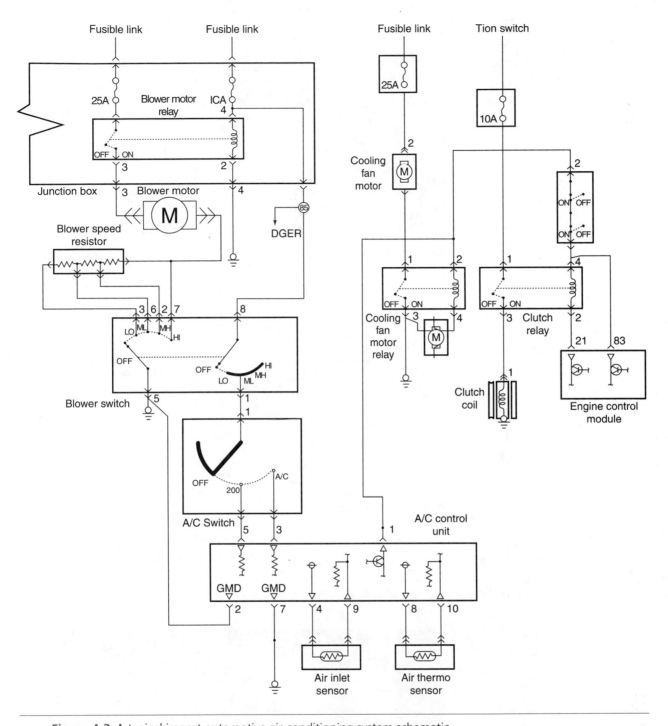

Figure 4-3 A typical import automotive air conditioning system schematic.

a. The reading should be the same as at point "A." If there are 0 volts, the fuse is defective and must be replaced.

 CAUTION: Before replacing the fuse, turn off air conditioner and ignition switch.

b. Disconnect the DVOM lead from point "C" and attach it to point "D." Note the reading, then disconnect the DVOM.

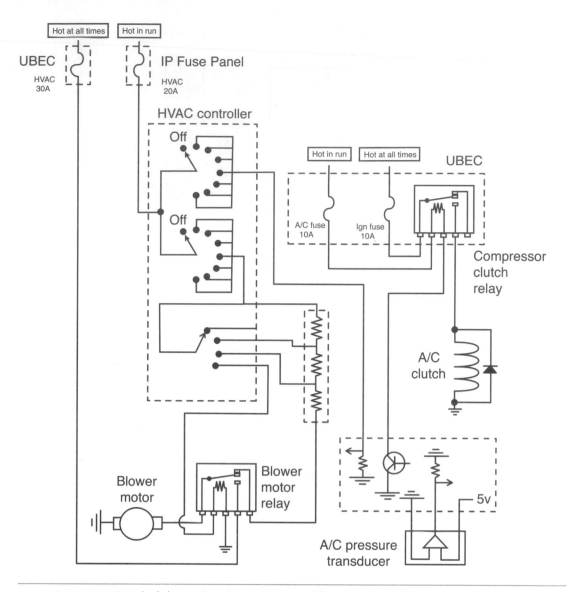

Figure 4-4 A typical domestic automatic air conditioning system schematic.

Figure 4-5 Testing battery voltage.

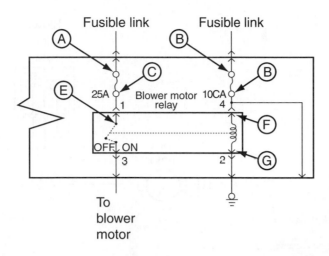

Figure 4-6 Test points for fuses and blower motor relay.

   c. The reading should be the same as at point "B." If there are 0 volts, the fuse is defective and must be replaced. (See CAUTION above.)

   d. If 12 volts or more are noted at points "C" and "D," proceed with step 3.

3. Connect the DVOM lead to point "E." Voltage reading should be the same as at point "A." Disconnect the DVOM.

   a. If there are 0 volts, look for a broken or disconnected wire.

   b. If less than 12 volts, look for a loose or corroded terminal.

   c. If 12 volts or more are noted, proceed with step 4.

4. Connect the DVOM to point "F" and note the voltage. Disconnect the DVOM.

   a. If 0 volts are noted, the switch is open. Skip to step 5.
      NOTE: It should be closed.

   b. If 12 volts or more are noted, the relay is operating properly. Proceed with further testing.

5. Connect the DVOM lead to point "G." The voltage reading should be the same as at point "B." Disconnect the DVOM.

   a. If there are 0 volts, look for a broken or disconnected wire.

   b. If less than 12 volts, look for a loose or corroded terminal.

   c. If 12 volts or more are noted, proceed with step 6.

6. Connect the DVOM to point "H." Note the voltage and disconnect the DVOM.

   a. If 12 volts or more are noted, the relay coil circuit to ground is open. Check for disconnected or defective ground wire.

   b. If there are less than 12 volts, check for loose or corroded ground wire connection.

   c. If 0 volts are noted, replace the relay.

Testing the Blower Motor (Figure 4-7). Turn the ignition switch to the ON position, the air conditioning switch to A/C, and the blower motor switch to high speed.

   1. Connect the DVOM to point "A" of the blower motor.

      a. The voltage should be the same as measured at point "F" of the previous test, 12 volts or more.

      b. If 0 volts are noted, look for a disconnected wire or connection.

      c. If less than 12 volts, look for a loose or corroded wire or connector.

      d. If there are 12 volts or more, proceed with step 2.

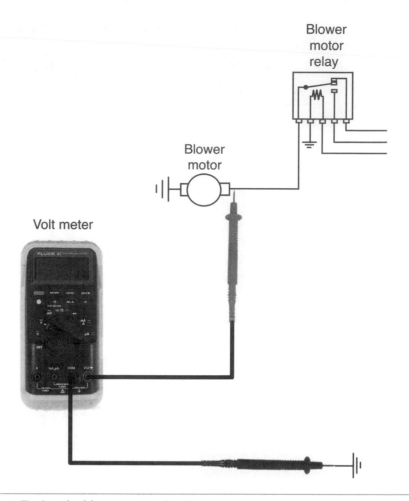

Figure 4-7 Testing the blower motor circuit.

2. Connect the DVOM to point "B" of the blower motor. Note the voltage and disconnect the meter.

   a. If the meter reads any voltage, check for a loose or corroded wire or terminal.

   b. If the motor is not running, turn off the A/C switch and disconnect the wires from the blower motor.

   c. Conduct a resistance test (step A) or current test (step B).

      A. Resistance test (Figure 4-8)

         i. Select the x1 ohm scale on the DVOM.

         ii. Connect the DVOM to the blower motor terminals.

            NOTE: DO NOT connect the DVOM to the disconnected wires.

         iii. Note the motor resistance. A high resistance indicates a defective motor and a low resistance indicates a good motor.

      B. Current test (Figure 4-9)

         i. Connect a jumper wire from the motor positive (+) terminal to the battery positive (+) terminal.

         ii. Connect a 20-amp. fused jumper wire from the motor negative (–) terminal to an ammeter having a minimum current carrying capacity of 20 amps.

         iii. Connect a jumper wire from the other side of the ammeter to chassis ground.

A defective ground wire accounts for a high percentage of electrical failures.

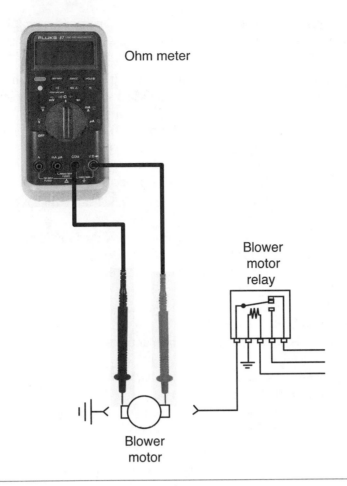

Figure 4-8 Blower motor resistance test.

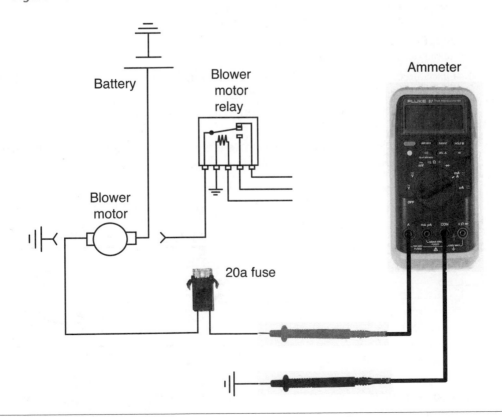

Figure 4-9 Blower motor current test.

NOTE: DO NOT connect this wire to the battery negative (–) terminal.

  iv. If the motor does not run or if the fuse blows, it is defective and must be replaced.

  v. If the motor runs, observe the current. It should be within 10 percent (±) of specifications. If not within specifications, replace the motor.

Testing the Blower Motor Speed Resistor (Figure 4-10). Though blower motor speed resistor resistance values differ between vehicle makes, they are typically as suggested in this procedure. Before conducting this test, disconnect all wires from the resistor.

NOTE: DO NOT connect the DVOM to the disconnected wires.

  1. Select the x1 ohm scale of the DVOM. Connect one lead of the DVOM to terminal 2 of the blower motor resistor.

    a. Touch the other DVOM test lead to terminal 4 of the blower motor resistor. The resistance should be 1.8–2.1Ω.

    b. Touch the other DVOM test lead to terminal 1 of the blower motor resistor. The resistance should be 0.88–1.0 Ω.

    c. Touch the other DVOM test lead to terminal 3 of the blower motor resistor. The resistance should be 0.31–0.35Ω.

  2. If the resistance is not as specified by the manufacturer, the blower motor resistor should be replaced.

Testing the Blower Motor Switch (Figure 4-11). Methods of testing the blower motor switch, depending on vehicle make and model, are outlined in the manufacturer's service manual. Procedures given here assume reasonable access to the control.

Test 1:

  1. Select the x1 ohm scale of the DVOM. Connect one test lead to terminal 5 of the blower motor switch.

    a. Connect the other test lead to terminal 3 of the blower motor switch and, while observing the meter, turn the switch to LO.
    NOTE: The meter should have gone from infinity (∞) to a very low resistance, such as 0Ω.

    b. Disconnect the test lead and connect it to terminal 6 of the blower motor switch and, while observing the meter, turn the switch to ML.
    NOTE: The meter should have gone from infinity (∞) to a very low resistance, such as 0Ω.

    c. Disconnect the test lead and connect it to terminal 2 of the blower motor switch and, while observing the meter, turn the switch to MH.

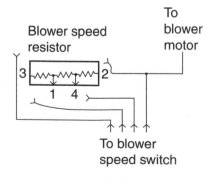

Figure 4-10 Test points for blower motor speed resistor.

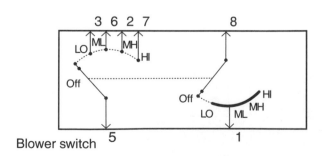

Figure 4-11 Test points for blower motor switch.

NOTE: The meter should have gone from infinity (∞) to a very low resistance, such as 0Ω.

 d. Disconnect the test lead and connect it to terminal 7 of the blower motor switch and, while observing the meter, turn the switch to HI.

  NOTE: The meter should have gone from infinity (∞) to a very low resistance, such as 0Ω.

Test 2:

 1. Connect one test lead to terminal 1 of the blower motor switch.

  a. Connect the other test lead to terminal 8 of the blower motor switch.

  b. While observing the meter, turn the switch to LO.

  NOTE: The meter should go from infinity (∞) to a very low resistance, such as 0Ω.

  c. While observing the meter, turn the switch to ML, MH, then to H.

  NOTE: The meter should remain on a very low resistance, such as 0Ω.

If the blower motor speed control switch failed either of the above tests, it must be replaced.

Testing the A/C Switch (Figure 4-12). Methods of testing the A/C switch, depending on vehicle make and model, are outlined in the manufacturer's service manual. Procedures given here are typical and assume reasonable access to the switch.

 1. Gain access to the A/C switch and disconnect all wires.

 2. Select the x1 ohm scale of the DVOM.

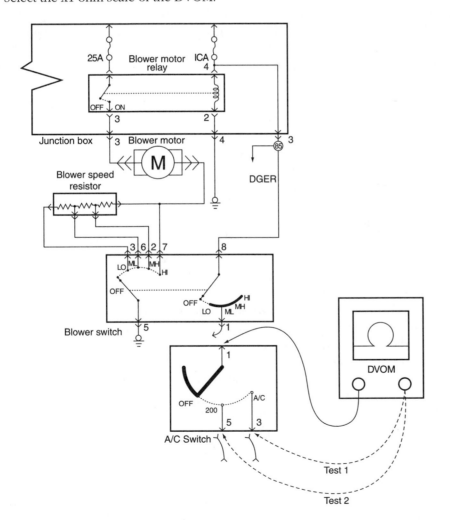

Be sure that the test lamp is not defective; that it will light when it is expected to do so.

Figure 4-12 Testing the A/C Switch.

3. Connect one test lead of the DVOM to terminal 1 of the A/C switch.
   a. Connect the other test lead to terminal 5 of the A/C switch.
   b. While observing the meter, turn the switch to ECO.
      NOTE: The meter should go from infinity (∞) to a very low resistance, such as 0Ω.
4. Disconnect the test lead from terminal 5 and connect it to terminal 3. While observing the meter, turn the switch to A/C.
   NOTE: The meter should go from infinity (∞) to a very low resistance, such as 0Ω.
5. If the A/C switch failed either of the above tests, it must be replaced.

The Compressor Clutch Does Not Work

If the compressor clutch is the only component that does not work, or if it works intermittently, the problem may be the clutch. It is more likely, however, that it is in the clutch electrical circuit (Figure 4-13).

Testing the Clutch Circuit. The following procedure should be considered a typical procedure only. For specific procedures, follow the manufacturer's instructions outlined in the appropriate service manual for safely testing an electrical circuit.

1. Touch the test lamp leads to the battery terminals to check the integrity of the fuse and bulb (Figure 4-14).
2. Disconnect the **clutch coil** from the wiring harness.
3. Turn the ignition switch to ON and place the air conditioning system controls in any COOL position.
4. Connect a fused test lamp from ground to the positive (+) terminal of the disconnected wiring harness.
   a. If the test lamp does not light, the fuse, clutch relay, or wiring may be defective. Troubleshooting procedures are similar to those outlined for the blower motor circuit. Repair or replace components as necessary.
      NOTE: The problem could be in the powertrain control module (PCM). Consult the manufacturer's service manual for troubleshooting procedures.
   b. If the test lamp lights, proceed with step 5.
5. Turn the control back and forth from minimum (min) to maximum (**MAX**) cooling several times while observing the test lamp.
   a. If the test lamp flickers or goes out the PCM may be defective. Further testing of the PCM is indicated.
   b. If the test lamp remains on, proceed with step 6.
6. Remove the clutch and clutch coil and bench test the individual parts as follows:
   a. Visually inspect the clutch **rotor** and **armature** assembly.
   b. If the rotor and/or armature is heavily scored, as in Figure 4-15, or shows signs of overheating, replace the assembly.
7. Bench test the clutch coil.
   a. Connect a jumper wire from the clutch coil frame (Figure 4-16A) or ground wire (Figure 4-16B) to the battery ground (–) cable.
   b. Connect a fused test lamp from the battery positive (+) terminal to the clutch coil lead wire.
   c. If the test lamp does not light, the clutch coil is defective (open) and must be replaced.
   d. If the test lamp lights, the clutch coil is not defective.
      NOTE: If the clutch coil is shorted, the test lamp will also light. If suspected of being shorted, hold the resistance test as outlined in Job Sheet 12.

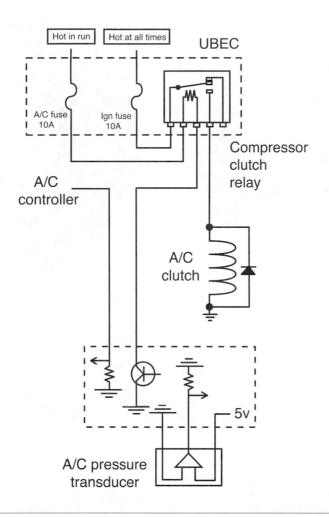

Figure 4-13 A typical clutch electrical circuit.

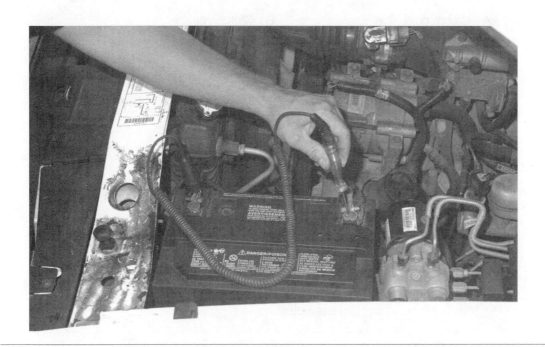

Figure 4-14 Touch fused test lamp leads to vehicle battery to test for integrity.

Figure 4-15 Inspecting clutch rotor and armature surfaces. (Courtesy of BET, Inc.)

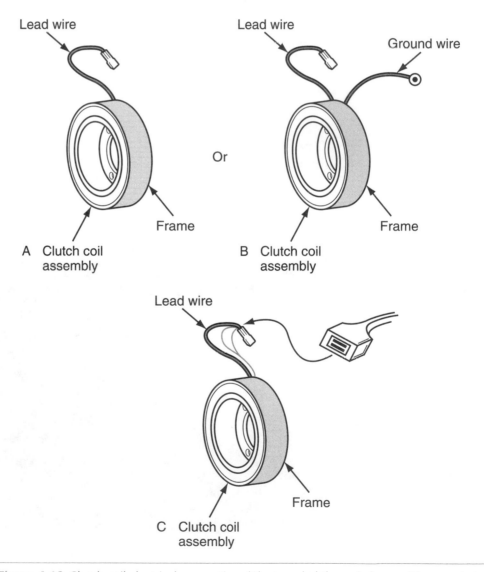

Figure 4-16 Clutch coil electrical connection: (A) grounded through frame, (B) grounded through ground wire, (C) grounded through connector.

e. Reinspect the ground wire, rotor, and armature to determine the problem.

f. Correct as necessary.

Defective Components

Following is an overview of what the technician may expect if a component of the air conditioning system is found to be defective.

Evaporator

A defective evaporator produces an insufficient supply of cool air. This symptom is often the result of a leak in the evaporator core. Other causes include:

❏ Dirt- or debris-plugged core (clean the core)

❏ Cracked or broken case (replace and/or caulk the case)

❏ Leaking seal or O-ring (replace the seal and/or O-ring)

Compressor

A malfunctioning compressor will be indicated by one, or more, of the following symptoms:

❏ Noise, indicating premature failure

❏ Seizure, usually due to oil loss

❏ Leakage, due to defective seals, gaskets, or O-rings

❏ Low suction and low high-side pressure caused by an undercharge of refrigerant or a restriction in the low side of the system

❏ High suction and low discharge pressure usually caused by a defective compressor valve plate and/or gasket assembly.

Some noise is to be expected from most compressors and may be considered normal during regular operation. Irregular rattles and noises, however, are an indication of early failure due to loss of oil and/or broken parts.

When the air conditioning system control calls for cooling and the compressor is inoperative, verify that 12 volts are present at the clutch terminals. To check for compressor seizure:

1. Turn off the engine.

2. De-energize the compressor clutch.

3. Try to rotate the drive plate by hand (Figure 4-17).

NOTE: Compressors that have not been used for a long period of time may "stick." If this is the case, turning it four or five times in both directions should free it up. If it will not rotate, or takes great effort, the compressor may have an internal defect. Also, if there is no resistance to turning, the shaft may be broken internally.

If the compressor clutch is slipping it may be due to an internal compressor problem or it may be due to an incorrect clutch air gap. Before condemning the compressor, determine if the air gap is correct. Low voltage or a defective clutch coil may also cause clutch slipping problems. First check to ensure that there are at least 12 volts available at the clutch coil. Next, disconnect the clutch coil and one side of the diode, if applicable, then check its resistance. Refer to the manufacturer's specifications for specific values; however, the resistance should generally be between 3 and 4 ohms (3–4Ω). Refer to Job Sheet 12 for procedures.

Low discharge pressure can be caused by poor internal compressor sealing such as the valve plate and/or valve plate gasket set. It can also be caused by a restriction in the compressor or in the

Classroom Manual
Chapter 5, page 115

Inspect the compressor for visual indications of a loss of oil.

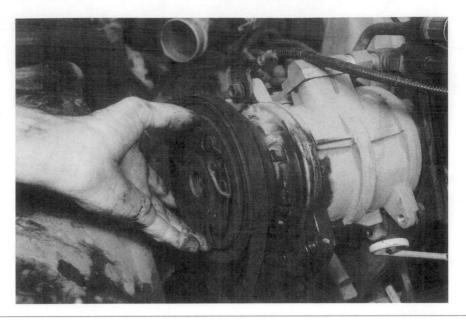

Figure 4-17 Rotate the drive plate (armature) by hand.

low side of the system. Yet another cause of low discharge pressure is an insufficient charge of refrigerant. All possibilities should be explored before diagnosing the problem as a defective compressor.

Condenser

Classroom Manual
Chapter 5, page 115

Look for a "frost patch" indicating a restriction.

There are three possible malfunctions of a condenser:

1. A leak due to rust and corrosion or by being struck by a sharp object or stone.
2. A restriction if a tube has been bent when struck with an outside object, such as a stone, with insufficient force to cause a leak but sufficient to kink or collapse a tube.

☑ **SERVICE TIP:** A restricted condenser may also result in excessive compressor discharge pressure. A partial restriction can cause a temperature change and even frost or ice to form immediately after the restriction. In this case, the restriction is serving as a metering device.

3. Restricted airflow through the condenser caused by dirt, debris, or foreign matter. When the airflow through the condenser is restricted or blocked, high discharge pressures will result.

NOTE: Though not easily detected, the outlet tube of the condenser may be slightly cooler than the inlet tube. Carefully feel the temperature by hand. It should go from a hot inlet at the top to a warm outlet at the bottom. A proper heat exchange should result in an even gradient across the surface of the condenser. Some technicians prefer the use of a laser-sighted, digital readout (DRO), infrared (IR) thermometer (Figure 4-18) for taking temperature measurements. This type thermometer provides an immediate and accurate indication of the surface temperature of any object, in °F or °C, simply by pointing the thermometer at the object.

Orifice Tube

Orifice tube failures are often indicated by low suction and discharge pressures. This condition results in insufficient cool air from the evaporator.

The common cause of orifice tube failure is a restriction. A less common cause is a clogged inlet screen due to contamination, corrosion particles, or refrigerant desiccant loose in the refriger-

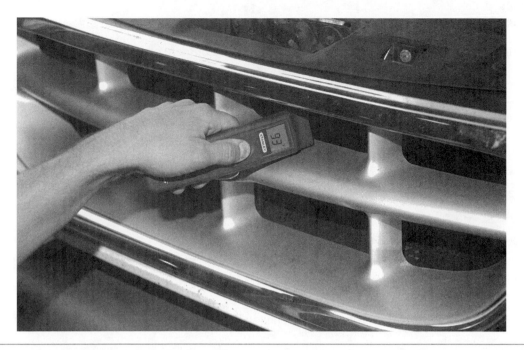

Figure 4-18 Using a laser-sighted, digital thermometer for measuring temperature gradient.

ant system due to a defective accumulator. Regardless of the cause, the recommended repair is to replace the orifice tube (Figure 4-19) and the defective parts that caused the problem.

Thermostatic Expansion Valve

Most thermostatic expansion valve (**TXV**) failures are indicated by the same symptoms as orifice tube failures. Many TXV failures, however, are due to a malfunction of the power element. This failure usually results in a closed valve that will not allow refrigerant to enter the evaporator. Regardless of the cause of failure, except for a clogged inlet screen, replacement of the valve is the recommended repair.

The inlet screen of the TXV can also become plugged due to contamination, corrosion particles, or refrigerant desiccant loose in the system due to a defective receiver-drier (Figure 4-20).

Classroom Manual
Chapter 5, page 117

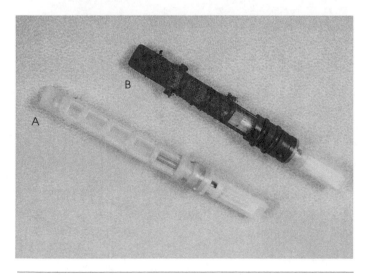

Figure 4-19 An (A) orifice tube with a (B) clogged screen.

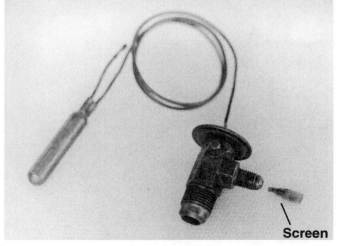

Figure 4-20 Inlet screen of a thermostatic expansion valve (TXV).

Refrigerant Lines

Refrigerant line restrictions are generally indicated by one or more of the following symptoms:

❑ Suction line: A restriction of the suction line causes low suction pressure, low discharge pressure, and little or no cooling. The evaporator is starved of refrigerant.

❑ Discharge line: A restricted discharge line generally causes the pressure relief valve to open to release excess pressure to the atmosphere. Pressure relief valves are generally self-reseating whenever the pressure drops to a predetermined safe level.

❑ Liquid line: A liquid line restriction has the same general symptoms as a suction line restriction—low suction pressure, low discharge pressure, and little or no cooling from the evaporator.

Causes of Failure

Following are some of the common causes of failure found in an automotive air conditioning system:

❑ Leaks; undercharge of refrigerant
❑ High pressure; overcharge of refrigerant in the system, air in the system, excess oil in the system
❑ Poor connections
❑ Restrictions
❑ Contaminants
❑ Moisture
❑ Defective component

Leaks

Service valves and protective caps (Figure 4-21) are among the most common causes of refrigerant leaks. The primary purpose of the cap and O-ring is to serve as a dirt seal. As much as a pound (0.45 kg) of refrigerant per year can escape from the service valve if the cap is missing or the O-ring is defective. Leak testing the service valve is often neglected because the service hoses are generally connected to them (Figure 4-22). Service valves should be leak tested with the caps and the service hoses removed. System integrity should not depend on the sealing power of a protective

Figure 4-21 Service valve with protective cap removed. (Courtesy of BET, Inc.)

Figure 4-22 Service valves with hoses connected are often neglected when leak testing.

cap. The primary purpose of the cap is to keep debris out of the service valve. If found to be leaking, the service valve should be repaired or replaced as applicable. Often, a new Schrader assembly is sufficient to stop most leaks.

An insufficient refrigerant charge, for any reason, will cause oil to become trapped in the evaporator. Oil also leaks out with refrigerant at the point of a leak. Any oil loss due to any reason can result in compressor seizure.

The compressor circulates a small amount of oil through the system with the refrigerant. Oil pumped out of the compressor in small quantities is mixed with the refrigerant in the condenser. This oil enters the evaporator with the refrigerant, and if the evaporator is properly flooded with refrigerant, passes to the compressor through the low-pressure line. Some of the oil passes to the compressor in small droplets. Most of the oil, however, is swept along the walls of the refrigerant lines by the velocity of the refrigerant vapor. This oil is returned to the compressor as a mist. If the evaporator is starved of refrigerant, oil will not return to the compressor in sufficient quantity to keep it properly lubricated. The major cause of premature compressor failure is a lack of lubricant. The tendency of a customer to have refrigerant added to the system "every few months or so" is a sure sign that the compressor is doomed. If the system is leaking refrigerant, it is a good bet that it is also leaking lubricant and compressor failure is sure to follow.

High Pressure

As refrigerant pressure increases in an air conditioning system, its temperature also increases. The resulting high temperature quickly accelerates the failure of a contaminated system. An increase in temperature of only 15°F (8°C) doubles the chemical reaction rate in the system. High temperature starts a chain of harmful reactions even in a clean system. Contamination, resulting from high temperature, may cause seizure of the compressor bearings.

High heat may also cause the refrigerant in the system to decompose or break down. High heat can cause synthetic rubber parts to become brittle and susceptible to cracking and breaking.

> **CAUTION:** High heat may cause components to get hot enough to cause severe burns.

As temperature and pressure in an air conditioning system increase, stress and strain on compressor discharge valves increase. If this condition is not corrected, the discharge reeds in the valve plates may fail.

High pressure and the accompanying high temperature can be caused by air in the air conditioning system. Air can enter the system through careless or incomplete service procedures. Systems that have been opened to the atmosphere during service procedures must be properly evacuated. If the system is not properly evacuated, the results, most surely, will be an air-contaminated system.

A system with air contamination does not operate at full efficiency. Air in the air conditioning system can cause oil to oxidize. Oxidized oil forms gums and varnishes that coat the inside walls of the tubes, reducing the efficiency of the heat transfer process. Still more damaging, air usually carries moisture into the system in the form of humidity.

When an air conditioning system is operated with a low-side pressure below **atmospheric pressure** (14.696 psig at sea level), air will be drawn into the system through the leak. This occurs when a noncycling system is low on refrigerant; the low side often operates below atmospheric pressure, in a vacuum. If a system contaminated in this way is recharged without proper evacuation procedures, high temperature and pressure conditions will result. Air, a noncondensable gas, has a tendency to collect in the condenser during the off cycle.

Air in the System

Air in the system can cause high head pressures. To bleed off excess air (any air is excess), connect a manifold and gauge set as outlined in Chapter 3 of this manual and follow these procedures:

Classroom Manual
Chapter 4,
pages 81–90

The top of the condenser is often the highest point of the system; air, lighter than refrigerant, seeks the highest point.

1. Hang the manifold and gauge set on a hood bracket so that it becomes the highest part of the system.

   NOTE: Air is lighter than refrigerant and should rise to the highest part of the system, the manifold hoses.

2. Turn off the air conditioner and engine for a time sufficient to **stabilize** the system.
3. Momentarily "crack" the low-side hose at the low-side gauge fitting and allow a slight amount of air to escape.
4. Repeat step 3 with the high-side hose.
5. Start the engine and turn on the air conditioner.
6. Note the high-side gauge.

If the pressure has been lowered, the problem was air in the system. If the pressure was lowered, but is not yet low enough, the procedure may be repeated, starting with step 2. If, however, the pressure was not lowered, look elsewhere for the problem.

The method outlined in this procedure is not recommended as a cure. If air is in the system, moisture is probably in the air conditioning system as well. The proper cure for this problem is to first recover the refrigerant, then evacuate and recharge the air conditioning system.

NOTE: If, during any refrigerant service, the evacuation step is omitted or if the recovery tank is not evacuated before recovery, air can be introduced into the air conditioning system.

Connections

Clamps used with hose fittings must be installed properly to ensure system integrity (Figure 4-23). To reduce the likelihood of leaks or early failure, inspect the fitting flanges for nicks, scores, or burrs. Hoses should be installed over the flange, with the end of the hose at the flange stop. The hose clamps must be tightened to the torque specified in the appropriate manufacturer's shop manual.

If there is a compression fitting, avoid overtightening. Overtightening can cause O-ring damage, resulting in a leak or early failure. Before assembly, inspect the fitting for burrs, which may cut the O-ring. It is important that the proper O-ring be used for the type of refrigerant and the type of fitting. It is also important to follow the manufacturer's recommendations for selecting O-rings.

When a connection is made with a compression fitting, place the gaskets or O-ring over the tube before inserting it into the connection (Figure 4-24). Use a torque and backing wrench to ensure a proper connection. Again, follow the manufacturer's specifications for proper torque requirements.

Restrictions

Most restrictions are caused by dirt, foreign matter, or corrosion. Corrosion is generally due to excess moisture in the system. Contaminants can lodge in **filters** and screens and can block the

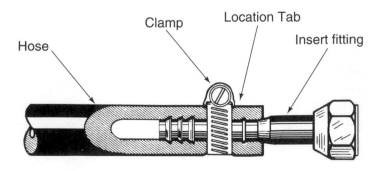

Figure 4-23 Detail of a typical hose fittings.

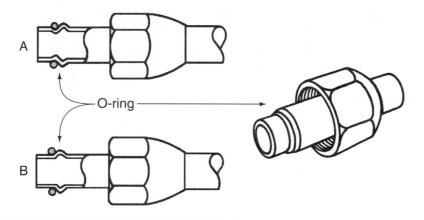

Figure 4-24 Place the O-ring over the fitting before inserting it into the connection.

flow of refrigerant through the system. Filters are found in the receiver-drier and suction line accumulator usually as a means to hold the desiccant in place. Screens are generally found:

- ❏ At the metering device inlet
- ❏ In the receiver-drier or accumulator
- ❏ At the compressor inlet

A restriction in the system can cause a "starved" evaporator. This can result in reduced cooling, poor oil return, and eventually, if not corrected, compressor seizure.

Supplemental aftermarket liquid line filters are available for installation air conditioning systems that have been contaminated (Figure 4-25). The filter should be installed in the system:

1. After repeated metering device plugging
2. When a seized compressor has been replaced

The liquid line filter contains a screen and a filter pad. It does not contain a desiccant. The fine-mesh screen catches larger particles and holds the filter in place. The filter catches smaller particles and filters the refrigerant oil.

The filter is installed in the liquid line between the condenser outlet and the evaporator inlet. Filters are available with or without an expansion tube orifice. The filter without an orifice is generally preferred. This type can be installed anywhere in the liquid line, preferably close to the

A contaminated system may have both a strainer and a drier.

Figure 4-25 A supplemental liquid line filter/drier.

metering device. A filter with an orifice is required when the installation is to be made in the low-pressure side of the system beyond the original expansion tube location. This installation, which is usually found on General Motors vehicles, requires that the original expansion tube be removed from the system.

Contamination

Contamination by foreign matter has many sources, including:

- ❏ Failed desiccant
- ❏ Preservative oils
- ❏ Lint
- ❏ Soldering or brazing fluxes
- ❏ Loose corrosion flakes

Any of these materials in the air conditioning system can cause:

- ❏ Compressor bearings to seize
- ❏ Metering device failure
- ❏ Corrosion of metal parts
- ❏ Decomposition of refrigerant
- ❏ Breakdown of the oil

Corrosion and the byproducts of corrosion can clog metering device screens, ruin compressor bearings, and accelerate the failure of compressor discharge valves. Moisture is the primary cause of corrosion in the air conditioning system.

In fact, the greatest enemy of an air conditioning system is moisture. When combined with the metals found in the system, moisture causes the formation of iron hydroxide and aluminum hydroxide. When combined with refrigerant, moisture can form three acids:

Classroom Manual
Chapter 7,
page 151

1. Carbonic (H_2CO_3)
2. Hydrochloric (HCl)
3. Hydrofluoric (HF)

CAUTION: Avoid contact with hydrochloric and hydrofluoric acids; both are very poisonous.

Carbonic acid is a weak solution generally found in solutions of carbon dioxide in water.

Moisture also causes metering devices to freeze up. As the operating temperature of the evaporator is reduced to the freezing point, moisture collects in the metering device orifice and freezes. This, in turn, restricts the flow of refrigerant into the evaporator. The result is an erratic or no cool condition of the evaporator.

High temperature and foreign matter are responsible for many refrigerant system difficulties. In most cases, it is the presence of moisture that accelerates these conditions. The acids that result from the combination of high pressure, moisture, and refrigerant cause damaging corrosion.

Functional Testing

A functional test may be performed to determine the operating conditions of the air conditioning control head (Figure 4-26) as well as the air distribution system. The functional test consists of checking the operation of the blower, **heater**, and air conditioning control assembly mode lever. This test also includes comparing mode lever and switch positions in relation to air delivery.

Insufficient Cooling: Cycling Clutch Orifice Tube (CCOT)

Classroom Manual
Chapter 7, page 164

The quick test procedure in CCOT systems helps to determine if the air conditioning system is properly charged with refrigerant. This test can only be performed if the ambient temperature is above 70°F (21°C).

Figure 4-26 A typical air conditioning system control head.

The quick test can simplify system diagnosis by verifying the problem of insufficient refrigerant. This quick test also eliminates a low refrigerant charge as a source of the problem. The quick test is performed as follows:

1. Start the engine. Allow it to warm at normal **idle speed**.
2. Open the hood and doors.
3. Select the NORM mode.
4. Move the lever to the full COLD position.
5. Select HI blower speed.
6. Feel the temperature of the evaporator inlet after the orifice tube (Figure 4-27).
7. Feel the temperature of the accumulator surface when the compressor is engaged.

 SERVICE TIP: If the compressor is cycling, wait until the clutch is engaged.

   a. Both surfaces (steps 6 and 7) should be at the same temperature. If they are not the same, check for other problems.
   b. If the inlet of the evaporator is cooler than the suction line accumulator surface or if the inlet has frost accumulation, a low refrigerant charge is indicated.
8. If a low refrigerant charge is indicated, add 4 oz. (120 mL) of refrigerant and repeat steps 6 and 7.
9. Add 4 oz. (120 mL) at a time until both surfaces feel the same temperature.

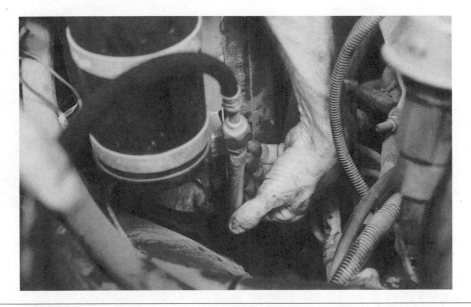

Figure 4-27 Feel the evaporator inlet after the orifice tube.

✓ **SERVICE TIP:** Allow the system sufficient time, five minutes or so, to stabilize between each addition of refrigerant.

NOTE: It is normal for an accumulator to sweat if the system is properly charged with refrigerant. It means that the evaporated refrigerant is absorbing heat from the ambient air surrounding the accumulator and suction line. This heat, added to the heat adsorbed in the evaporator, is called superheat and does not cause a change in pressure. Overall sweating, then, does not indicate that there is a restriction in the accumulator.

▲ **WARNING:** Adding refrigerant in this manner should only be done if the system is known to be sound, free of leaks. Adding refrigerant to a leaking system does more harm than good.

Diagnosing Orifice Tube Systems

If the indication is "no cooling" or "insufficient cooling" from the air conditioning system, inspect the air conditioning and cooling systems for defects as follows:

Preliminary Checks (All Models)

Inspect the system components before connecting the manifold and gauge set. Procedures for diagnosing the General Motors Cycling Clutch Orifice Tube (CCOT) and Ford Fixed Orifice Tube (FFOT) systems follow. Diagnosing thermostatic expansion valve systems is covered later in this chapter.

General Motors CCOT Diagnosis

1. Check to verify that the temperature door strikes both stops when the lever is moved rapidly from hot to cold and cold to hot.
2. Check for a loose, missing, or damaged compressor drive belt.
3. Check for loose or disconnected wiring or connectors.

NOTE: Be sure fuses or circuit breakers are not defective.

4. Check to see if the cooling fan (Figure 4-28) is running continuously in all air conditioning modes.

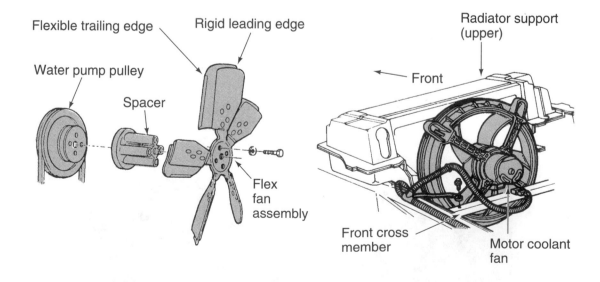

Figure 4-28 A typical engine cooling fan; (A) engine driven; (B) electrically driven. (From Erjavec, *Automotive Technology,* © 1992 by Delmar Publishers Inc.)

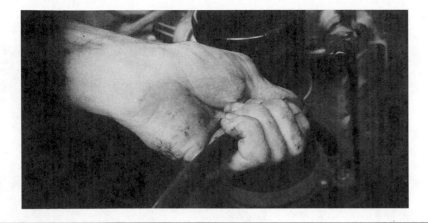

Figure 4-29 Feel the liquid line before the orifice tube.

Often, the technician can hear a vacuum leak—a hissing sound.

NOTE: Make repairs as necessary, and recheck cooling.

5. If items checked in steps 1 through 4 are satisfactory, and the system still does not cool:
   a. Set the temperature lever to full (MAX) cold.
   b. Move the selector lever to NORMal A/C.
   c. Set the blower switch to HI.
   d. Open the doors and hood.
   e. Warm the engine at 1,500 rpm.
6. Perform a visual check for compressor clutch operation.
   a. If the clutch does not engage, proceed with the "Compressor Clutch Test."
   b. If the clutch engages or cycles, feel the liquid line before the orifice tube (Figure 4-29).
   c. If the tube is warm, proceed with step 7.
   d. If the tube is cold, check the high-side tubing for a restriction.

NOTE: A restriction will be marked with a drop in temperature or frost spot. If the tubing is restricted, repair, evacuate, and recharge the system.

7. If the system is equipped with the Harrison V-5 compressor, proceed with "Diagnosing Harrison V-5 Systems." For all other models, feel the evaporator inlet and outlet tubes.
   a. If the inlet is colder than the outlet, the system may be undercharged. Check for and repair any leaks; evacuate, charge, and retest the system. If no leaks are found, proceed with "System Charge Test."
   b. If the outlet tube is colder than the inlet tube or both tubes are the same temperature, proceed with "Pressure Switch Test."

Classroom Manual
Chapter 6, page 134

Clutch Test

If the clutch is not operational, check the condition of the wiring and the adjustments of the throttle or vacuum cut-out switches. Connect a jumper wire with an in-line fuse (Figure 4-30) between the positive (+) battery terminal and the clutch coil lead. If the clutch does not energize, proceed with step 1. If the clutch energizes, proceed with step 2.

1. Connect a jumper wire from the clutch coil to an engine ground (-). If the clutch does not energize, remove and repair it as necessary.
2. If the clutch operates but cooling is not sufficient, allow the system to operate for a few minutes and check the low-side pressure at the accumulator access fitting.
   a. If the pressure is above 50 psig (345 kPa), proceed with step 3.
   b. If the pressure is below 50 psig (345 kPa), proceed with step 4.
3. Connect a jumper wire across the pressure switch connector (Figure 4-31).
   a. If the compressor operates, the switch is defective. Replace the switch and retest the system.

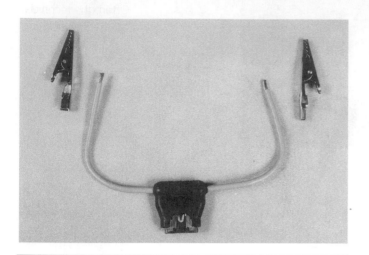

Figure 4-30 A typical jumper wire. The in-line fuse guards against an accidental short circuit.

Figure 4-31 Jump out the pressure switch.

    b. If the compressor does not operate, check for an open or short circuit between the switch and clutch.

4. Connect a high-pressure gauge and check the high-side pressure.

    a. If the high-side pressure is above 50 psig (345 kPa), discharge the system and check for a plugged orifice tube or a restriction in the high side.

    b. If the high-side pressure is below 50 psig (345 kPa), the refrigerant charge is lost. Leak test, repair, evacuate, recharge, and retest the system.

Clutch Diode. A strong electromagnetic field is generated when electrical power is applied to the clutch. When this power is disconnected, the magnetic field collapses and creates high-voltage spikes. These spikes, harmful to the delicate electronic circuits, must be eliminated.

A diode, across the clutch coil, provides a path to ground for the electrical spikes as power is interrupted. This diode is usually taped inside the wiring harness across the 12-volt and ground leads (Figure 4-32).

Typical procedures for testing a diode are given in Photo Sequence 3. The diode may also be tested using a digital volt ohmmeter (DVOM) following the procedures outlined in the instructions included with the meter.

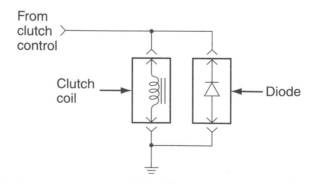

Figure 4-32 A diode in the clutch circuit prevents electrical spikes.

Photo Sequence 3
Bench Testing the Compressor Clutch Coil and Diode

The following procedure may be followed by technicians that use an analog or digital ohmmeter. Many digital volt ohmmeters (DVOM), however, have a "short cut" method for testing a diode. Follow instructions included with the DVOM. If either the clutch coil or the diode fails the test it is defective and must be replaced.

P3-1 Before bench testing an inoperative clutch coil check for voltage present at the connector with the coil disconnected and controls set for COOL. NOTE: IF 12-volts are available, proceed with the bench test, step PS4-2. If 12 volts are not present, the problem is probably neither the clutch coil nor the diode.

P3-2 Carefully cut and remove the tape to expose the diode leads to the clutch coil.

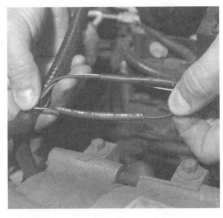

P3-3 Isolate the diode by disconnecting one lead form the clutch coil.

P3-4 Connect an ohmmeter to the clutch coil leads or from the coil lead to ground, as applicable. Note the resistance. NOTE: 0Ω indicates that the coil is shorted. Infinite () Ω indicates that the coil is open.

P3-5 Connect an ohmmeter to the diode. Note the resistance.

P3-6 Reverse the ohmmeter leads to the diode and, again, note the resistance. NOTE: There should be low ohms (due to internal resistance of the diode) when the ohmmeter is connected in one direction and infinite ohms resistance when connected in the other direction.

System Charge Test

This test may be performed if cooling is not adequate. While performing the system charge test, watch the high-side gauge for any indication of overcharging such as excessively high pressure. Discontinue the test if the system pressure exceeds that expected for any given ambient temperature condition.

1. Add a "**pound**" **of refrigerant**. Check the clutch cycle rate.
   a. If the clutch cycles more than 8 times per minute (less than every 7 seconds), discharge the system and check for a plugged orifice tube or some other restriction.
   b. If the clutch cycles less than 8 times per minute (more than every 8 seconds), proceed with step 2.
2. Feel the accumulator inlet and outlet tubes (Figure 4-33).
   a. If the inlet tube is warmer or the same temperature as the outlet tube, add 3–4 oz. (89–118 mL) more refrigerant.
   b. If the inlet tube is colder than the outlet tube, add 3–4 oz. (89–118 mL) more refrigerant.
3. Feel the inlet and outlet tubes of the accumulator.
   a. If the inlet tube is, again, warmer or if it is the same temperature as the outlet tube, add 3–4 oz. (89–118 mL) more refrigerant.
   b. If the inlet tube is still colder than the outlet tube, add 3–4 oz. (89–118 mL) more refrigerant.
4. Again, feel the accumulator inlet and outlet tubes.
   a. If the inlet tube is still warmer or at the same temperature as the outlet tube, add 3–4 oz. (89–118 mL) more refrigerant.
   b. If the inlet tube is still colder than the outlet tube, recover the refrigerant and check for a clogged or restricted orifice tube.

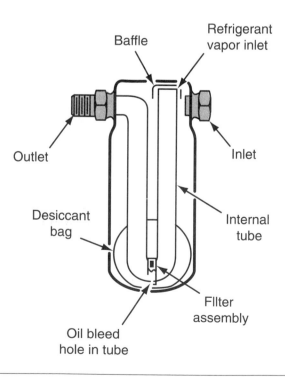

Figure 4-33 A typical accumulator showing inlet and outlet tubes.

WARNING: While performing this test, watch for any indication of overcharging such as excessively high discharge pressure. If high pressures occur, discontinue the test.

 CAUTION: When adding refrigerant, *never* exceed the recommended capacity of the system.

Pressure Switch Test

This test may be performed whenever the inlet and outlet tube temperatures are acceptable, but cooling is not sufficient.

1. Using a manifold and gauge set (Figure 4-34), check to determine if the clutch cycles on between 41–51 psig (283–352 kPa), and cycles off between 20–28 psig (138–193 kPa).
   a. If cycling is correct, proceed with "Performance Test."
   b. If the clutch cycles at pressures too low or too high, replace the pressure cycling switch.
2. If the clutch runs continuously, disconnect the evaporator blower motor wire (Figure 4-35). The clutch should cycle off between 20–28 psig (138–193 kPa).
   a. If the pressure drops to below 20 psig (138 kPa), replace the pressure cycling switch.
   b. If the clutch cycles off, proceed to "Performance Test."

Performance Test

To conduct the performance test:

1. Set the temperature lever to full cold, selector to MAX A/C, and place the blower motor switch in the HI position.
2. Start the engine. Allow it to run at 2,000 rpm.
3. Close the doors and windows.

Classroom Manual
Chapter 7, page 164

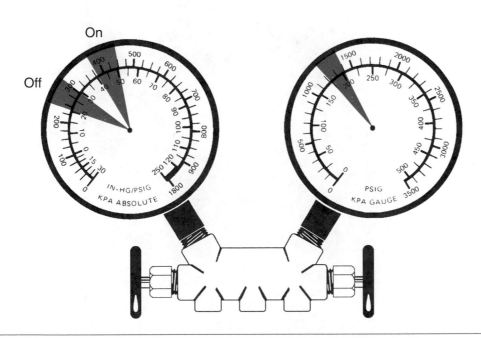

Figure 4-34 The clutch should cycle (A) on and (B) off within a specified range.

Figure 4-35 Disconnect the blower motor.

4. Place an auxiliary fan in front of the condenser.
5. Allow the system to stabilize; 5–10 minutes.
6. Place a thermometer in the register nearest the evaporator and check the temperature (Figure 4-36). The temperature should be 35–45°F (1.7–7.2°C) with an ambient temperature of 80°F (27°C).
7. If the outlet temperature is high, check the compressor **cycling time**.
   a. If the clutch is energized continuously, discharge the system and check for a missing orifice tube, plugged inlet screen, or other restriction in the suction line.

An overcharge of refrigerant is generally accompanied by excessive high-side pressure.

Classroom Manual
Chapter 7, page 164

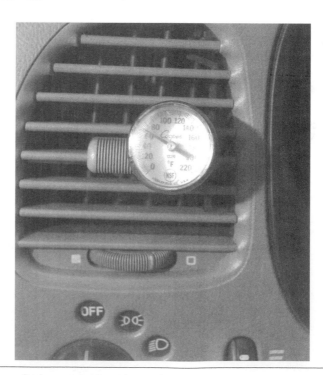

Figure 4-36 Measure discharge air at center register.

b. If the clutch cycles on and off or remains off for an extended period, discharge the system and check for a plugged orifice tube. Replace the tube, and evacuate, charge, and retest the system.

Diagnosing Harrison V-5 Compressor Systems

1. Connect the gauge set to the system.
2. Use a jumper wire to bypass the cooling fan switch.
3. Start and run the engine at about 1,000 rpm.
4. Set the selector lever to NORM A/C and the blower motor switch to the HI position.
5. Measure the discharge air temperature at the center register as shown in Figure 4-36.
6. If the temperature is less than 60°F (16°C), proceed with step 8; if it is more than 60°F (16°C), check the pressure at the accumulator.
   a. If this pressure is 35–50 psig (241–345 kPa), proceed with step 8 (Figure 4-37).
   b. If the pressure is greater than 50 psig (345 kPa), proceed to step 11.
7. If the accumulator pressure in step 6 was less than 35 psig (241 kPa), add a "pound" can of refrigerant.
   a. If the pressure is now more than 35 psig (241 kPa), leak-test the system.
   b. If the pressure is still low, discharge the system and examine the orifice tube for a restriction. If it is plugged or otherwise defective, replace the orifice tube, and evacuate, charge, and retest the system.
8. Set the selector lever to the DEF mode. Disconnect the engine cooling fan and allow the compressor to cycle on the high pressure cut-out switch.
   a. If a compressor knocking noise is noted on clutch engagement, the system oil charge is high. If this is the case, discharge the system, flush all components, and charge with the appropriate amount of refrigerant and oil.

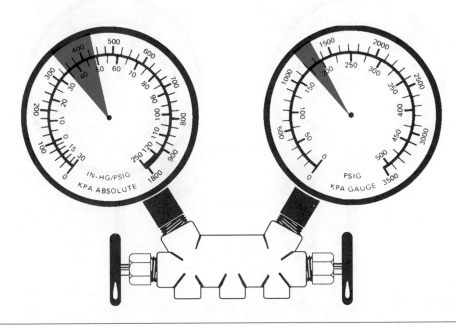

Figure 4-37 Pressure between 35–50 psig (241–345 kPa).

b. If no compressor noise was **observed** in step 8a, set the selector lever to MAX cooling. Adjust the blower control to its LO setting.

9. Idle the engine for 5 minutes at 1,000 rpm. If the pressure at the accumulator is now 29–35 psig (200–241 kPa), the system is operating properly (Figure 4-38).

10. If the pressure at the accumulator is below 28 psig (193 kPa), discharge the system. Replace the compressor control valve, and evacuate, charge, and retest the system.

CAUTION: Replace any lubricant that may have been removed from the air conditioning system during the refrigerant recovery process plus any that may have been lost due to a leak. One of the major causes of compressor failure is lack of lubricant.

a. If the pressure is above 36 psig (248 kPa), discharge the system. Replace the compressor control valve, and evacuate, charge, and retest the system.

b. If step 10a did not prove effective—the pressure is still above 36 psig (248 kPa)—replace the compressor.

11. In step 6, if the pressure at the accumulator (Figure 4-39) was above 50 psig (345 kPa) and below 160 psig (1,103 kPa), discharge the system and check for a missing orifice tube.

a. If the orifice tube is not missing, replace the compressor control valve. Evacuate, charge, and retest the system.

b. If the condition is not corrected with a new control valve, replace the compressor.

12. If the pressure in step 6 was higher than 160 psig (1,103 kPa), the system is overcharged. Discharge, evacuate, charge, and retest the system.

Diagnosing Ford's FOT System

Proper diagnosis of Ford's Fixed Orifice Tube (FFOT) system may be accomplished by observing the clutch cycle rate, time on plus time off, and system pressures. Low and high pressure will vary somewhat between the low and high points as the clutch cycles on and off.

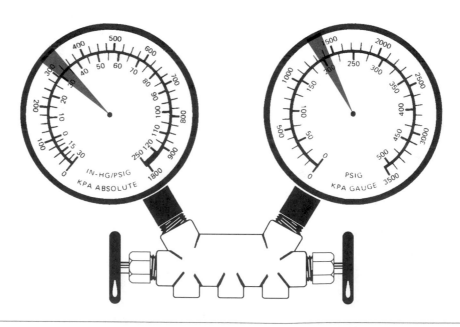

Figure 4-38 Pressure at the accumulator of 29–35 psig (200–241 kPa).

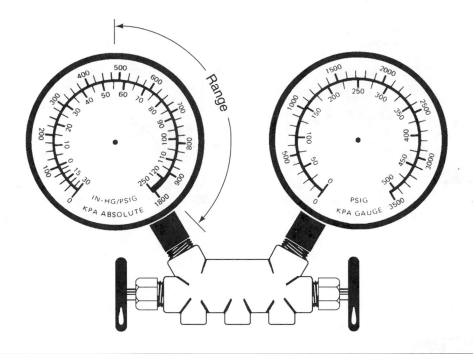

Figure 4–39 Pressure range of 50–160 psig (345–1,103 kPa).

Prepared charts (Figure 4-40) are used to compare the system pressures and clutch cycle rate to determine if the system pressures and clutch cycle rate are as specified.

Most Ford lines, with an electric cooling fan, use an electronic module to control the fan and clutch circuits. For electrical troubleshooting and repairs, refer to the manufacturer's appropriate wiring diagrams for specific model year vehicles. The following diagnosis assumes that the cooling fan and clutch electrical circuits are functioning properly.

FFOT is an acronym for Ford Fixed Orifice Tube.

The accuracy of clutch cycle timing for the FFOT system depends on the following conditions:

1. The in-car temperatures must be stabilized at 70–80°F (21–27°C).
2. MAX air conditioning with **RECIR** (recirculating) air must be selected.
3. MAX blower speed must be selected.
4. The engine should be running at 1,500 rpm for a minimum of 10 minutes.

The lowest pressure noted on the low-side gauge, as observed as the clutch is disengaged, is the low-pressure setting of the clutch cycling pressure switch. Conversely, the pressure recorded when the clutch first engages is the high-pressure setting for the clutch cycling pressure switch.

Compressor clutch cycling will not normally occur if the ambient temperature is above 100°F (38°C), and in some instances above 90°F (21°C), depending on conditions such as relative humidity and engine speed. Also, clutch cycling does not usually occur when the engine is operating at curb idle speed.

If the system contains no refrigerant or is extremely low on refrigerant, the clutch may not engage. If, on the other hand, a clutch cycles frequently, it is an indication that the system is undercharged or the orifice tube is restricted.

FFOT System Diagnosis

If poor or insufficient cooling is noted and the system does not have an electrodrive engine cooling fan, proceed to step 2. If it is equipped with an electrodrive engine cooling fan, as shown in Figure 4-28, check to determine if the clutch energizes. If the clutch does not energize, check the electrical clutch circuit.

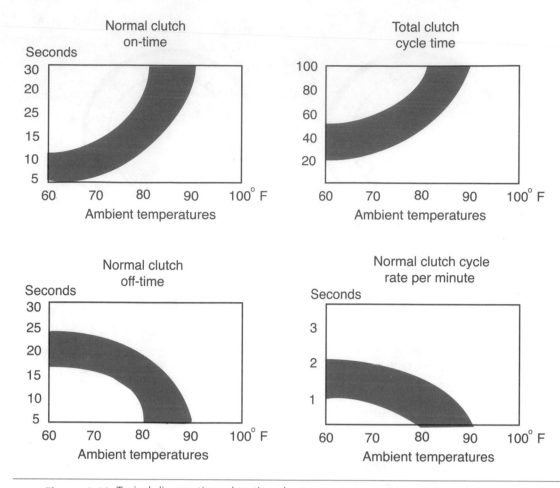

Figure 4-40 Typical diagnostic and testing charts.

1. If the clutch energizes, determine if the cooling fan operates when the clutch is engaged.
   a. If it does not, check the cooling fan electrical circuit.
   b. If the fan operates properly, proceed to step 2.
2. Check for a loose, missing, or damaged compressor drive belt.
3. Inspect for loose, disconnected, or damaged clutch, clutch cycling wires, or connectors.
4. Check the resistor connections, if equipped.
5. Check the connections of all vacuum hoses (Figure 4-41).
6. Check for blown fuses and proper blower motor operation.
7. Be sure all vacuum motors and temperature doors provide full travel.
8. Inspect all control electrical and vacuum connections.

NOTE: Repair all items as necessary and recheck the system.

9. If cooling is still inadequate, refer to the pressure-cycle time charts.
   a. Hook up a manifold gauge set.
   b. Set the selector lever to MAX A/C and the blower switch to HI.
   c. Set the temperature lever to full cold.
   d. Close all doors and windows.
10. Insert a thermometer in the center grill outlet.
   a. Allow the engine to run for 10–15 minutes at approximately 1,500 rpm with the compressor clutch engaged.
   b. Check and note the discharge temperature.
   c. Check and record the outside ambient temperature.

Ambient: surrounding air. Ambient temperature refers to surrounding air temperature.

Figure 4-41 A vacuum leak can be a source of trouble.

11. With a watch, time the compressor on and off time. Compare the findings with the appropriate chart.
   a. If the clutch does not cycle rapidly, proceed with step 15.
   b. If the clutch cycles rapidly, bypass the clutch cycling switch with a jumper wire. The compressor should now operate continuously.
12. Feel the evaporator inlet and outlet tubes.
   a. If the inlet tube is warm or if the outlet tube is colder after the orifice tube, leak-test the system. Repair leaks, and evacuate, charge, and retest the system.
   b. If no leaks are found, add approximately 4 oz. (113 g) of refrigerant.
13. Again, feel the inlet and outlet tubes.
   a. If the inlet tube is colder, add 4 oz. (113 g) of refrigerant.
   b. Once more, check the inlet and outlet tubes. Continue to add refrigerant in 4-oz. (113-g) increments until the tubes feel equal in temperature and are about 28–40°F. (–2–4°C).
   c. If, in step 12, the inlet tube was equal to outlet tube (approximately 28–40°F [(–2–4°C)], add 8–12 oz. (226–339 g) of refrigerant (Figure 4-42).
   d. Check the outlet discharge temperature for a minimum of 50°F (10°C).
14. If, in step 11, the outlet tube temperature was equal to the inlet tube temperature, 28–40°F (–2–4°C), replace the clutch cycling switch and retest the system.
15. Feel the evaporator inlet and outlet tubes.
   a. If the inlet tube is warm or if the outlet tube is colder after the orifice tube, perform steps 12 and 13 to restore the system.
   b. If the inlet and outlet tubes are at the same temperature, 28–40°F (–2 –4°C), or if the outlet tube after the orifice tube, is slightly colder than the inlet tube, check for normal system pressure requirements.
16. If the compressor cycles within limits, the system is functioning properly.
   a. If the compressor cycles on high or low pressures, on above 52 psig (359 kPa) and off below 21 psig (145 kPa), replace the clutch cycling switch and retest the system.
   b. If the compressor runs continuously, disconnect the blower motor wire. Check for the compressor cycling OFF at 21–26 psig (145–179 kPa) suction pressure (Figure 4-43).
   c. If so, reconnect the blower motor wire. The system is functioning properly.
   d. If the suction pressure fell below 21 psig (145 kPa) when the blower motor wire was disconnected, replace the clutch cycling switch and retest the system.

Ambient air contains moisture in the form of "humidity."

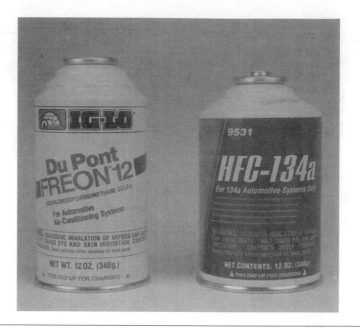

Figure 4-42 Typical "pound" cans actually contain 12 oz. (339 g) of refrigerant

Poor Compressor Performance

Refer to the FFOT chart (Figure 4-40). Some of the other problems relating to poor compressor performance include:

❑ Clutch slippage
❑ Loose drive belt
❑ Clutch coil open
❑ Dirty control switch contacts
❑ High resistance in clutch wiring
❑ Blown fuse or open circuit breaker

Classroom Manual
Chapter 6, page 109

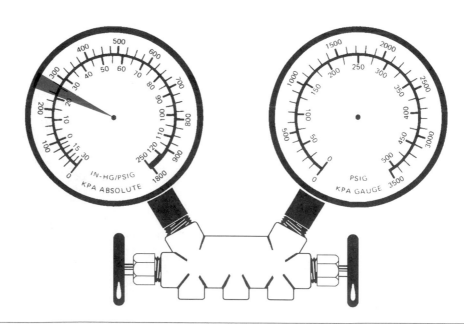

Figure 4-43 Compressor cycles off at 21–26 psig (145–179 kPa).

Additional problems associated with compressors include:

- ❏ Cycling switch
- ❏ Clutch seized
- ❏ Accumulator refrigerant oil bleed hole plugged
- ❏ Refrigerant leaks

Diagnosing Thermostatic Expansion Valve Systems

The following procedure may be followed to diagnose the thermostatic expansion valve (TXV) system performance:

1. Connect the manifold gauge set to the system.
2. Start the engine and adjust the engine speed to fast idle, 1,000–1,200 rpm.
3. Place a large fan in front of the condenser to substitute for normal ram airflow (Figure 4-44).
4. Operate the air conditioner to "stabilize" the system:
   a. Adjust all controls for MAX cooling.
   b. Operate the system for 10–15 minutes.
5. Observe and note the gauge readings.

Classroom Manual
Chapter 5,
pages 117–119

Interpreting gauge readings will become "second nature" to the technician.

☑ **SERVICE TIP:** Gauge readings vary and their application is often a matter of "professional opinion." Gauge readings should, then, only be used as a guide in diagnostic procedures. Proceed with step 5 for abnormally low low-side gauge readings. Go to step 9 for abnormally high low-side gauge readings.

6. If the low-side gauge reading is abnormally low, place a warm rag (125°F [52°C]) around the valve body (Figure 4-45).
7. Observe the low-side gauge.
   a. If the pressure rises, there is moisture in the system. Correct as required.
   b. If the pressure does not rise, proceed with step 7.
8. Remove the remote bulb and place it in a warm (125°F [52°C]) rag (Figure 4-46).

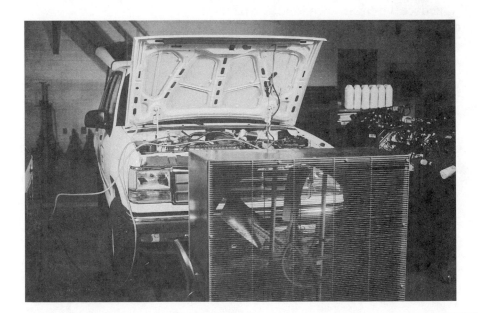

Figure 4-44 Place a fan in front of the car to provide ram air.

Figure 4-45 Place a warm rag around the TXV body.

9. Observe the low-side gauge.
   a. If the pressure rises, the remote bulb was probably improperly placed. Reposition the bulb, tighten it, and retest the system.
   b. If the pressure does not rise, proceed with step 9.
10. If the low-side gauge reading, step 4, was abnormally high, remove the remote bulb from the evaporator outlet tube and place it in an ice water (H_2O) bath.
11. If the low-side pressure drops to normal or near normal, the problem may be:
   a. Lack of insulation of the remote bulb. Reinstall, insulate, and retest.
   b. Improperly placed remote bulb. Reposition the remote bulb, insulate, and retest.

Rock salt (sodium chloride, NaCl) in ice water (H_2O) will produce a freezing (32°F [0°C]) liquid temperature for testing purposes.

☑ **SERVICE TIP:** Replace the thermostatic expansion valve (TXV) if it fails the bulb warming/cooling test. If the problem in step 4 was low pressure, check the inlet screen for foreign matter before replacing the valve.

12. Conclude the test.
   a. Turn off all air conditioning controls.
   b. Reduce the engine speed and turn off the engine.
   c. Remove the manifold and gauge set.

Figure 4-46 Place the remote bulb in a warm rag.

The customer complains that the air conditioner/heater blower is working on only two speeds. It runs on LO and MED but does not run on HI.

An inspection of the schematic for the particular make/model vehicle reveals that the blower motor speed for HI is not controlled by the blower resistor. Voltage for HI is supplied through another circuit intended to prevent high blower speed when the headlamps are ON. Since there is no high speed with the headlamps OFF, this circuit is suspect. Further inspection reveals a blown fuse in the high-speed blower circuit. The fuse is replaced and the blower operation for HI is restored.

Terms to Know

Armature	Filter	"Pound" (of refrigerant)
Atmospheric pressure	Functional test	RECIR
Clutch coil	Ground	Rotor
Cycling time	Heater	Stabilize
Diagnosis	Idle speed	Tension gauge
Duct	MAX	TXV
Evaporator core	Observe	

ASE-Style Review Questions

1. Refer to Figure 4-47, page 132.
   Technician A says that the compressor clutch diode in the clutch electrical circuit is in series with the compressor clutch coil.
   Technician B says that a blown HVAC 20A fuse in the IP fuse block will prevent blower motor operation in all speeds EXCEPT high.
   Who is right?
   A. A only   **C.** Both A and B
   B. B only   **D.** Neither A nor B

2. A test lamp has been connected from the blower motor housing to the "hot" side of the fuse or circuit breaker:
   Technician A says if the lamp does not light, a defective ground wire is indicated.
   Technician B says that if the lamp lights, a defective blower motor is indicated.
   Who is right?
   A. A only   **C.** Both A and B
   B. B only   **D.** Neither A nor B

3. *Technician A* says that low suction and discharge pressure may indicate a low refrigerant charge.

   Technician B says that low suction and discharge pressure may indicate a restriction in the system.
   Who is right?
   A. A only   **C.** Both A and B
   B. B only   **D.** Neither A nor B

4. *Technician A* says a liquid line restriction has the same general symptoms as a suction line restriction.
   Technician B agrees, adding "An undercharge of refrigerant also has the same symptoms."
   Who is right?
   A. A only   **C.** Both A and B
   B. B only   **D.** Neither A nor B

5. *Technician A* says that air is a noncondensable gas.
   Technician B says that air collects in the evaporator during the OFF cycle of the air conditioning system.
   Who is right?
   A. A only   **C.** Both A and B
   B. B only   **D.** Neither A nor B

6. *Technician A* says that moisture in the system can cause poor or no cooling.
Technician B says that moisture in the system can cause harmful acids.
Who is right?
- **A.** A only
- **B.** B only
- **C.** Both A and B
- **D.** Neither A nor B

7. *Technician A* says a restriction in the suction line will cause excessive high-side pressure.
Technician B says a restriction will result in a pressure change, usually marked by frost or ice.
Who is right?
- **A.** A only
- **B.** B only
- **C.** Both A and B
- **D.** Neither A nor B

8. *Technician A* says if the orifice tube is missing, the low-side pressure will be low.
Technician B says if the orifice tube is plugged with debris, low-side pressure will be high.
Who is right?

- **A.** A only
- **B.** B only
- **C.** Both A and B
- **D.** Neither A nor B

9. *Technician A* says the Harrison V-5 compressor control valve sometimes gives trouble.
Technician B says that the V-5 control valve is serviceable.
Who is right?
- **A.** A only
- **B.** B only
- **C.** Both A and B
- **D.** Neither A nor B

10. *Technician A* says the minimum flow test on a TXV is made by wrapping the remote bulb in a warm rag.
Technician B says a maximum flow test may be made on a TXV by immersing the remote bulb in ice water (H_2O).
Who is correct?
- **A.** A only
- **B.** B only
- **C.** Both A and B
- **D.** Neither A nor B

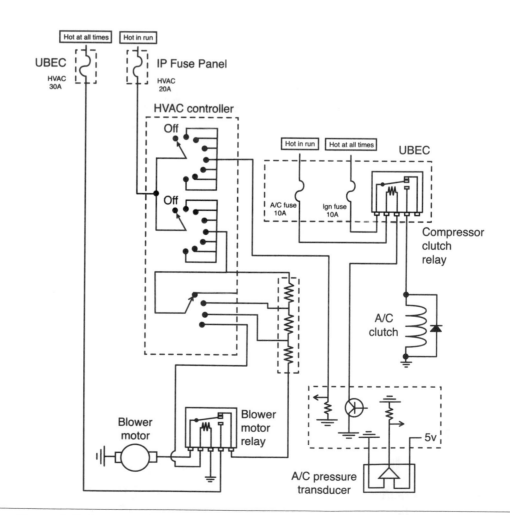

Figure 4-47 For ASE Style Review Question #1.

ASE Challenge

1. The refrigerant most harmful to the ozone is:
   - A. HCFC-22
   - C. HFC-134a
   - B. CFC-12
   - D. CHG-12

2. All of the following may cause a low-pressure gauge to go into a vacuum, EXCEPT:
   - A. Clogged screen in the metering device
   - B. Clogged screen in the accumulator
   - C. Clogged screen in the receiver-drier
   - D. Thermostatic expansion valve stuck closed

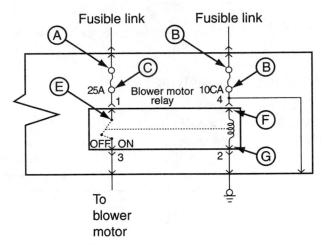

4. The blower motor does not operate. The voltage at point F on the illustration above is 12 volts. The least likely cause of this problem is a defective:
   - A. 10-amp. fuse
   - B. 25-amp. fuse
   - C. Blower motor relay coil.
   - D. Blower motor relay contacts

5. A slipping compressor clutch may be due to:
   - A. Incorrect clutch air gap
   - B. Internal compressor problem
   - C. Either A or B
   - D. Neither A nor B

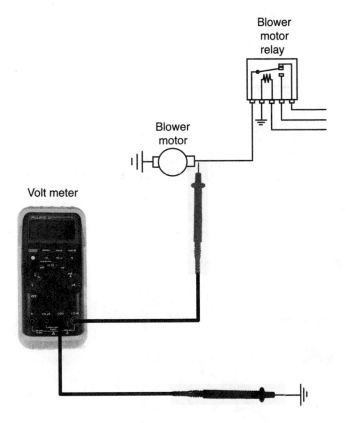

3. The blower motor does not operate although there are 12-volts available at the motor terminal in the illustration above. The most likely problem is:
   - A. "Blown" fuse or open circuit breaker
   - B. Defective relay ground
   - C. Defective blower motor relay
   - D. Loose or defective motor ground wire

Table 4-1 ASE TASK

Inspect A/C condenser for air flow restrictions. Clean and straighten fins.

Problem Area	Symptoms	Possible Causes	Classroom Manual	Shop Manual
RESTRICTED AIRFLOW	Excessive (high) head pressure	1. Dirt, debris, or bugs on condenser	116	166
		2. Defective (mechanical) fan clutch	260	416–419
		3. Defective cooling fan	260	416
		4. Fins bent	116	405–407

Table 4-2 ASE TASK

Inspect, test, and replace A/C system condenser and mountings.

Problem Area	Symptoms	Possible Causes	Classroom Manual	Shop Manual
NOISE	Rattles on road	Loose or broken condenser mounting	116	166

Table 4-3 ASE TASK

Inspect, test, and replace fan (both electrical and mechanical), fan clutch, fan belts, fan shroud, and air dams.

Problem Area	Symptoms	Possible Causes	Classroom Manual	Shop Manual
NOISE	Rattles on road	1. Loose fan shroud	225–226	416
		2. Worn fan clutch	260	416–419
		3. Loose fan blades(s)	260	416
		4. Defective electrical fan mounting	262–263	417–420
	Squeal, squeak, or chatter	1. Defective fan clutch	260	416–420
		2. Defective electric fan motor	262	417–420

JOB SHEET 10

Name _____ Date _____

Inspect the V-Belt Drive

Upon completion of this job sheet you should be able to visually check the compressor drive V-belt and check its tension.

ASE Correlation

This job sheet is related to the ASE Heating and Air Conditioning Test's content area: *Refrigeration System Components Diagnosis and Repair, 1. Compressor and Clutch*, Task: *Inspect, adjust, and replace A/C compressor drive belts, pulleys, and tensioners.*

Tools and Materials

A vehicle with V-belt drive
Service manual
Belt tension gauge

Describe the vehicle being worked on:

Year _____ Make _____ Model _____

VIN _____ Engine type and size _____

Procedure

1. List the different belts, and their purpose. _____
   _____

2. Visually inspect the belts and describe the condition of each._____
   _____

3. Check the tension of the A/C belt. According to specifications, the tension should be _____. The tension is _____.

4. Based on the above inspection, what is your recommendation?_____
   _____

5. What procedure would you follow to adjust the belt tension?_____
   _____

✔ **Instructor's Check** _____

JOB SHEET 11

Name _____ Date _____

Inspect the Serpentine Drive Belt

Upon completion of this job sheet you should be able to visually check the compressor serpentine drive belt and check its tension.

ASE Correlation

This job sheet is related to the ASE Heating and Air Conditioning Test's content area: *Refrigeration System Component Diagnosis and Repair, 1. Compressor and Clutch,* Task: *Inspect, adjust, and replace A/C compressor drive belts, pulleys, and tensioners.*

Tools and Materials

A vehicle with serpentine-belt drive
Service manual
Belt tension gauge

Describe the vehicle being worked on:

Year _____ Make _____ Model _____
VIN _____ Engine type and size _____

Procedure

1. If more than one belt, list the different belts and their purpose. _____
   _____

2. Visually inspect the belt(s) and describe their condition. _____
   _____

3. Check the tension of the belt. According to specifications, the tension should be _____. The tension is _____.

4. Based on the above inspection, what is your recommendation? _____
   _____

5. What procedure would you follow to adjust the belt tension? _____
   _____

☑ **Instructor's Check** _____

JOB SHEET 12

12

Name _____ Date _____

Inspect and Test the Clutch Coil and Diode

Upon completion of this job sheet you should be able to inspect and test the compressor clutch coil and diode assembly.

ASE Correlation

This job sheet is related to the ASE Heating and Air Conditioning Test's content area: *Refrigeration System Component Diagnosis and Repair, 1. Compressor and Clutch,* Task: *Inspect, test, service, and replace A/C compressor clutch components or assembly.*

Tools and Materials

An air-conditioned vehicle
Service manual
Digital volt ohmmeter (DVOM)

Describe the vehicle being worked on:

Year _____ Make _____ Model _____

VIN _____ Engine type and size _____

Procedure

1. Gain access to the compressor clutch area. Describe its location._____

   _____

2. Visually inspect wiring and describe the condition. _____

   _____

3. With the engine not running, but ignition switch ON and A/C controls to MAX cooling, check the available voltage at the clutch coil connector. According to specifications, the voltage should be _____. The voltage is _____.

4. Turn the ignition OFF. Disconnect the clutch coil and isolate the diode, if equipped. Check the resistance of the clutch coil. According to specifications, the resistance should be _____. The resistance is _____.

5. With the diode isolated, connect it to an ohmmeter. What is the resistance? _____ Reverse the leads on the ohmmeter. What is the resistance? _____

6. Based on the above inspection and tests, what is your recommendation?_____

   _____

☑ **Instructor's Check** _____

JOB SHEET 13

13

Name _____ Date _____

Inspect the Condenser

Upon completion of this job sheet you should be able to visually check the air conditioning system condenser for proper air flow.

ASE Correlation

This job sheet is related to the ASE Heating and Air Conditioning Test's content area: *Refrigeration System Components Diagnosis and Repair, 2. Evaporator, Condenser, and Related Components,* Task: *Inspect A/C condenser for air flow restrictions.*

Tools and Materials

An air-conditioned vehicle

Describe the vehicle being worked on:

Year _____ Make _____ Model _____

VIN _____ Engine type and size _____

Procedure

1. Visually inspect the air conditioning system and describe its overall condition.

   _____

2. In a well-ventilated area, start the engine, place the transmission in park and turn on the air conditioning system to MAX cooling. If a standard transmission, place in NEUTRAL and chock the wheels. Describe your procedure for accomplishing this step. _____

   _____

   █ **CAUTION:** Take care not to come into contact with moving parts, such as fan blades or belts, as well as heated metal, such as the exhaust manifold, when performing steps 3 through 5.

3. Carefully place your hand on the condenser near the refrigerant inlet. Describe what you feel. _____

   _____

4. Carefully place your hand on the condenser near the refrigerant outlet. Describe what you feel. _____

   _____

5. When moving your hand across the condenser, refrigerant inlet to outlet, do you feel a change in temperature? If so, describe the change._____

   _____

6. Based on the above inspection, what is your recommendation?_____

   _____

7. What procedure would you follow to increase the air flow across the condenser?

_____

☑ **Instructor's Check** _____

Servicing System Components

Upon completion and review of this chapter you should be able to:

❑ Identify and compare the differences between English and metric fasteners.

❑ State the purpose of good safety practices when servicing an automotive air conditioning system.

❑ Diagnose air conditioning system malfunctions based on customer complaints.

❑ Identify the different types of automotive air conditioning systems.

❑ Remove and replace automotive air conditioning system components, such as hoses and fittings, metering devices, driers and accumulators, compressors, condensers and evaporators, and pressure switches.

English and Metric Fasteners

The servicing of automotive air conditioning systems seems to become more and more complex each year. Although basic theories do not change, refrigeration and electrical control are redesigned or modified year to year. To add to the confusion, domestic automotive manufacturers use metric nuts and bolts on many components and accessories. Both English and **metric fasteners** can be found on the same automobile.

Some metric fasteners closely resemble **English fasteners** in size and appearance. The automotive service technician must be very careful to avoid mixing these fasteners. English and metric fasteners are not interchangeable. For example, a metric 6.3 (6.3-mm) capscrew may replace by design an English 1/4-28 (1/4-in. by 28 threads per in.) capscrew. Note in Figure 5-1 that the diameters of the two fasteners differ by only 0.002 in. (0.05 mm). The threads differ by only 2.6 per in. (1 per cm). There are 28 threads per in. (11 threads per cm) for a 1/4-28 capscrew, and 30.6 threads per in. (12 threads per cm) for a 6.3-mm capscrew.

While the differences are minor, an English 1/4-28 nut will not hold on a metric 6.3 capscrew. Mismatching of fasteners can cause component damage and/or early failure. Such component failure can result in personal injury.

Safety

It must be recognized that the skills and procedures of those performing **service procedures** vary greatly. It is not possible to anticipate all of the conceivable ways or conditions under which service procedures may be performed. It is, therefore, not possible to provide precautions for every possible hazard that may result.

The following precautions are basic and apply to any type of automotive service:

❑ Wear safety glasses or goggles for eye protection when working with refrigerant (Figure 5-2). This is important while working under the hood of the vehicle.

❑ If the engine is to be operated, set the parking brake.

    a. Place the gear selector in **PARK** if the vehicle is equipped with an automatic transmission.

    b. Place the transmission in **NEUTRAL** if the vehicle is equipped with a manual transmission.

❑ Unless required otherwise for the service procedure, be certain that the ignition switch is turned to the OFF position.

❑ If the engine is to be operated, be certain that the vehicle is in a well-ventilated area or that provisions are made to vent the exhaust gases.

Basic Tools

Mechanic's basic tool set (English and metric)

Fender cover

Flare nut wrench set

Torque wrench

Hacksaw with 32 tpi blade

Springlock coupling tool set

Single-edge razor blade

FOT remove-replace tool

Tube cutter

Calibrated container

Internal-external snapring plier

Pressure/temperature switch socket

Both English and metric fasteners may be found on an assembly.

Classroom Manual
Chapter 1,
pages 9–12

Follow all safety precautions when servicing an automotive air conditioner.

English Series				Metric Series			
Size	Diameter		Threads Per Inch	Size	Diameter		Threads Per Inch (prox)
	in	mm			in	mm	
#8	0.164	4.165	32 or 36				
#10	0.190	4.636	24 or 32				
1/4	0.250	6.350	20 or 28	M6.3	0.248	6.299	25
				M7	0.275	6.985	25
5/16	0.312	7.924	18 or 24	M8	0.315	8.001	20 or 25
3/8	0.375	9.525	16 or 24				
				M10	0.393	9.982	17 or 20
7/16	0.437	11.099	14 or 20				
				M12	0.472	11.988	14.5 or 20
1/2	0.500	12.700	13 or 20				
9/16	0.562	14.274	12 or 18	M14	0.551	13.995	12.5 or 17
5/8	0.625	15.875	11 or 18				
				M16	0.630	16.002	12.5 or 17
				M18	0.700	17.780	10 or 17
3/4	0.750	19.050	10 or 16				
				M20	0.787	19.989	10 or 17
				M22	0.866	21.996	10 or 17
7/8	0.875	24.765	9 or 14				
				M24	0.945	24.003	8.5 or 12.5
1	1.000	25.400	8 or 14				
				M27	1.063	27.000	8.5 or 12.5

Figure 5-1 A comparison of English and metric fasteners.

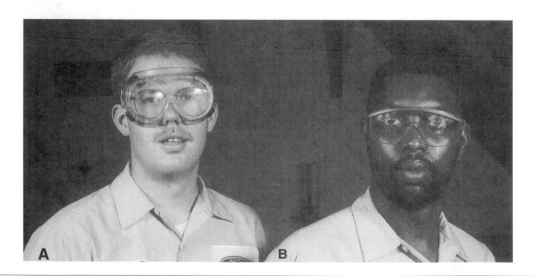

Figure 5-2 Wear safety (B) glasses or (A) goggles when servicing air conditioning components.

Figure 5-3 Carefully disconnect the battery cable.

❏ Avoid loose clothing. Roll up long shirt sleeves. Tie long hair securely behind the head. Remove rings, watches, and loose-hanging jewelry.

❏ Keep clear of all moving parts when the engine is running. Engine-driven cooling fans have been known to separate. A loose fan blade can cause serious injury.

❏ Keep hands, clothing, tools, and test leads away from the engine cooling fan. Electric cooling fans may start without warning even when the ignition switch is in the OFF position.

❏ Avoid personal contact with hot parts such as the radiator, exhaust manifold, and high-side refrigerant lines.

❏ Disconnect the battery when required to do so (Figure 5-3). Follow the recommendations of the manufacturer; Chrysler, for example, requires that the negative (–) cable be disconnected to disable the air bag. General Motors requires the positive (+) cable to be disconnected, and Ford requires both cables be disconnected, first the negative (–), then the positive (+).

❏ Batteries normally produce explosive gases. DO NOT allow flames, sparks, or any lighted substances to come near the battery. Always shield your face and protect your eyes when working near a battery.

NOTE: When the battery has been reconnected after being disconnected volatile memory information—such as radio station presets, clock, seat, mirror, and window memory as well as customer input keyless entry codes—will be lost. Also, when the battery is reconnected some abnormal drive symptoms may occur for the first 10 miles (18 kilometers) or so.

❏ If in doubt, ASK; do not take a chance. If there is no one to ask, consult an appropriate service manual. Again, do not take chances.

CAUTION: The technician must exercise extreme caution and pay heed to every established safety practice when performing these or any automotive air conditioning service procedures.

Diagnostic Techniques

Before attempting to service an automotive air conditioning system, be certain that the diagnosis is based on sound reasoning. Consider the following:

❏ Did you listen carefully to the customer's complaint?
❏ Does your diagnosis of the problem have merit based on the customer complaint?
❏ What type of system are you working on?
  a. Cycling or **noncycling clutch**?
  b. CFC-12, HFC-134a, or unknown refrigerant?
❏ Do you have the proper tools, equipment, and parts to service the air conditioning system?
  a. What tools are required?
  b. What equipment is required?
  c. What parts are required?

Listen carefully to the customer's complaint of the problem. If you don't understand, ask questions. Suppose, for example, that the customer complains of a moaning sound. The word *moaning* means different things to different people, so ask questions. Take a test drive with your customer so the noise can be identified.

Assume that you test drive the vehicle and hear nothing. When you return to the shop and tell the customer you heard nothing, the response may be, "Well, you did not start off very fast at the traffic light." Further discussion will reveal that the noise is only heard when pulling away from a traffic light after stopping.

The diagnosis to this problem is a relatively simple one. You are looking for a problem that only occurs during heavy acceleration. The problem can be a defective vacuum check valve or split vacuum hose (Figure 5-4), anything that may cause a vacuum loss during heavy acceleration.

NOTE: The loss of a vacuum signal at the control head will generally cause domestic vehicles to "fail safe" in either the heat or defrost mode.

Noises sound like different things to different people.

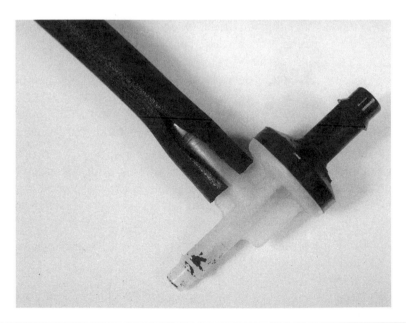

Figure 5-4 A split vacuum hose can cause a system malfunction.

Figure 5-5 A typical temperature control.

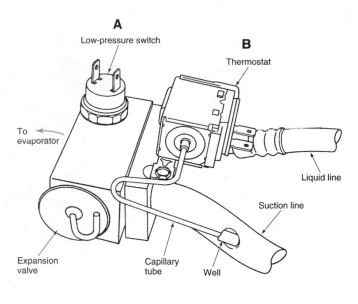

Figure 5-6 Typical pressure controls: (A) low pressure; (B) thermostatically controlled.

What Type System

There are basically two methods of temperature control for automotive air conditioning systems:

❏ Cycling clutch
❏ Noncycling clutch

Cycling Clutch

The cycling clutch system relies on two methods for temperature control:

1. Temperature cycling switch. The temperature cycling switch (Figure 5-5) is a temperature-sensitive switch that cycles the compressor clutch on and off at **predetermined** temperature levels.
2. Pressure cycling switch. The pressure cycling switch (Figure 5-6), as its name implies, is sensitive to system pressure and turns the compressor clutch on and off at predetermined pressure levels.

Noncycling Clutch

The noncycling clutch system relies on a variable displacement (VD) compressor to control the in-car temperature. The amount of refrigerant permitted to flow through the system is controlled by the compressor's ability to alter the stroke of the pistons as required by varying system conditions.

The only purpose of the clutch in a noncycling system, then, is to disengage the compressor when the air conditioner is not in use and to engage the compressor when the driver calls for cooling.

What Type Refrigerant

Several refrigerants have been approved by the Environmental Protection Agency (EPA) for use in automotive air conditioning systems. Only two types, however, are approved by the automotive industry for use. The use of a refrigerant not approved by industry may void manufacturer's warranties on the system as well as on replacement components.

The two industry approved refrigerants (Figure 5-7) are:

Classroom Manual
Chapter 4,
pages 91–93

Use the proper
refrigerant.

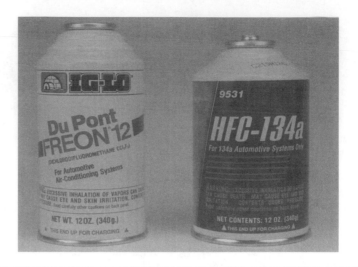

Figure 5-7 "Pound" cans of two approved refrigerants.

1. CFC-12
2. HFC-134a

Other refrigerants approved by the EPA for automotive use include:

- ❏ FRIGC FR-12
- ❏ Freeze-12
- ❏ Free Zone (also, RB-276)
- ❏ GHG-HP
- ❏ GHG-X4 (also, Autofrost and Chill-It)
- ❏ GHG-X5
- ❏ Hot Shot (also, Kar Kool)
- ❏ Ikon-12
- ❏ R-406A (also, GHG)

Refrigerant CFC-12

A refrigerant, known as R-12 or CFC-12, was used in automotive air conditioning systems through the early 1990s. Because of environmental concerns, its production and use has been phased out. Certain system changes, however, have to be made in order to use the new refrigerant. There is no drop-in refrigerant available that is approved for automotive use.

Refrigerant HFC-134a

Refrigerant-134a (HFC-134a) is at present the automotive industry's refrigerant of choice to replace CFC-12 in automotive service. It is not, however, a drop-in replacement. Certain system modifications must be made before the new refrigerant can be used in an old system. This new refrigerant was first used in automotive applications in the early 1990s. By 1995, all new cars manufactured contained the new refrigerant.

Do not contaminate recovery system equipment or cylinders.

Other Refrigerants

In 1994, the EPA established the Significant New Alternatives Policy (SNAP) Program to review alternatives to ozone-depleting substances. Under authority of the 1990 Clean Air Act (CAA), the EPA also examines potential substitute refrigerants as to their flammability, effects on global warming, and toxicity. As of this writing, the agency has determined that ten "new" refrigerants, including HFC-134a, are acceptable for use as a CFC-12 replacement in motor vehicle air conditioning systems.

They are all, however, "acceptable subject to use conditions." All alternate refrigerants, except HFC-134a, are "blends," which means that they contain more than one component in their composition.

"Acceptable subject to use conditions" indicates that the EPA believes these refrigerants, when used in accordance with the use conditions, to be safer for human health and for the environment than the CFC-12 they are meant to replace. This designation, however, is not intended to imply that the refrigerant will work as satisfactorily as CFC-12 in any specific system. Also, it is not intended to imply that the refrigerant is perfectly safe regardless of how it may be used.

The EPA does not test refrigerants and therefore does not specifically approve or endorse any one refrigerant over any others. The agency reviews all of the information about a refrigerant submitted by its manufacturer and independent testing laboratories. The EPA does not determine what effect, if any, a "new" refrigerant may have on vehicle warranty.

Some refrigerant manufacturers use the term *drop-in* to imply that their refrigerant will perform identically to CFC-12 and that no modification is required for its use. The term also implies that the alternate refrigerant can be used alone or mixed with CFC-12. The EPA believes the term *drop-in* confuses and obscures at least two important regulatory points:

1. Charging one refrigerant into a system before extracting the old refrigerant is a violation of the SNAP use conditions and is, therefore, illegal.
2. Certain components may be required by law, such as hoses and compressor shutoff switches. If these components are not present, they must be installed. Five blends, for example, contain HCFC-22 and require barrier hoses.

It may also be noted that system performance is affected by such variables as outside temperature, relative humidity, and driving conditions. It is therefore not possible to ensure equal performance of any refrigerant under all of these conditions.

The service facility must have service and recovery equipment specifically designed for each type of refrigerant that is to be serviced. This means that at least two systems are required: one for CFC-12 and one for HFC-134a. A third set is required if **contaminated** systems are to be serviced, and a fourth set is required if a blend refrigerant is to be used.

Each new alternate refrigerant must be used with a unique set of fittings attached on the service ports, all recovery and recycling equipment, on can taps and other charging equipment, and on all refrigerant containers. A unique label must be affixed over the original label to identify the type refrigerant as well as lubricant used in the air conditioning system.

Proper Tools, Equipment, and Parts

Having and using the proper tools and equipment is an essential part of performing a successful repair procedure. A screwdriver, for example, makes a very poor chisel; an adjustable wrench is a very poor hammer.

Tools

It is important that tools be used for the purpose for which they are designed. For example, a flare nut wrench, not an open-end wrench, should be used on flare nuts (Figure 5-8). A flare nut wrench should not be used to remove a bolt or nut; this service requires either an open, box, combination, or socket wrench.

Equipment

It is equally important that the proper equipment be used for a particular service procedure. A vacuum pump, for example, can be constructed using an old refrigerator compressor (Figure 5-9). It is neither as attractive nor as efficient as a commercially available vacuum pump, however.

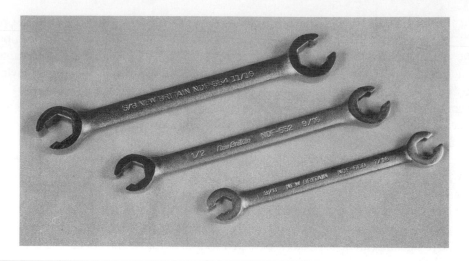

Figure 5-8 A flare nut wrench set.

Figure 5-9 A typical refrigeration compressor.

Parts

It is not practical to stock all of the parts that may be required in the course of doing business. The local parts distributors are responsible for that. If, however, a particular customer relies on your shop for a certain service, an adequate stock is suggested. For example, if a local off-road equipment repair facility relies on you for hoses, it would be wise to stock an adequate supply of various types of hose fittings and the several sizes of bulk hose necessary to supply the customer's needs.

Service Procedures

The service procedures given in this chapter are typical and are to be used as a guide only. Due to the great number of variations in automotive air conditioning system configurations, it is impossible to include all specific and detailed information in this text. When specific and more detailed information is required, the service technician must consult the appropriate manufacturer's service manual for any particular year and model vehicle. General service manuals are available that cover most service procedures in detail for automobiles of a specific year, make, and model. One such manual, covering only the past four years, has little theory and is more than 3-in. (76-mm) thick. Just one manufacturer's shop service manual, the 1993 Buick Riviera, devotes 99 pages to air conditioning service.

The information given in this manual is given only as a guide for the student technician to perform basic service procedures that are normally required. Proper service and repair procedures are vital to the safe, reliable operation of the system. Most important, proper service procedures and techniques are essential to providing personal safety to those performing the repair service and to the safety of those for whom the service is provided.

Be sure to follow the manufacturer's recommendations. To **disarm** the air bag restraint system, for example, Chrysler had suggested disconnecting the battery ground (–) cable; General Motors had suggested disconnecting the positive (+) battery cable; and Ford had suggested disconnecting both cables—first the ground (–) cable, then the positive (+) cable.

Since the computers are disabled and often must "relearn" a program, disconnecting the battery is now discouraged. General Motors suggests the following typical procedures:

1. Turn off the ignition switch.
2. Remove the restraint system (air bag) fuse from the fuse panel.
3. Remove the left-side sound insulator.
4. Disconnect the connector position assurance (CPA) and yellow two-way connector found at the base of the steering column.

There is no set universal procedure. The best approach is to refer to the specific service manual appropriate for the make and model vehicle being serviced to ensure that proper procedures are followed.

Preparation

There are certain basic procedural steps that must be taken before attempting to perform any service procedure. Whenever applicable, this procedure will be referenced in the various service procedures that follow.

1. Place the ignition switch in the OFF position.
2. Place a fender cover on the car to protect the finish.
3. Disconnect the battery cable. Follow procedures as outlined in the appropriate manufacturer's service manual.
4. Recover the system refrigerant. Use the proper recovery equipment and follow the instructions outlined in Chapter 7 of this manual.
5. Locate the component that is to be removed for repair or replacement.
6. Remove access panel(s) or other hardware necessary to gain access to the component.

Servicing Refrigerant Hoses and Fittings

There are many types of fittings used to join refrigerant hoses to the various components of the air conditioning system. Some of these fittings are (Figure 5-10):

- ❑ Male and female **SAE** flare
- ❑ Male and female upset flange, commonly called **O-ring**
- ❑ Male and female **spring lock**
- ❑ Male barb
- ❑ Peanut
- ❑ Beadlock

Consult the manufacturer's service manual for specific guidelines when performing any major automotive air conditioning service.

Special Tools
Battery pliers
Recovery equipment

Take extreme care when disconnecting a battery.

Classroom Manual
Chapter 5,
pages 123–124

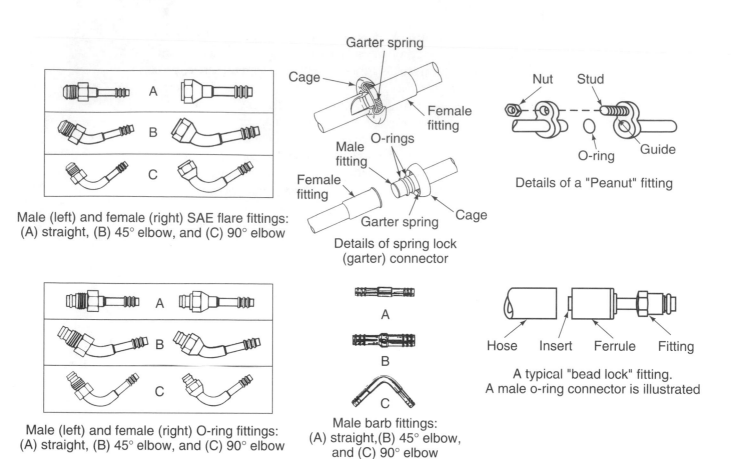

Male (left) and female (right) SAE flare fittings: (A) straight, (B) 45° elbow, and (C) 90° elbow

Garter spring

Cage

Female fitting

O-rings

Male fitting

Female fitting

Garter spring

Cage

Details of spring lock (garter) connector

Nut   Stud

O-ring   Guide

Details of a "Peanut" fitting

Male (left) and female (right) O-ring fittings: (A) straight, (B) 45° elbow, and (C) 90° elbow

Male barb fittings: (A) straight, (B) 45° elbow, and (C) 90° elbow

Hose   Insert   Ferrule   Fitting

A typical "bead lock" fitting. A male o-ring connector is illustrated

Figure 5-10 Various types of fittings used in automotive air conditioning service

Special Tools

Flare nut wrench set

1/4-in. drive socket set

Torque wrench

Hacksaw with 32-tpi blade

Pliers

Spring lock coupling tool set

Single-edge razor blade

Use caution when using a razor blade. Always cut away from the body.

This procedure will be given in five parts:

1. Installing an insert **(barb) fitting**
2. Repairing a hose using an **insert fitting**
3. Repairing a damaged factory fitting
4. Servicing spring lock fittings
5. O-ring service

Part 1. Insert Barb Fitting

NOTE: Barb-type fittings should not be used for barrier-type hoses. If in doubt as to the type of hose, use a beadlock fitting for the repair. Follow the procedure given in Photo Sequence 4 for beadlock hose assembly.

1. **a.** Measure and mark the required length of replacement hose, or
   b. Determine how much hose must be cut ahead of the damaged fitting.
2. Use a single-edge razor blade to cut the hose.

NOTE: Industrial single-edge razor blades are available at most commercial cleaning supply houses or hardware stores.

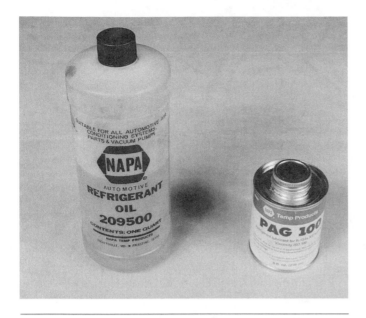

Figure 5-11 Small containers of mineral and PAG oil.

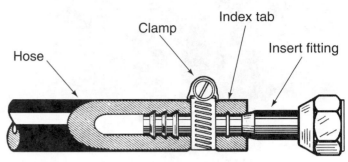

Figure 5-12 Barb hose fitting detail.

3. Trim the end of the hose to be used to ensure that the cut is at a right angle (square).

4. Apply clean refrigeration oil to the inside of the hose to be used.

> ⚠️ **WARNING:** Use the proper oil for the refrigerant used in the system; i.e., mineral oil for CFC-12 and PAG for HFC-134a (Figure 5-11).

5. Be sure that the fitting is free of all nicks and burrs.

6. Coat the fitting liberally with clean refrigeration oil. See warning following step 4.

7. Slip the insert fitting into the refrigeration hose in one constant, deliberate twisting motion.

8. Install the hose clamp.

9. Tighten the hose clamp to a **torque** of 30 ft.-lb. (40.6 N·m).

NOTE: The hose clamp should be placed at the approximate location of the fitting barb closest to the nut end of the fitting (Figure 5-12).

Part 2. Hose Repair Using Insert Fitting

1. Determine how much of the damaged hose must be cut out and mark the hose.

2. Follow steps 2 through 9 of Part 1.

Part 3. Repairing a Damaged Factory Fitting

1. Remove the hose from the car and place the damaged fitting securely in a vise or other holding device.

2. Use a hacksaw to cut through the ferrule (Figure 5-13).

3. Use pliers to "peel" off the ferrule (Figure 5-14).

4. Slice the hose with a single-edge razor blade and remove the hose (Figure 5-15).

5. Cut off the damaged end of the hose.

6. Apply clean refrigeration oil to the inside of the hose. See warning following step 4 of Part 1.

7. Coat the fitting liberally with clean refrigeration oil. See warning following step 4 of Part 1.

Figure 5-13 Use a hacksaw to cut through the ferrule. (Courtesy of BET, Inc.)

Figure 5-14 Use pliers to peel off the ferrule. (Courtesy of BET, Inc.)

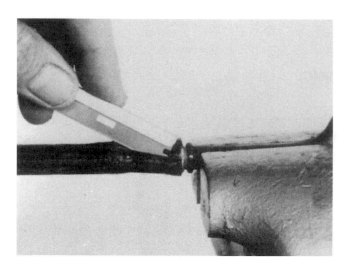

Figure 5-15 Slice the hose with a single-edge razor blade. (Courtesy of BET, Inc.)

8. Slip the hose onto the fitting with one constant, deliberate twisting motion.
9. Install a hose clamp and tighten to 30 ft.-lb. (40.6 N·m).

Part 4. Servicing Spring Lock Fittings

For simplicity, this procedure is given in two parts: To Separate, steps 1 through 4, and To Join, steps 5 and 6.

To Separate

1. Install the special tool onto the coupling so it can enter the cage to release the garter spring (Figure 5-16).
2. Close the tool and push it into the cage to release the female fitting from the garter spring (Figure 5-17).
3. Pull the male and female coupling fittings apart (Figure 5-18).
4. Remove the tool from the disconnected spring lock coupling.

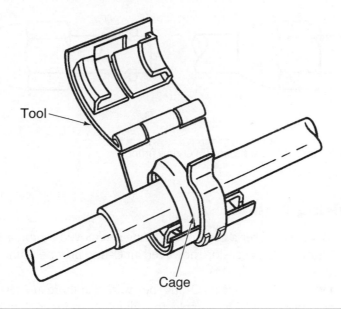

Figure 5-16 Installing the special tool to separate the spring lock coupling

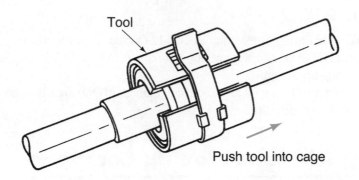

Figure 5-17 Close the tool and push it into the cage to release the coupling.

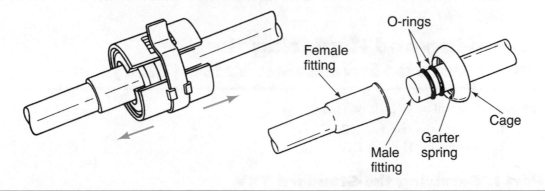

Figure 5-18 Pull the coupling apart.

To Rejoin

5. Lubricate two new O-rings with clean refrigeration oil and install them on the male fitting.

NOTE: The O-ring material is of a special composition and size. To avoid leaks, use the proper O-rings. Also, see the warning following step 4 of Part 1.

6. Insert the male fitting into the female fitting and push them together to join.

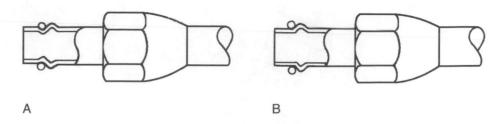

A   B

Figure 5-19 Captive (A) and standard (B) O-ring fittings

Part 5. Servicing O-rings

O-rings must be replaced whenever a component fitting is removed for any reason. They do not usually leak if not disturbed. On occasion, however, an O-ring may be found to be leaking and must be replaced.

If it becomes necessary to replace an O-ring, be aware that there are several different types available for different applications. For example, in addition to spring lock fittings, there are two different types of O-ring fittings used on CFC-12 systems (Figure 5-19):

1. Captive
2. Standard

Although HFC-134a O-rings are similar to CFC-12 O-rings, they are made of a different material. Most CFC-12 and HFC-134a O-rings are not compatible. When replacing them, it is important to use the proper O-ring for the fitting.

The inside diameter (ID) as well as the outside diameter (OD) are also important considerations to ensure a leak-free connection.

Use the proper
O-ring.

Replacing Air Conditioning Components

The following procedures may be considered typical for step-by-step instructions for the replacement of air conditioning system components. For specific replacement details, however, refer to the manufacturer's shop service manual for the particular year, model, and make of the vehicle.

Removing and Replacing the Thermostatic Expansion Valve (TXV)

This procedure will be given in two parts:

1. Servicing the Standard TXV
2. Servicing the H-Block TXV

Special Tools

Flare nut wrench set

Classroom Manual
Chapter 5,
pages 117

Part I. Servicing the Standard TXV

1. Follow the procedures outlined in "Preparation."
2. Remove the insulation tape from the remote bulb.
3. Loosen the clamp to free the remote bulb.
4. Disconnect the external equalizer, if the TXV is so equipped.
5. Remove the liquid line from the inlet of the TXV.
6. Remove and discard the O-ring, if equipped.
7. Inspect the inlet screen (Figure 5-20).
   a. If it is clogged, clean and replace the screen. Skip to step 14.

Photo Sequence 4
Assemble Bead Lock Hose Assembly

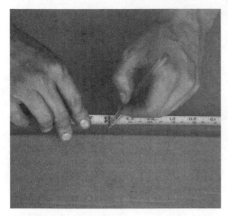

P4-1 Measure hose and mark proper length.

P4-2 Cut hose ensuring that the end is "squared."

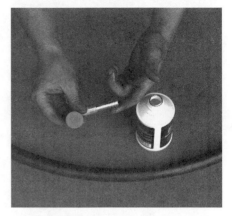

P4-3 Liberally lubricate Bead Lock fitting with mineral oil.

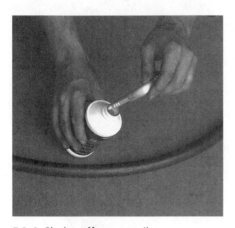

P4-4 Shake off excess oil.

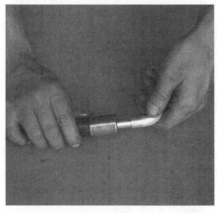

P4-5 Insert hose with ferrule into the fitting.

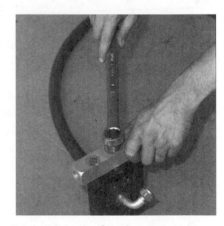

P4-6 Crimp the ferrule.

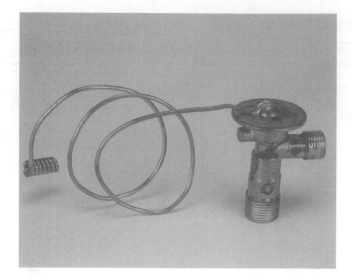

Figure 5-20 Inspect the inlet screen of the thermostatic expansion valve.

Take care not to damage the capillary tube or remote bulb.

    b. If it is not clogged, proceed with step 8.

8. Remove the evaporator inlet fitting from the outlet of the TXV.

9. Remove and discard the O-ring, if equipped.

10. Remove the holding clamp (if provided on the TXV) and carefully lift the TXV from the evaporator.

11. Carefully locate the new TXV in the evaporator.

12. Insert new O-ring(s) on evaporator inlet, if equipped.

13. Attach evaporator inlet to TXV outlet. Tighten to the proper torque.

14. Install new O-rings on the liquid line fitting, if so equipped.

15. Attach the liquid line to the TXV inlet. Tighten to the proper torque.

16. Reconnect the external equalizer tube, if equipped.

17. Position the remote bulb and secure it with a clamp (Figure 5-21).

18. Tape the remote bulb to prevent it from sensing ambient air.

19. Proceed with step 14 of Part 2.

Figure 5-21 Position the remote bulb and secure it with a clamp. (Courtesy of BET, Inc.)

Part 2. H-Valve TXV

1. Follow the procedures outlined in "Preparation."
2. Disconnect the wire connected to the pressure cut-out or pressure differential switch, as applicable.
3. Remove the bolt from the line sealing plate found between the suction and liquid lines.
4. Carefully pull the plate from the H-valve.
5. Cover the line openings to prevent the intrusion of foreign matter.
6. Remove and discard the plate to the H-valve **gasket**.
7. Remove the two screws from the H-valve.
8. Remove the H-valve from the evaporator plate.
9. Remove and discard the H-valve to evaporator plate gasket.
10. Install a new H-valve with gasket (Figure 5-22).
11. Replace two screws. Torque to 170-230 in.-lb. (20–26 N·m).
12. Replace the line plate to H-valve gasket.
13. Hold the line assembly in place and install the bolt. Torque to 170–230 in.-lb. (20–26 N·m).
14. Replace the access panels and any hardware previously removed.
15. Reconnect the battery following the instructions given in the manufacturer's shop manual.
16. Leak-test, evacuate, and charge the system with refrigerant as outlined in Chapter 7 of this manual.
17. Hold performance test, if required.

Be careful not to mar or nick the mating surface(s).

Classroom Manual
Chapter 5,
page 117

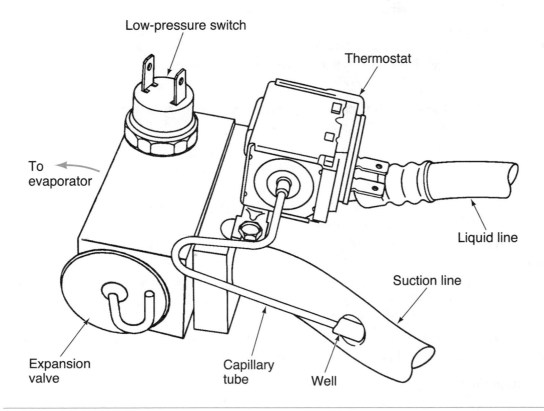

Figure 5-22 A typical H-valve.

Removing and Replacing the Fixed Orifice Tube (FOT)

The fixed orifice tube (FOT) is also known as:

❏ Cycling Clutch Orifice Tube (CCOT)
❏ Fixed Orifice Tube/Cycling Clutch (FOTCC)
❏ Variable Displacement Orifice Tube (VDOT)

Although orifice tubes may look alike, they are not interchangeable.

It should be noted that orifice tubes are not interchangeable. An orifice tube used on Ford car lines may not be used on General Motors car lines. The same service tool may be used, however, to remove and replace any of them. When replacing the orifice tube, it is most important that the correct replacement part be used.

Some car lines have a nonaccessible orifice tube in the liquid line. Its exact location, anywhere between the condenser outlet and the evaporator inlet, is determined by a circular depression or three indented notches in the metal portion of the liquid line. An orifice tube replacement kit is used to replace this type orifice tube and 2.5 in. (63.5 mm) of the metal liquid line.

This service procedure is given in two parts:

1. Servicing the Accessible FOT
2. Replacing the Nonaccessible FOT

Part 1. Servicing the Accessible FOT

The procedure (Photo Sequence 5) is given in two parts: removing the FOT, steps 1 through 9, and replacing the FOT, steps 10 through 14.

Special Tools

Flare nut wrench set
FOT removal tool
Extractor tool
Torque wrench
Tube cutter

Removing the FOT

1. Perform the procedures outlined in "Preparation."

NOTE: It may not be necessary to disconnect the battery for this procedure.

2. Using the proper flare nut wrenches, remove the liquid line connection at the inlet of the evaporator to expose the FOT.
3. Remove and discard the O-ring(s) from the liquid line fitting, if equipped.
4. Pour a very small quantity of clean refrigeration oil into the FOT well to lubricate the seals.

⚠ **WARNING:** Use an oil that is proper for use with the system refrigerant; i.e., mineral oil for CFC-12 and PAG for HFC-134a.

5. Insert the FOT removal tool onto the FOT (Figure 5-23).
6. Turn the T-handle of the tool slightly clockwise (cw) only enough to engage the tool onto the tabs of the FOT.
7. Hold the T-handle and turn the outer sleeve or spool of the tool clockwise to remove the FOT. Do not turn the T-handle.

NOTE: If the FOT breaks during removal, proceed with step 8. If it does not break, proceed with step 10.

8. Insert the extractor into the well and turn the T-handle clockwise until the threaded portion of the tool is securely inserted into the brass portion of the broken FOT (Figure 5-24).
9. Pull the tool. The broken FOT should slide out.

NOTE: The brass tube may pull out of the plastic body. If this happens, remove the brass tube from the puller and reinsert the puller into the plastic body. Repeat steps 8 and 9.

Installing the FOT

10. Liberally coat the new FOT with clean refrigeration oil.
11. Place the FOT into the evaporator cavity and push it in until it stops against the evaporator tube inlet dimples.

160

Photo Sequence 5
Typical Procedure for Replacing a Fixed Orifice Tube

This service procedure presumes that all of the refrigerant has been properly removed from the air conditioning system.

P5-1 Using the proper open end or flare nut wrenches, remove the liquid line from the evaporator inlet fitting.

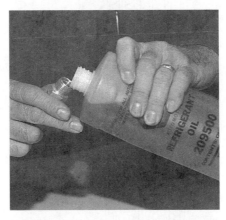

P5-2 Pour a small quantity of refrigeration lubricant into the orifice tube well to lubricate the O-rings.

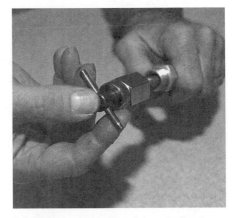

P5-3 Insert an orifice tube removal tool and turn the T-handle slightly clockwise to engage the orifice tube.

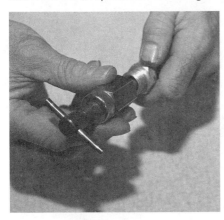

P5-4 Hold the T-handle and turn the outer sleeve clockwise to remove the orifice tube. Do not turn the T-handle. **NOTE:** If the orifice tube broke, proceed with P5-5. If not broken, proceed with P5-6.

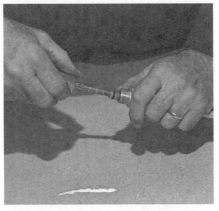

P5-5 Insert the broken orifice tube extractor into the orifice tube well and turn the T-handle clockwise several turns until the tool has been threaded into the orifice tube.

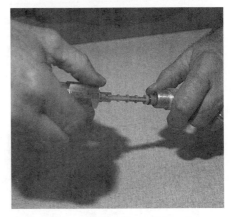

P5-6 Pull the tool. The orifice tube should slide out.

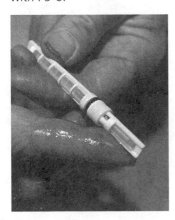

P5-7 Coat the new orifice tube liberally with clean refrigeration lubricant.

P5-8 Slide the new orifice tube into the evaporator until it stops against the tube inlet dimples.

P5-9 Slide a new O-ring onto the evaporator or liquid line, as applicable.

P5-10 Connect the liquid line to the evaporator and tighten nut with proper wrenches.

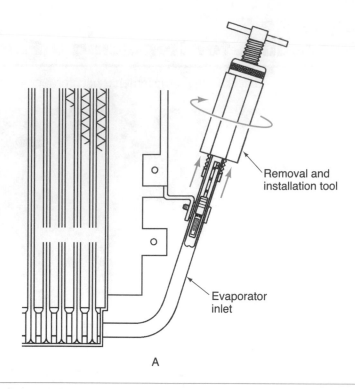

A

Figure 5-23 Insert the FOT removal tool.

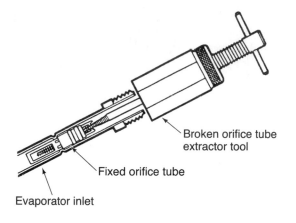

Figure 5-24 Insert the broken FOT extractor tool.

12. Install a new O-ring, if equipped.

13. Replace the liquid line and tighten to recommended torque.

14. Replace the accumulator as outlined in this chapter.

Part 2. Servicing the Nonaccessible FOT

1. Follow the procedures outlined in "Preparation."

NOTE: It may not be necessary to disconnect the battery.

2. Remove the liquid line from the evaporator inlet.
Remove and discard the O-rings, if equipped.

3. Remove the liquid line from the condenser outlet.
Remove and discard the O-rings, if equipped.

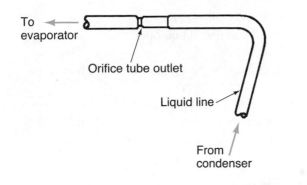

Figure 5-25 Locate the FOT.

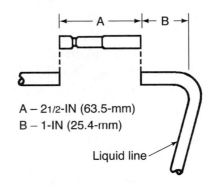

A – 2-1/2-IN (63.5-mm)
B – 1-IN (25.4-mm)

Figure 5-26 Remove 2.5 in. (63.5 mm) of the liquid line.

 WARNING: Note how the liquid line was routed, steps 2 and 3, so it can be replaced in the same manner.

4. Locate the orifice tube. The outlet side of the orifice tube can be identified by a circular depression or three notches (Figure 5-25).
5. Use a tube cutter to remove a 2.5-in. (63.5-mm) section of the liquid line (Figure 5-26).

 SERVICE TIP: Allow at least 1 in. (25.4 mm) of exposed tube at either side of any bend. Also, do not use excessive pressure on the feed screw of the tube cutter to avoid distorting the liquid line. A hacksaw should not be used if a tube cutter is available. If a hacksaw must be used, however, flush both pieces of the liquid line with clean refrigeration oil to remove contaminants such as metal chips.

> Be sure all flushing residue is removed from the tubes before reassembly.

6. Slide a **compression nut** onto each section of the liquid line.
7. Slide a **compression ring** onto each section of the liquid line with the taper portion toward the compression nut.
8. Lubricate the two O-rings with clean refrigeration oil of the proper type and slide one onto each section of the liquid line. See the warning following step 4 of part 1.
9. Attach the orifice tube housing, with the orifice tube inside, to the two sections of the liquid line (Figure 5-27).
10. Hand tighten both compression nuts. Note the flow direction indicated by the arrows. The flow should be toward the evaporator.
11. Hold the orifice tube housing in a vise or other suitable fixture to tighten the compression nuts.

 SERVICE TIP: Be sure that the hose bends are in the same position as when removed for ease in replacing the liquid line.

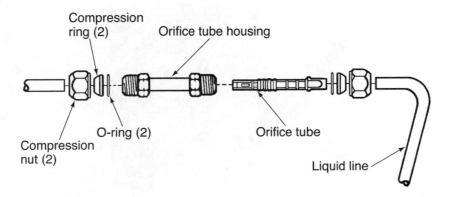

Figure 5-27 The replacement orifice tube assembly.

12. Tighten each compression nut to 65–70 ft.-lb. (87–94 N·m) torque.
13. Insert new O-rings on both ends of the liquid line, if equipped.
14. Install the liquid line:
    a. Attach the condenser end of the liquid line to the condenser and tighten to the proper torque.
    b. Repeat step 14a with the evaporator end of the liquid line.
15. Leak-test, evacuate, and charge the system as outlined in Chapter 7 of this manual.
16. Repeat or continue performance testing.

Removing and Replacing the Accumulator

Special Tools

Flare nut wrench set

Calibrated container

Torque wrench

1. Follow procedures outlined in "Preparation."
2. Disconnect the electrical connection on the pressure control switch.
3. Remove the accumulator inlet fitting.
4. Remove and discard the O-ring.
5. Remove the accumulator outlet fitting.
6. Remove and discard the O-ring.
7. Remove the bracket attaching screw and bracket.
8. Remove the accumulator from the vehicle (Figure 5-28).
9. Remove the pressure switch.
10. Remove and discard the O-ring from the pressure switch.
11. Pour the oil from the accumulator into a calibrated container.

Do not reuse oil that was removed from the system.

12. With a new O-ring, install the pressure switch on the new accumulator.
13. Add a like amount of oil, as removed in step 11, to a new accumulator.

 WARNING: Use the proper oil for the refrigerant in the system; e.g., mineral oil for CFC-12 and PAG for HFC-134a.

14. Position the new accumulator.
15. Using new O-rings, attach the evaporator outlet line to the accumulator inlet and finger tighten.
16. Using new O-rings, attach the suction line to the accumulator outlet and finger tighten.
17. Replace the retainer bracket, removed in step 7.
18. Using appropriate flare nut wrenches, torque the inlet and outlet fittings, steps 15 and 16, as specified.

Classroom Manual
Chapter 5,
pages 122–123

Figure 5-28 Remove the accumulator from the vehicle.

19. Reattach the electrical connector to the pressure switch.
20. Connect the battery, if removed in step 1, according to the manufacturer's recommendations.

NOTE: For steps 20 through 22, refer to Chapter 7.

21. Leak-test the system.
22. Evacuate the system.
23. Charge the system.
24. Hold performance test, if indicated.

Classroom Manual
Chapter 5,
pages 122

Removing and Replacing the Compressor

1. Follow all of the required procedures as outlined in "Preparation."
2. Remove the inlet and outlet hoses or service valves from the compressor.

NOTE: The suction and discharge lines of many systems are connected to the compressor with a common manifold.

3. Remove the clutch lead wire.
4. Loosen and remove the belt(s).
5. Remove the mounting bolt(s) from the compressor bracket(s) and brace(s).
6. Remove the compressor from the vehicle (Figure 5-29).
7. Position the new or **rebuilt** compressor.
8. Install the mounting bolt(s) into the compressor bracket(s) and brace(s).
9. Position and reinstall the belt(s).
10. Replace the clutch lead wire.
11. Using new gaskets and/or O-rings, replace the suction and discharge lines in the reverse order as removed in step 2.

NOTE: Steps 12 through 14 procedures are found in Chapter 7 of this manual.

12. Leak-test the system.
13. Evacuate the system.
14. Charge the system with refrigerant.
15. Hold performance test, if indicated.

Remove any other wires, such as superheat switch wire, from the compressor, if equipped.

Classroom Manual
Chapter 6,
pages 129

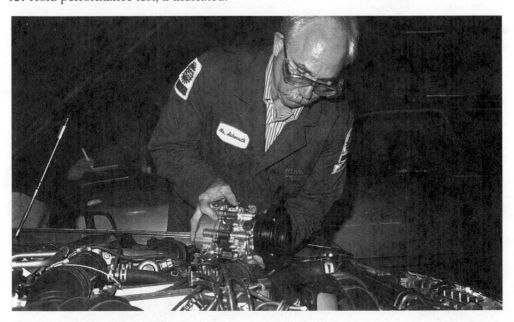

Figure 5-29 Remove the compressor from the vehicle.

Removing and Replacing the Condenser

Special Tools

Flare nut wrench set

Torque wrench

Do not reuse
O-rings.

Classroom Manual
Chapter 5,
pages 115

1. Follow the procedures outlined in "Preparation."

NOTE: It is not generally necessary to disconnect the battery for this procedure.

2. Remove the hood hold-down mechanism and any other cables or hardware that inhibit access to the condenser.
3. Remove the hot-gas line at the top of the condenser.
4. Remove and discard any O-rings.
5. Remove the liquid line at the bottom of the condenser.
6. Remove and discard any O-rings.
7. Remove and retain any attaching bolts and/or nuts holding the condenser in place.
8. Lift the condenser from the car.
9. Install the new condenser by reversing the procedure given in steps 2, 3, 5, and 7.

NOTE: Be sure to install new O-rings, if equipped, on the hot gas and liquid lines.

10. Leak-test the system.
11. Evacuate the system.
12. Charge the system with refrigerant.

Removing and Replacing the Evaporator

Special Tools

Flare nut wrench set

Torque wrench

Classroom Manual
Chapter 5,
pages 120

This procedure will be given in four parts:

1. Aftermarket
2. Factory or Dealer Installed (Domestic)
3. Factory or Dealer Installed (Import)
4. Rear Heating/Cooling Unit

Prior to servicing the evaporator, the air conditioning system refrigerant must be removed by the recovery process. This procedure should be performed prior to step 1 outlined in either part. On completion of this service, the air conditioning system should be properly evacuated and charged with the appropriate refrigerant as outlined in Chapter 7 of this manual.

Part 1. Aftermarket

1. Remove the liquid line from the metering device.
2. Remove the suction line from the evaporator.
3. Remove and discard O-rings, if equipped.
4. Remove the electrical lead wire(s). Also, disconnect the ground wire.
5. Remove the mounting hardware; remove the evaporator from the vehicle.
6. Check to see if there is a measurable amount of lubricant in the evaporator.

NOTE: Add an equivalent amount of proper, clean, and fresh lubricant to the replacement evaporator.

7. To install the evaporator reverse steps 1, 2, 4, and 5. Install new O-rings, if applicable.

Part 2. Factory or Dealer Installed (Domestic) (Figure 5-30)

1. Remove the liquid line from the metering device inlet.

NOTE: If an H-valve metering device, skip to step 5.

2. Remove and discard the O-ring, if equipped.
3. Remove the suction line from the evaporator.
4. Remove and discard the O-ring, if equipped.

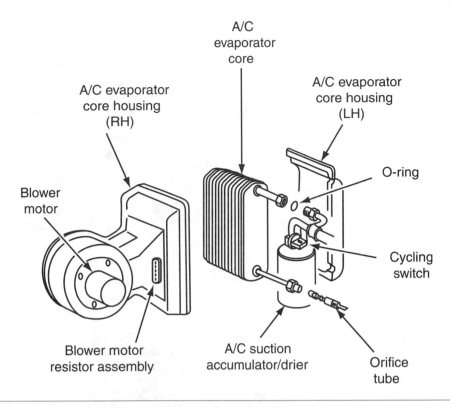

A/C evaporator core

A/C evaporator core housing (RH)

A/C evaporator core housing (LH)

Blower motor

O-ring

Cycling switch

Blower motor resistor assembly

A/C suction accumulator/drier

Orifice tube

Figure 5-30 Exploded view of an evaporator assembly.

NOTE: If not H-valve system, skip steps 5 and 6 and proceed with step 7.

5. Remove the suction/liquid lines from the H-valve.
6. Remove and discard the gasket or O-rings, as applicable.
7. Remove the accumulator if equipped, following procedures as outlined in this chapter.
8. Remove any mechanical linkages or vacuum lines from the evaporator controls.
9. Remove mounting bolts and hardware, as applicable, from the evaporator housing.
10. Separate the housing to gain access to the evaporator core.
11. Carefully lift the evaporator assembly from the vehicle.

NOTE: Do not force the assembly.

12. If there is a measurable amount of lubricant in the evaporator, add an equivalent amount of proper, clean, and fresh lubricant to the replacement evaporator.
13. Install a replacement evaporator by reversing steps 1, 3, and 7–11 or 5 and 7–11, as applicable.

H-valve systems are not equipped with an accumulator.

Use new gaskets and O-rings when required.

Part 3. Factory or Dealer Installed (Import) (Figure 5-31)

1. Remove the liquid line and suction line from the evaporator.
2. Immediately cap the fittings (evaporator and hose) to keep moisture and debris out of the system.
3. Remove and discard O-rings or gaskets, as applicable.
4. Remove any obstructions, such as dash panels or glove box to gain access to the cooling unit.
5. Remove mechanical linkages, electrical wires, or vacuum lines from the cooling unit.
6. Remove mounting bolts and hardware, as applicable, from the cooling unit.
7. Lift the cooling unit assembly from the vehicle. There should be no need to force the assembly.

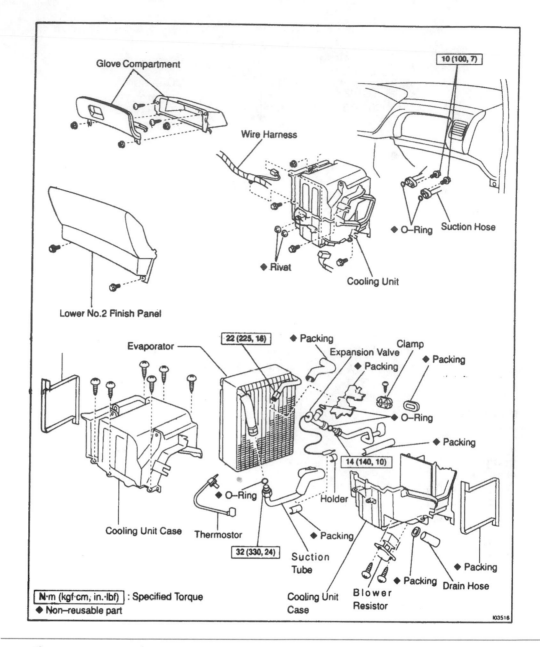

Figure 5-31 Typical import cooling unit. (Reprinted with permission)

8. Separate the housing to gain access to the evaporator core.

9. Install a replacement evaporator core by reversing steps 1 and 4–8.

NOTE: Use new O-rings or gaskets to replace those removed in step 3.

10. If there was a measurable amount of lubricant in the evaporator add an equivalent amount of proper, clean, and fresh lubricant to the replacement evaporator.

Part 4. Rear Heating/Cooling Unit (Figure 5-32)

There are many variations on procedures to service the rear heating/cooling unit. The manufacturer's service manual should be followed for any particular procedure. The following procedure is given as typical only.

1. Drain the coolant from the radiator. It may not be necessary to drain all of the coolant, however.

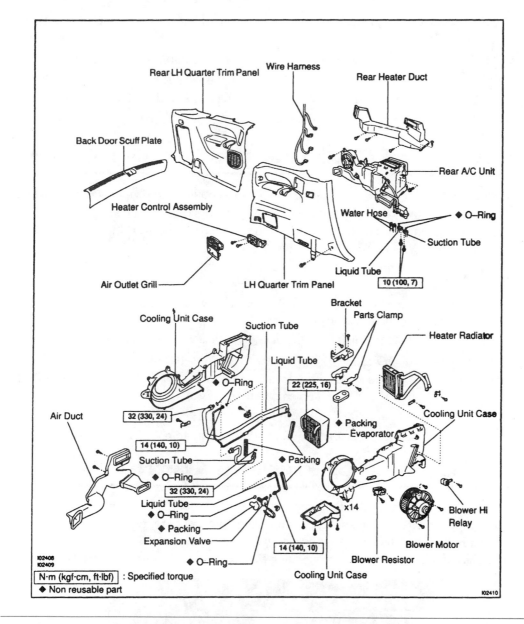

Figure 5-32 Typical rear heating/cooling unit. (Reprinted with permission)

2. Remove the liquid line and suction line from the evaporator.
3. Immediately cap the fittings (evaporator and hose) to keep moisture and debris out of the system.
4. Remove and discard O-rings or gaskets, as applicable.
5. Remove the coolant hoses from the heater core.
6. Remove the heater grommet, if applicable.
7. Remove any obstructions, such as seats, panels, trim, or controls to gain access to the heating/cooling unit.
8. Remove any mechanical linkages, electrical wires, or vacuum lines from the heating/cooling unit.
9. Remove mounting bolts and hardware, as applicable, from the unit.
10. Lift the heating/cooling assembly from the vehicle. There should be no need to use force.

11. Separate the case to gain access to the evaporator and heater cores as well as to the metering device.

NOTE: If replacing the evaporator and there was a measurable amount of lubricant in the evaporator, add an equivalent amount of proper, clean, and fresh lubricant to the replacement.

12. Install a replacement component by reversing steps 1, 2, and 5–11.

NOTE: Use new O-rings or gaskets to replace those removed in step 4.

Removing and Replacing the Receiver-Drier

Special Tools

Flare nut wrench set
Torque wrench

The receiver-drier contains the desiccant in a thermostatic expansion valve (TXV) system.

Classroom Manual
Chapter 5, page 116

1. Recover the air conditioning system refrigerant as outlined in Chapter 7 of this manual.
2. Remove the low- or high-pressure switch wire, if applicable.
3. Remove the inlet and outlet hoses (liquid lines) from the receiver-drier.
4. Remove and discard O-rings or gaskets, if applicable.
5. Loosen and/or remove the mounting hardware.
6. Remove the receiver-drier from the vehicle (Figure 5-33).
7. Measure the amount of lubricant in the receiver-drier. Add the same amount of clean, fresh lubricant to the replacement receiver-drier, step 9.
8. Remove the pressure switch from the drier, if applicable. Discard the gasket or O-ring.
9. Install a new receiver-drier. Reverse the order of removal, steps 2, 3, and 5–7.

NOTE: Most receiver-driers are marked with an arrow (→) or the word(s) **IN** and **OUT** to denote the direction of refrigerant flow. Remember, flow is away from the condenser and toward the metering device.

10. Leak-test, evacuate, and charge the air conditioning system.

For a Superheat or Pressure Switch

Special Tools

Internal snap ring plier, if applicable
Appropriate wrench for the type pressure/temperature switch

Determine the location of the switch to be replaced. If it is the superheat switch on the rear head of a Harrison compressor, follow the applicable steps in the procedure for replacing the compressor.

If it is a pressure switch on the accumulator, follow the appropriate steps in the procedure for replacing the accumulator. If it is a drier pressure switch, follow the appropriate steps in the proce-

Figure 5-33 Remove the receiver-drier from the vehicle.

dure for replacing the drier. If it is anywhere else in the system, the following general procedures may apply:

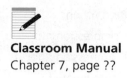

Classroom Manual
Chapter 7, page ??

1. Recover the refrigerant as outlined in Chapter 7 of this manual.
2. Using the proper tool, remove the defective component.
3. Remove and discard the gasket and/or O-ring.
4. Place a new gasket and/or O-ring on the new component.
5. Install the new component, again using the proper tool.
6. Leak-test, evacuate, and charge the system with refrigerant.
7. Hold performance test, if indicated.

In conclusion:

❏ Use only components and parts designated for a particular system: CFC-12 or HFC-134a.
❏ Use new gaskets and/or O-rings when replacing a component.
❏ Liberally coat all components with clean refrigeration oil before reassembly.
❏ For reassembly, reverse the removal procedure.

CASE STUDY

A customer complains of a chattering noise that occurs only when driving home.
Technician: "The noise occurs only on the way home, never on the way to work?"
Customer: "That's right, only on the way home."
Technician: "What time do you go home?"
Customer: "Usually seven or eight o'clock. I've even tried to take different routes. The chatter always happens about five miles from work."
Technician: "Do you operate your air conditioner in the evening?"
Customer: "I seldom turn it off. I like to avoid the road fumes whenever I can."
Technician: "Are you sure that it is the air conditioner?"
Customer: "Yes. When it makes a noise, I turn the air conditioner off. The noise stops."
Technician: "Only on the way home from work. Does it ever happen when you're on the way home from a movie or the grocery store?"
Customer: "When we go out later, I take my wife's car."
Technician: "Why?"
Customer: "I have a bad generator (alternator) and use my little battery charger to keep the battery up. An overnight charge is just enough to get me to work and back."

Based on what you have learned, you know that low voltage will cause the clutch to chatter. By the time the customer drove approximately five miles toward home with his headlamps on, the battery voltage was reduced to a level that caused this chatter. The problem is a defective alternator; it is not an air conditioning problem at all.

Get to know your customers. The more you communicate, the more effective you are when performing troubleshooting and diagnosis of needed service. Be it a groan, chatter, a squeak, or a bang, the customer will eventually reveal the problem. Often, the customer will hear a noise that you may not identify as being a problem. The customer hears it as a noise that is different than those experienced since owning the vehicle. Questioning the customer will often reveal when, why, and how.

Other problems may cause an air conditioning system to malfunction.

Terms to Know

Barb fitting	Insert fitting	Rebuilt
Compression nut	Metric fastener	SAE
Compression ring	Neutral	Service procedures
Contaminated	Noncycling clutch	Spring lock
Disarm	O-ring	Torque
English fastener	Park	
Gasket	Predetermined	

ASE-Style Review Questions

1. Before installation,
   Technician A says that all gaskets and O-rings should be coated with Permatex.
   Technician B says they should be coated with Form-a-gasket.
   Who is right?
   - **A.** A only
   - **B.** B only
   - **C.** Both A and B
   - **D.** Neither A nor B

2. *Technician A* says that the hot gas discharge hose runs from the condenser to the drier.
   Technician B says that the hot gas discharge line also runs from the drier to the metering device.
   Who is right?
   - **A.** A only
   - **B.** B only
   - **C.** Both A and B
   - **D.** Neither A nor B

3. *Technician A* says that it is permissible, according to the EPA, to release refrigerant from the hoses of a manifold and gauge set because it is such a small amount.
   Technician B says that, to minimize refrigerant loss, shut-off valves are required by the EPA within 12 in. (3.04 cm) of the service end of the manifold and gauge set hoses.
   Who is right?
   - **A.** A only
   - **B.** B only
   - **C.** Both A and B
   - **D.** Neither A nor B

4. *Technician A* says that desiccant is found in the receiver-drier.
   Technician B says that desiccant is found in the accumulator.
   Who is right?
   - **A.** A only
   - **B.** B only
   - **C.** Both A and B
   - **D.** Neither A nor B

5. *Technician A* says that the expansion tube is located at or before the evaporator inlet.
   Technician B says that the expansion tube is located at or after the receiver-drier.
   Who is right?
   - **A.** A only
   - **B.** B only
   - **C.** Both A and B
   - **D.** Neither A nor B

6. Another name for the fixed orifice tube (FOT).
   Technician A says it is expansion tube.
   Technician B says that it is low-side tube.
   Who is right?
   - **A.** A only
   - **B.** B only
   - **C.** Both A and B
   - **D.** Neither A nor B

7. *Technician A* says that the accumulator should be replaced if there is moisture in the system.
   Technician B says that the accumulator must be replaced if the fixed orifice tube (FOT) is replaced.
   Who is right?
   - **A.** A only
   - **B.** B only
   - **C.** Both A and B
   - **D.** Neither A nor B

8. When removing the accessible orifice tube:
   Technician A says both the T-handle and the outer sleeve are turned.
   Technician B says that either the T-handle or the outer sleeve may be turned.
   Who is right?
   - **A.** A only
   - **B.** B only
   - **C.** Both A and B
   - **D.** Neither A nor B

9. *Technician A* says that all orifice tubes are the same size, but they are not interchangeable. *Technician B* says that all orifice tubes are not the same size, but they are interchangeable. Who is right?
   A. A only
   B. B only
   C. Both A and B
   D. Neither A nor B

10. *Technician A* says that mineral oil may be used to lubricate O-rings used on an HFC-134a system. *Technician B* says that PAG and ester lubricants may also be used on O-rings for an HFC-134a system. Who is right?
    A. A only
    B. B only
    C. Both A and B
    D. Neither A nor B

ASE Challenge

1. The phrase *subject to use conditions* means that:
   A. Only EPA-approved refrigerant may be used in a vehicle
   B. Proper lubricant for the refrigerant must be used
   C. Fittings and labels must be in place to identify refrigerant used
   D. All of the above

2. Refrigerant superheat is being discussed.
   Technician A says that superheated refrigerant is at a temperature above its boiling point.
   Technician B says that superheated refrigerant is caused by an overcharge of refrigerant.
   Who is right?
   A. A only
   B. B only
   C. Both A and B
   D. Neither A nor B

3. An accumulator contains all of the following EXCEPT:
   A. Oil bleed hole
   B. Orifice tube
   C. Desiccant
   D. Fine-mesh screen

4. Most vendors will not honor the warranty on a new or rebuilt compressor if the _____ is not replaced at the time of service.
   A. Accumulator or receiver-drier
   B. Expansion valve or orifice tube
   C. O-ring seals or gaskets
   D. Oil or lubricant

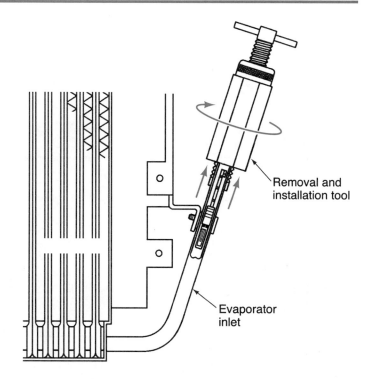

Removal and installation tool

Evaporator inlet

5. The illustration above shows:
   A. A spring lock coupling being removed
   B. An O-ring seal or gasket being removed
   C. Both A and B
   D. Neither A nor B

Table 5-1 ASE TASK

Diagnose A/C system problems that cause the protection devices (pressure, thermal, and control modules) to interrupt system operation; determine needed repairs.

Problem Area	Symptoms	Possible Causes	Classroom Manual	Shop Manual
COMPRESSOR CLUTCH CYCLES FREQUENTLY	System cycles on low pressure	1. Undercharge of refrigerant	86	110
		2. Restriction in low side of system	87	110
		3. Defective evaporator fan motor	84	110
		4. Defective control	89	111
	System cycles on high pressure	1. Overcharge of refrigerant	87	77–78
		2. Restriction in high side of system	87	77–78
		3. Defective (mechanical) fan clutch	89	111
		4. Defective (electrical) cooling fan motor	89	111

Table 5-2 ASE TASK

Inspect and replace A/C system pressure protection devices.

Problem Area	Symptoms	Possible Causes	Classroom Manual	Shop Manual
INTERMITTENT, POOR, OR NO COOLING	Clutch cycles frequently on pressure control	1. Defective low-pressure control	225	354
		2. Defective high-pressure control	226	354

Table 5-3 ASE TASK

Inspect, repair, or replace A/C system mufflers, hoses, lines, filters, fittings, and seals.

Problem Area	Symptoms	Possible Causes	Classroom Manual	Shop Manual
NOISE	Dull pulsating sound	Improper routing of hose(s)	89	151
NO OR POOR COOLING	Loss of refrigerant	1. Defective hose(s)	89	151
		2. Damaged fitting	168–169	151
		3. Leaking seal (O-ring)	207	151

Table 5-4 ASE TASK

Inspect and replace receiver-drier (in TXV A/C systems).
Inspect and replace accumulator-drier in orifice tube A/C systems.

Problem Area	Symptoms	Possible Causes	Classroom Manual	Shop Manual
POOR OR NO COOLING	Moisture in the system	1. Saturated desiccant in receiver-drier	115–117	166
		2. Saturated desiccant in accumulator	122–123	164

Table 5-5 ASE TASK

Inspect, test, and replace expansion valve.
Inspect, test, and replace orifice tube.

Problem Area	Symptoms	Possible Causes	Classroom Manual	Shop Manual
POOR OR NO COOLING	Lower than normal low-side pressure	1. Defective or restricted expansion valve	117	156
		2. Defective or restricted orifice tube	120	160
	Higher than normal low-side pressure	Defective thermostatic expansion valve	117	156

Table 5-6 ASE TASK

Inspect, test, or replace evaporator.
Inspect, clean, and repair evaporator housing and water drain.
Inspect, test, and replace evaporator pressure control systems and devices.

Problem Area	Symptoms	Possible Causes	Classroom Manual	Shop Manual
POOR OR NO COOLING	Loss of refrigerant	Leaking evaporator core	120–122	166, 286
	Reduced airflow	1. Dirty evaporator fins	120–122	166
		2. Cracked/broken housing	120–122	286
	Improper low-side pressure	Adjust pressure control (if equipped)	89	111

Table 5-7 ASE TASK

Inspect and replace A/C system high-pressure relief device.

Problem Area	Symptoms	Possible Causes	Classroom Manual	Shop Manual
LOSS OF REFRIGERANT	High-pressure relief not seating properly	Defective high-pressure relief valve	226	354

JOB SHEET 14

14

Name _____ Date _____

Determining the Type of Air Conditioning System

Upon completion of this job sheet you should be able to identify the type of air conditioning system: cycling clutch or noncycling clutch.

ASE Correlation

This job sheet is related to the ASE Heating and Air Conditioning Test's content area: *A/C System Diagnosis and Repair,* Task: *Identify system type and conduct performance test of the A/C system; determine needed repairs.*

Tools and Materials

An air conditioned vehicle
Manufacturer's service manual

Describe the vehicle being worked on:

Year _____ Make _____ Model _____

VIN _____ Engine type and size _____

Procedure

1. Visually inspect the air conditioning system and describe its overall condition.

   _____

   _____

2. In a well-ventilated area, start the engine, place the transmission in park and turn on the air conditioning system to MAX cooling. If a standard transmission, place in NEUTRAL and chock the wheels. Describe your procedure for accomplishing this step.

   _____

   _____

3. Move the cold control from one extreme to the other.

   a. Does the compressor cycle off and on? Describe your findings.

   _____

   _____

   b. Does the heater control valve change positions? Describe your findings.

   _____

   _____

4. What is the type system: cycling clutch or noncycling clutch?

   _____

   _____

5. How did you make this determination?

_____

_____

6. Locate the temperature control information in the shop manual. Where is this information located?

_____

_____

7. Does the information presented in step 6 verify your determination of step 5?

_____

_____

✔ **Instructor's Check** _____

JOB SHEET 15

15

Name _____ Date _____

Determining the Refrigerant Type

Upon completion of this job sheet you should be able to identify the type refrigerant used in the air conditioning system.

ASE Correlation

This job sheet is related to the ASE Heating and Air Conditioning Test's content area: *Refrigerant Recovery, Recycling, Handling, and Retrofit*, Task: *Identify and recover A/C system refrigerant*.

Tools and Materials

An air conditioned vehicle
Manufacturer's shop manual

Describe the vehicle being worked on:

Year _____ Make _____ Model _____
VIN _____ Engine type and size _____

Procedure

1. Visually inspect the air conditioning system and describe its overall condition.

   _____
   _____

2. Inspect the low-side service fitting. Is it a CFC-12, HFC-134a, or other type fitting? Describe.

   _____
   _____

3. Inspect the high-side service fitting. Is it a CFC-12, HFC-134a, or other type fitting? Describe.

   _____
   _____

4. Are there any decals under the hood to identify the refrigerant type? Describe your findings.

   _____
   _____

5. What type of refrigerant is in the system?

   _____
   _____

6. Does the shop manual verify your findings?

   _____
   _____

7. What special precautions would you take when recovering this refrigerant?

_____

_____

✔ **Instructor's Check** _____

JOB SHEET 16

16

Name _____ Date _____

Identifying Hose Fittings

Upon completion of this job sheet you should be able to identify the type of hose fittings used in the air conditioning system.

ASE Correlation

This job sheet is related to the ASE Heating and Air Conditioning Test's content area: *Refrigeration System Component Diagnosis and Repair, 2. Evaporator, Condensor, and Related Components,* Task: *Inspect, repair, or replace A/C system mufflers, hoses, lines, filters, fittings, and seals.*

Tools and Materials

An air conditioned vehicle
Manufacturer's shop manual

Describe the vehicle being worked on:

Year _____ Make _____ Model _____

VIN _____ Engine type and size _____

Procedure

1. Visually inspect the air conditioning system and describe its overall condition.

   _____

   _____

2. Inspect the hose to condenser inlet fitting. Describe its type.

   _____

   _____

3. Inspect the high-side liquid line to evaporator inlet fitting. Describe its type.

   _____

   _____

4. Are the two hose fittings (steps 2 and 3) interchangeable? Explain.

   _____

   _____

5. Why do you think that a barb-type fitting should not be used with a barrier hose?

   _____

   _____

6. Locate hose fittings in the shop manual. Does the shop manual verify your findings?

   _____

   _____

7. What special precautions would you take when servicing hoses and fittings?

_____

_____

☑ **Instructor's Check** _____

JOB SHEET 17

17

Name _____ Date _____

Replacing the Receiver-Drier

Upon completion of this job sheet you should be able to identify a receiver-drier and describe how to replace it.

ASE Correlation

This job sheet is related to the ASE Heating and Air Conditioning Test's content area: *Refrigeration System Component Diagnosis and Repair, 2. Evaporator, Condensor and Related Components*, Task: *Inspect and replace receiver-drier or accumulator-drier.*

Tools and Materials

An air conditioned vehicle with a defective receiver-drier
Refrigerant recovery station
Set of mechanic's hand tools
Manufacturer's shop manual

Describe the vehicle being worked on:

Year _____ Make _____ Model _____

VIN _____ Engine type and size _____

Procedure

1. Visually inspect the air conditioning system and describe its overall condition.

_____

_____

2. Locate the receiver-drier. Describe its location.

_____

_____

3. Troubleshoot the receiver-drier following procedures outlined in the shop manual. Is it defective? Explain your findings.

_____

_____

4. Are there decals under the hood to identify the refrigerant type? Describe your findings. What type of refrigerant is in the system?

_____

_____

5. Does the shop manual verify your findings?

_____

_____

6. What procedure would you follow to replace the receiver-drier?

_____

_____

☑ **Instructor's Check** _____

JOB SHEET 18

18

Name _____ Date _____

Replacing the Superheat or Pressure Switch

Upon completion of this job sheet you should be able to troubleshoot and replace a high-pressure release device.

ASE Correlation

This job sheet is related to the ASE Heating and Air Conditioning Test's content area: *Refrigeration System Component Diagnosis and Repair, 2. Evaporator, Condensor and Related Components*, Task: *Inspect and replace A/C system high-pressure relief device.*

Tools and Materials

An air conditioned vehicle with high-pressure relief valve
Refrigerant recovery station
Set of mechanic's hand tools
Manufacturer's shop manual

Describe the vehicle being worked on:

Year _____ Make _____ Model _____

VIN _____ Engine type and size _____

Procedure

1. Visually inspect the air conditioning system and describe its overall condition.

   _____

   _____

2. Locate the high-pressure relief valve. Describe its location.

   _____

   _____

3. Troubleshoot the high-pressure relief valve following procedures outlined in the shop manual. Is it defective? Explain your findings.

   _____

   _____

4. Can the relief valve be replaced without recovering the refrigerant? Explain your answer.

   _____

   _____

5. Are there decals under the hood to identify the refrigerant type? Describe your findings. What type of refrigerant is in the system?

   _____

   _____

6. Does the shop manual verify your findings?

_____

_____

7. What procedure would you follow to replace the high-pressure relief valve? _____

_____

_____

☑ **Instructor's Check** _____

Compressors and Clutches

Upon completion and review of this chapter you should be able to:

- Identify the various makes and models of compressors used in automotive air conditioning service.

- Check and correct the oil level in various models of compressors.

- Leak-test and replace shaft seals in various models of compressors.

- Leak test and correct shell and fitting leaks of various models of compressors.

- **Troubleshoot** and replace various types of compressors.

- Troubleshoot and make mechanical repairs to clutch coils and rotor assemblies.

Introduction

The compressor (Figure 6-1) is thought of as the heart of the automotive air conditioning system. Without the compressor, the system would not function. Actually, all five components of the system are essential if the system is to function properly. In addition to the connecting hoses and fittings, these five components are the compressor, condenser, metering device, evaporator, and accumulator or receiver-drier.

This chapter covers the compressor and clutch. The other components of the system are covered in detail in other chapters of this manual as well as the Classroom Manual. Refer to the index of either manual for further reference.

Compressor

There are over 12 manufacturers of compressors with hundreds of different models and configurations available for use on the modern automobile. Some that are still in use date back to 1961. Others, which are seldom found, have been discontinued generally because of size and weight.

Regardless of manufacture or model, the compressor must serve two important functions in the automotive air conditioning system. First, the compressor must create a low pressure in the system. This low pressure is to exist from the evaporator inlet metering device to the compressor inlet service fitting.

The second function of the compressor is to compress the refrigerant, changing it from a low-pressure and low-temperature vapor to a high-pressure and high-temperature vapor. This high-pressure and high-temperature vapor condition exists from the compressor outlet to the condenser inlet. It is in the condenser that the refrigerant vapor is changed to a liquid. While it is at slightly lower temperature, it is still at a high pressure until it again reaches the metering device, generally at the inlet of the evaporator.

Compressor Identification

When servicing a compressor, it is important to be able to identify its type, manufacturer, and model number. This is especially true for identifying replacement parts, such as a compressor **shaft seal**. It is, perhaps, equally important if it is necessary to replace the compressor. Following is a brief overview in alphabetical order of most of the compressors that are currently available, new or rebuilt.

Basic Tools

Basic technician's tool set

Refrigerant recovery equipment

Manifold and gauge set with hoses

Safety glasses/ goggles

Graduated container

Funnel

Fender covers

Proper refrigerant is also essential for an air conditioning system.

Classroom Manual Chapter 4, page 88

Classroom Manual Chapter 5, pages 112 and 120

Classroom Manual Chapter 6, pages 139

Compressors are designed to pump vapor only. Liquid may cause severe damage.

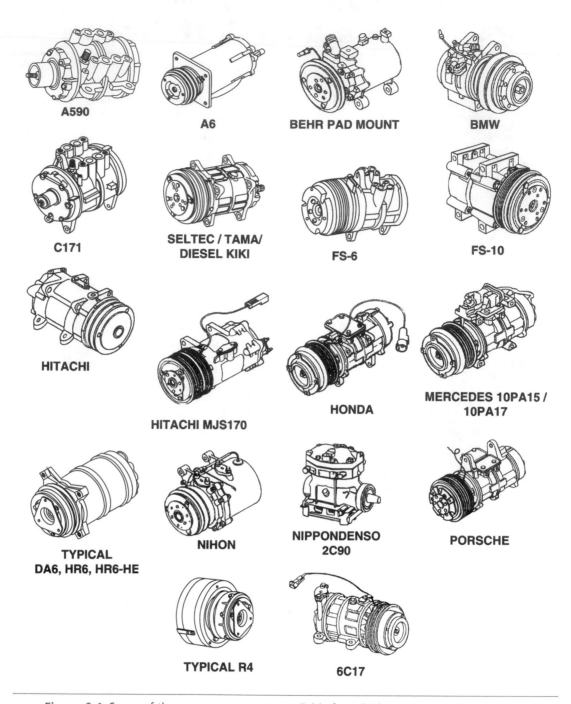

A590

A6

BEHR PAD MOUNT

BMW

C171

SELTEC / TAMA/ DIESEL KIKI

FS-6

FS-10

HITACHI

HITACHI MJS170

HONDA

MERCEDES 10PA15 / 10PA17

TYPICAL DA6, HR6, HR6-HE

NIHON

NIPPONDENSO 2C90

PORSCHE

TYPICAL R4

6C17

Figure 6-1 Some of the many compressors available for vehicle service.

Behr/Bosch

There are two types of Behr and Bosh compressors, both found on cars manufactured by BMW: the wing cell (rotary) and oscillating (swash) plate. In all, there are four basic styles; three rotary and one swash plate. They are not interchangeable in that their mounting provisions and hose connections vary by model and style. This compressor is generally identified by the **OEM decal** on the compressor body.

Calsonic

There are 12 identified models of the Calsonic compressor. They are found primarily on 1987 and later Infinity and Nissan car lines as well as on a few Isuzu and Saturn car lines. Extreme care must be taken when replacing a Calsonic compressor. To ensure proper application, the important considerations are:

- ❏ Mounting provisions
- ❏ Tube fittings and location
- ❏ Number of grooves and spacing of clutch
- ❏ Clutch coil terminal location

Basically, there are one single-groove, four 4-groove poly-, and one 6-groove poly-clutch-equipped Calsonic compressors available. A Calsonic compressor may sometimes be found as a replacement for a Diesel Kiki compressor. If it is to be used to replace a Diesel Kiki, the Calsonic must first be carefully compared to the Diesel Kiki to ensure compatibility before attempting replacement.

Chrysler

Chrysler no longer manufactures a compressor. Their popular C171 and A590 compressors are manufactured by Nippondenso. These two compressors are identified as follows:

Model A590

The model A590 compressor is found on 1985 and later Chrysler car lines. There are eight different models of this compressor: four with standard clutch coils and four with **high-torque clutch** coils. Also, there are two models with single-groove clutch rotors and two models with four-groove clutch rotors. High-torque clutch coils and rotors are discussed in more detail in the "Clutch" section later in this chapter. Though this compressor resembles the earlier C171 compressor, it is not interchangeable and therefore not a replacement for this compressor.

Model C171

The C171 compressor, manufactured by Nippondenso (Figure 6-2), is found on 1980 and later Chrysler car lines. There is one basic C171 compressor that has nine different clutch combinations with five different styles of clutch pulley:

- ❏ Two single-groove (four-cylinder or six-cylinder engine)
- ❏ Two double-groove (four-cylinder or V8 engine)
- ❏ One eight-groove poly (V8 engine)

There are three different types of clutch coil available:

- ❏ Single terminal (used up to 1983)
- ❏ Double terminal (1984 and later)
- ❏ High torque (double terminal)

High-torque clutches are used for systems operating under higher than normal head pressures, such as vans.

Delco Air

The Delco Air compressor is also known as the Delco, Delco Harrison, Frigidaire, Frigidaire Harrison, and Harrison. This line of compressors is covered as "Harrison" in this manual.

Figure 6-2 Typical Nippondenso six-cylinder compressor. (Courtesy of BET, Inc.)

Diesel Kiki

The Diesel Kiki compressor may be found on Isuzu, Mazda, and Volvo as well as some other applications. It is also a popular compressor for use in aftermarket applications. There are nine different types with four different clutch pulley arrangements. Occasionally, the Diesel Kiki will have other compressor line labels, such as Calsonic, Nihon Radiator, Celtec, or Zexel. The important consideration, regardless of name, is to refer to the "Type Code" and/or "Part Number" found on the sticker or tag on the compressor housing. Note the number of **belt** grooves on the compressor clutch rotor as well as its diameter. Also, measure the distance between the **mounting bosses**, center to center and outside to outside, to be sure of a proper match.

Ford/Lincoln/Mercury

The term *logo* refers to a company trademark, such as the Ford Motor Company "FORD" or General Motors' "GM."

The Ford Motor Company does not manufacture a compressor. Those under their logo, models FS6 and FX15, are manufactured for them by Nippondenso. Several other compressors may be found on Ford car line applications, such as the new Panasonic Vane Rotary that made its appearance in 1993. (See also Panasonic.)

Model FS6

The FS6 compressor, manufactured by Nippondenso for Ford, should not be confused with other look-alike Nippondenso compressors, such as models 6P148 and 10P15. There are only two types of model FS6 compressor used on 1980 and later Ford car lines: standard mount and cross bolt mount. The cross bolt mount compressor is only available with a six-groove poly-belt clutch pulley assembly, while the standard mount is available with either one or two V-groove, or five or six-groove poly-belt clutch pulley assembly.

▲ **WARNING:** Because of its design the FS6 compressor will not tolerate an **overcharge** of refrigerant. An overcharge of just 3.2 oz. (94.6 mL) will result in certain compressor failure.

The important considerations for selecting a replacement model FX6 compressor are:

❑ Type of mount.

❑ Clutch and pulley type, if standard mount.

❑ Mounting distance; front surface of the front head to the center of the pulley groove, single-groove pulley only. This will be 0.84 in. (21.3 mm) or 1.34 in. (34 mm).

Model FX15/FS10

The FX15/FS10 compressor, manufactured by Nippondenso, has been used on Ford car lines since 1988. Its radial design is unique in that it has five double-acting pistons driven by a swash plate. This is a somewhat difficult compressor to match up since Ford incorporated a "**running design change**" in the middle of 1991. In an effort to develop a more efficient compressor, one of the changes was to redesign the splined shaft that drives the clutch armature. The splined **shaft** was changed from 16 points to 21 points. Therefore, clutches used on early FX15/FS10 compressors will not fit later FX15/FS10 compressors and vice-versa. If it is necessary to replace the compressor, it is desirable to replace the compressor and clutch as an assembly rather than to attempt to match them. The primary consideration is the number of grooves on the clutch rotor: five, six, seven, or eight. The pulley diameter and coil location are also important considerations with a six-poly groove clutch.

Not all splined shafts are the same. Many have a different number of splines and/or are a different diameter.

Frigidaire

The Frigidaire compressor is also known as Delco, Delco Air, Delco Harrison, Frigidaire Harrison, and Harrison. This line of compressors is covered under the heading of "Harrison."

Harrison

There are four models of Harrison compressors in use today. They are model A6, model DA6/HR6/HR6HE, model R4, and model V5.

Classroom Manual
Chapter 6,
page 139

Model A6

The model A6 compressor (Figure 6-3) was claimed by many to be the most reliable and trouble-free compressor ever designed. Nonetheless, its heavy weight—over 35 lb. (15.9 kg)—forced its phase-out in the mid 1970s in favor of the lighter model R4. The A6 compressor, which was introduced in 1959, was still used as original equipment until well into the 1980s. The model A6 was also used by Audi beginning in 1978, Ford in 1972, and Jaguar in 1979. There are three factors in determining a proper replacement for the model A6 compressor.

First, identify the clutch. A double-groove clutch is the simplest because there is only one type. If it is a single-groove clutch, however, it will be one of the following four types:

❏ Small pulley diameter; less than 5.25 in. (133 mm). This is the most common application.

Figure 6-3 A Harrison six-cylinder compressor and V-belt clutch assembly. (Courtesy of BET, Inc.)

❑ Large pulley diameter (found mostly on trucks).

❑ Cadillac (1976 and earlier models).

❑ Off-road (farm equipment) with dust shield.

Next, identify the switch found in the rear head of the compressor.

❑ If it is a 1977 General Motors vehicle or any things other than a General Motors vehicle, regardless of year/model, use a nonswitch compressor.

❑ If it is a General Motors vehicle prior to 1977, use a compressor with a **superheat switch** in the rear head. The superheat switch, used on valves-in-receiver (VIR) systems, may be identified as having a large hole in the center of the cavity.

❑ If it is a General Motors vehicle, 1978 and later with a cycling clutch orifice tube (CCOT) system, use a compressor with a high-side pressure switch. A high-side pressure port is identified with a small off-center (offset) hole in the cavity.

Finally, identify the coil voltage and coil terminal location.

❑ 12 volts; generally for all automobiles and light trucks.

❑ 24 volts; generally for over-the-road and off-road vehicles.

❑ Facing the clutch coil, the terminal location will be either up to the left or up to the right.

Model DA6/HR6/HR6HE

The Harrison "Delco Air 6" (DA6), an extremely successful compressor, was first introduced in 1983. Recognizing that there were design deficiencies with the original model, General Motors superseded the DA6 (Figure 6-4) with a "Harrison Redesigned 6" (HR6). Further improvements over the years have produced the model HR6HE. This model number, HR6HE, is actually an acronym for the "Harrison Redesigned 6 (cylinder) High Efficiency" compressor. This compressor is found in many applications and its popularity is increasing each year. If it is necessary to replace this compressor, the following factors must be considered:

 WARNING: THE SUCTION PORT MUST BE FACING UP. If the suction port is in any other position, failure of the compressor will occur due to oil starvation.

❑ Identify the head configuration mounting type. There are nine configurations to ensure that the suction port is at its highest position when the compressor is properly installed.

Figure 6-4 The popular DA6 compressor by Harrison; shown with a poly-groove clutch assembly. (Courtesy of BET, Inc.)

- ❏ Identify the pulley diameter. Measure the pulley diameter to determine proper size. It will be from 4.21 in. (106.9 mm) to 5.6 in. (142.2 mm).
- ❏ Identify the coil terminal location. With the clutch facing you and the compressor in its operating position, the coil terminal will be at either 10 o'clock, 2 o'clock, or 6 o'clock.
- ❏ Identify the coil connector type: **pin type** or **spade type**.

⚠️ **WARNING:** General Motors offers both standard-amperage and high-amperage coils with either pin-type or spade-type connectors. It is not generally recommended that the two amperage types be interchanged.

- ❏ Identify the number of high-side switches in the rear head: none, one, or two.
- ❏ Identify the **mounting** distance from the clutch belt groove farthest from the compressor to the front mounting **flange** on the compressor. This will be 0.87 in. (22.1 mm), 0.92 in. (23.4 mm), or 1.23 in. (31.2 mm).

⚠️ **WARNING:** The measurements must be extremely accurate so the belt(s) will line up properly.

Model R4

In spite of the fact that the R4 compressor (Figure 6-5) is described by many as the "lemon of compressors," since it was introduced in 1975 it has become the largest selling compressor in the automotive air conditioning industry. The truth is, this compressor, manufactured almost entirely of aluminum (Al), simply will not tolerate poor workmanship during service procedures. The proper procedures for servicing a model R4 compressor is covered in this chapter.

⚠️ **WARNING:** The R4 compressor system oil charge is critical. Follow the instructions given in this and manufacturers' shop manuals for checking and adjusting the oil charge to ensure a long compressor service life.

To determine the proper replacement for an R4 compressor, follow these steps:
- ❏ Identify the style of the clutch
- ❏ Number of grooves: one V-groove or six-groove poly-groove

High-side switches include high pressure and superheat.

Classroom Manual
Chapter 6,
page 134

An undercharge of oil will result in early compressor failure; an overcharge in inadequate performance.

Classroom Manual
Chapter 4,
page 88

Figure 6-5 Harrison's R4 compressor with V-belt clutch assembly. (Courtesy of BET, Inc.)

Classroom Manual
Chapter 6,
page 139

American Motors
was acquired by
Chrysler Corpo-
ration.

❏ Diameter of clutch rotor. Measure the diameter of the grooves; not the rim of the pulley. The belt may be changed to "adjust" for a different diameter pulley.

❏ Mounting distance. Measure from the center of the front pulley groove to the front of the compressor case. This measurement must be accurate to ensure proper belt alignment.

❏ Clutch coil location: 3 o'clock or 6 o'clock.

❏ Switch-type access on rear head. A high-side-type switch was used on 1977 and later model vehicles. It may be identified by an offset hole found in the bottom of the switch cavity. A superheat-type switch was used on earlier models. It may be identified by a large hole found in the center of the bottom of the cavity.

⚠ **WARNING:** The high-side and superheat switches are not interchangeable, although each will fit into either cavity.

When an R-4 compressor has failed and a new or rebuilt compressor is installed, it is recommended that a filter with a molecular-sieve desiccant be installed in the liquid line of the air conditioning system. The molecular-sieve desiccant is capable of removing fine aluminum particles from the refrigerant to prevent clogging the inlet screen in the metering device. This filter, however, may itself become clogged and must be removed or replaced.

The R4 compressor is also used by American Motors, General Motors, Mercedes Benz, Peugeot, and Volvo as well as some other car lines.

Servicing the R4 Compressor

This service procedure is given in three sections: servicing the **seal** assembly, checking and adding oil, and servicing the clutch. The services outlined in this procedure assume that the compressor has been removed from the vehicle for repair.

Replacing the Seal on the R4 Compressor

Seal replacement on the Harrison R4 four-cylinder compressor is accomplished using a different procedure than is required for other types of compressors. See Photo Sequence 6 for a typical procedure for removing and replacing the R4 compressor seal assembly.

Special Tools

Clutch hub holder

Holding fixture

Hub and drive plate
remover/installer

Internal snap ring
pliers

Seal seat
remover/installer

O-ring
remover/installer

Seal protector

Shaft seal seat
remover/installer

Photo Sequence 6
Typical Procedure for Removing and Replacing the R4 Compressor Seal Assembly

P6-1 Using a 9/16-in. thinwall socket and hub holding tool, remove the shaft nut.

P6-2 Using the clutch hub and drive plate puller, remove the hub and drive plate. If the shaft key did not come out, remove the key.

P6-3 Remove the shaft seal seat retainer ring using the snap ring pliers.

P6-4 Remove the seal seat using the shaft seal seat remover.

P6-5 Using the shaft seal remover, remove the shaft seal.

P6-6 Using an O-ring remover, remove the shaft seal seat O-ring. Take care not to scratch the mating surfaces.

P6-7 Ensure that the inner bore of the compressor is free of all foreign matter. Flush the area with clean refrigeration oil.

P6-8 Place the seal seat O-ring on the installer tool and slide the O-ring into place. Remove the tool.

P6-9 Liberally coat the shaft seal with refrigeration oil and place it on the shaft seal installer tool. Slide the shaft into place in the bore. Rotate the seal clockwise until it seats on the flats provided. Rotate the tool counterclockwise and remove it.

P6-10 Place the shaft seal seat on the remover/installer tool. Slide the shaft seal seat into position and remove the tool. (Repeat Figure P1-4.)

P6-11 Install the shaft seal seat snap ring. Note that the beveled edge of the snap ring must face the outside of the compressor.

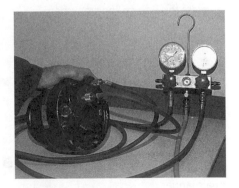

P6-12 Install the test fitting and connect a manifold and gauge set to the test fitting ports.

Photo Sequence 6
Typical Procedure for Removing and Replacing the R4 Compressor Seal Assembly (continued)

P6-13 Connect the manifold to a pressure source and open the high- and low-side manifold hand valves to pressurize the compressor.

P6-14 With a leak detector, check the shaft seal area for escaping refrigerant. Refer to Shop Manual Chapter 7 and Classroom Manual Chapter 7. If a small leak is detected, rotate the crankshaft a few turns to seat the seal; then recheck the seal area for leaks. If the leak is heavy or if it persists, the seal must be removed and checked for defects.

P6-15 Place the drive key into the clutch plate keyway. About 3/16 in. (4.8 mm) of the key should be allowed to protrude over the end of the keyway. Align the key with the keyways of the drive plate and compressor crankshaft. Then slide the drive plate into position.

P6-16 Using a hub and drive plate installer, press this part on the crankshaft.

P6-17 A clearance of 0.030 in. ±0.010 in. (0.76 mm ±0.25 mm) should exist between the drive plate and the rotor.

P6-18 Replace the shaft nut.

Classroom Manual
Chapter 6,
page 139

Oil must be discarded in a manner consistent with the Environmental Protection Agency (EPA) guidelines.

WARNING: Careful handling of all seal parts is important. The **carbon seal face** and the steel **seal seat** must not be touched with the fingers because of the **etching** effect of the acids normally found on the fingers.

Checking and Adding Oil to the Harrison R4

The design of the R4 compressor requires a different oil checking procedure than that used for other types of compressors. The Harrison R4 compressors are factory charged with 5.5 to 6.5 fluid oz. (163–192 mL) of refrigeration oil.

It is not recommended that the oil level be checked as a matter of routine. Generally, the compressor oil should be checked only where there is evidence of a major loss, such as that caused by a broken refrigeration line, a serious leak, or damage from a collision. The oil should also be checked if the compressor is to be repaired or replaced.

To check the oil charge, the compressor must be drained. The amount of oil drained from the compressor is noted. The old oil is then discarded.

Procedure

1. Clean the external surface of the compressor so that it is free of oil and grease.

2. Position the compressor with the shaft end up over a **graduated container**.

3. Drain the compressor (Figure 6-6). Allow it to drain for at least ten minutes. Measure and note the amount of oil removed, then discard the old oil.

4. Add new oil in the same amount as the oil drained.

   NOTE: If the replacement compressor is new, drain it as outlined in steps 2 and 3, then add new oil in the amount drained from the old compressor.

5. If the loss of refrigerant occurs over an extended period of time, add three fluid oz. (88.71 mL) of new oil. *Do not exceed a total of 6.5 oz. (192 mL) of oil.*

Service Notes. When the compressor is drained, if it shows signs that foreign matter is present or that the oil contains chips of metallic particles, the system should be flushed. The receiver-drier or accumulator, as applicable, should be replaced after the system is flushed. The compressor inlet and/or thermostatic expansion valve (TXV) or orifice tube inlet screens should be cleaned as well.

If the system is flushed, add a full 6 oz. (177 mL) of clean refrigeration oil to the compressor.

Servicing the Harrison R4 Compressor Clutch

The Harrison R4 compressor clutch requires special tools for service. This procedure is given based on the use of the proper service tools.

Removing the Clutch Plate and Hub Assembly

Follow the procedural sequence as outlined in Photo Sequence 6, P6-1 and P6-2.

An overcharge of oil will reduce system capacity.

Classroom Manual
Chapter 6, page 134

Figure 6-6 Drain compressor oil into a graduated container.

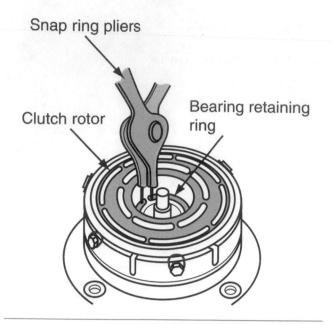

Figure 6-7 Remove the bearing and rotor retaining ring.

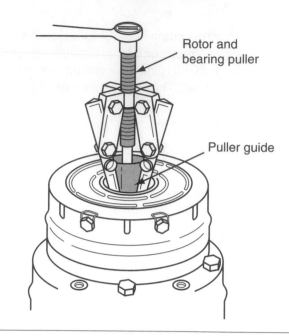

Figure 6-8 Remove the clutch rotor.

Clutch Rotor Bearing and Coil Pulley Rim Removal

1. Mark the location of the clutch coil terminals to ensure proper reassembly.

2. Remove the rotor and bearing assembly retaining ring using the **snap ring** pliers (Figure 6-7).

3. Install the rotor bearing and puller guide over the end of the compressor shaft. The guide should seat on the front head of the compressor. Then, using a puller, remove the clutch rotor and assembly parts (Figure 6-8).

Split the Clutch Rotor Bearing and Coil Pulley Rim

1. Using a cold chisel and hammer, bend the tabs of the six pulley rim mounting screw lockwashers flat (Figure 6-9).

2. Using a 7/16-in. 6-point box wrench, loosen and remove all six screws.

3. Separate the pulley rim from the rotor (Figure 6-10).

Diagnosis

1. Visually check the coil for loose connections or cracked insulation.

2. Briefly connect the coil to a 12V battery with an ammeter connected in series. If the coil draws more than 3.2 amperes at 12 volts, it should be replaced.

3. Inspect the clutch plate and **hub** assembly. Check for signs of looseness between the plate and hub. See the service tip following step 4.

4. Check the rotor and bearing assembly.

☑ **SERVICE TIP:** If the frictional surface of the clutch plate or rotor shows signs of warpage due to excessive heat, that part should be replaced. Slight scoring is normal; if either assembly is heavily scored, however, it should be replaced.

5. Check the bearing for signs of excessive noise, binding, or looseness. Replace the bearing if necessary.

Crayon or chalk temporarily marks component locations. A scribe or file is used to permanently mark the location.

Figure 6-9 Bending the locking tabs to permit removal of the capscrews. (Courtesy of BET, Inc.)

Figure 6-10 Separating the pulley rim from the rotor. (Courtesy of BET, Inc.)

Replacing the Bearing

1. Place the rotor and bearing assembly, split side down, atop two soft wood blocks (Figure 6-11).

2. Using the appropriate tool with a hammer, drive the bearing from the rotor (Figure 6-12). The bearing may also be removed with the arbor press.

3. Turn the rotor over with the frictional surface resting on a block of soft wood.

4. Using the appropriate tool with a hammer, drive the bearing into the rotor. To ensure

Figure 6-11 Place the rotor atop two soft wood blocks. (Courtesy of BET, Inc.)

Figure 6-12 Drive the old bearing from the rotor. (Courtesy of BET, Inc.)

Figure 6-13 Press the new bearing into the rotor. (Courtesy of BET, Inc.)

the alignment of the bearing outer surface into the rotor inner surface, the use of an arbor press (Figure 6-13) is recommended.

⚠ **WARNING:** Make sure that pressure is exerted on the outer bearing race during insertion. If pressure is exerted on the inner race by either method of insertion, premature failure of the bearing will result.

5. Use a prick punch to stake the bearing into the rotor (Figure 6-14).

Reassembling the Clutch Rotor Bearing and Coil Pulley Rim

1. With the coil in place, join the pulley rim to the rotor.
2. Replace and/or tighten the six retaining screws using a 7/16-in. 6-point box wrench.
3. Using a cold chisel and hammer, bend the tabs of the six mounting screw lockwashers up against a flat of each of the screws; one tab for each screw (Figure 6-15).

If it is slightly worn, a 12-point box wrench may slip and round off the hex head of a 6-point fastener.

Replacing the Clutch Rotor Bearing and Coil Pulley Rim

1. Position the assembly on the front head of the compressor.
2. Using the rotor assembly installer with a universal handle (Figure 6-16), drive the assembly into place.

⚠ **WARNING:** Before the assembly is fully seated, be sure that the coil terminals are in the proper location and the three protrusions on the rear of the coil housing align with the locator holes in the front head.

3. Install the retainer ring, using snap ring pliers.

Figure 6-14 Use a prick punch to stake the bearing into the rotor. (Courtesy of BET, Inc.)

Figure 6-15 Bend the tabs against the capscrews. (Courtesy of BET, Inc.)

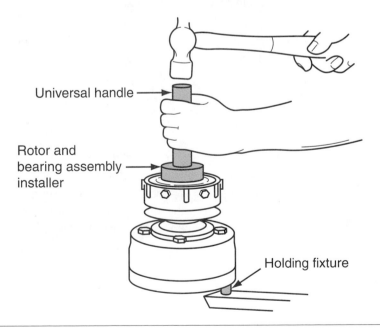

Universal handle

Rotor and bearing assembly installer

Holding fixture

Figure 6-16 Using the rotor and bearing assembly installer, drive the clutch rotor assembly into place on the front head of the compressor.

Figure 6-17 The key should protrude 3/16 in. (4.7 mm). (Courtesy of BET, Inc.)

Figure 6-18 Harrison's V5 compressor. (Courtesy of BET, Inc.)

Replacing the Clutch Plate and Hub Assembly

1. Clean the frictional surfaces of the clutch plate and rotor, if necessary.
2. Insert the key into the slot (keyway) of the hub. Do not insert the key into the compressor crankshaft slot (keyway). The key should protrude about 3/16 in. (0.1875 in. or 4.7 mm) below the hub (Figure 6-17).

> ⚠ **WARNING:** Improper placement of the key may result in damage to the clutch hub.

3. Place the clutch plate and hub assembly onto the compressor shaft by matching the key of the hub to the keyway of the shaft.
4. Using a clutch plate and hub installer, press this part on the crankshaft. *Do not hammer this part into position.*
5. Use a nonmagnetic feeler gauge to ensure an air gap of 0.020–0.040 in. (0.508–1.016 mm) between the frictional surfaces.
6. Replace the shaft nut and torque to 8–12 ft.-lb. (10.8–16.3 N·m).

Model V5

Classroom Manual
Chapter 6, page 145

The five-cylinder Harrison Model V5 (Figure 6-18), introduced by General Motors (GM) in selected car lines in 1985, was the first of the variable-displacement compressors. By design, a variable displacement compressor matches mobile air conditioning demand under all conditions without cycling the clutch off and on. Variable displacement is accomplished by a variable-angle swash plate and five axially located pistons.

The self-lubricating compressor collects up to 4 oz. (118 mL) of lubricant in the crankcase. A crankcase suction bleed is routed through the rotating swash plate, removes some of the lubricant, and reroutes it to the crankcase for component lubrication.

Displacement is controlled by a bellows-actuated control valve in the rear head of the compressor that senses suction pressure. The swash plate angle and compressor displacement are controlled by the suction pressure differential in the crankcase.

When the air conditioner capacity demand is high, the suction pressure will be above the control point and the control valve will maintain a bleed from crankcase to suction. In this condition, no pressure differential will exist, resulting in maximum displacement. On the other hand, when the capacity demand is low, suction pressure will reach the control point and the control valve will bleed discharge gas into the crankcase. The control valve also closes crankcase to suction plenum passage, which provides for smaller displacement.

Compressor swash plate angle is controlled by the force balance on the pistons. An increase of crankcase suction pressure differential (Δ_p) creates a total force on the pistons causing a movement about the swash plate pivot pin, thereby reducing the swash plate angle.

The following is a guide to identify a particular Harrison V5 compressor for service, parts, or replacement:

❏ Identify the type clutch: single V-groove or five- or six-groove poly-groove.

❏ Identify the switch or switch ports in the rear head: one or two.

❏ Note the location of the coil terminal when the suction port is at 12 o'clock: 12 o'clock or 2 o'clock.

❏ Measure the clutch pulley diameter.

❏ Measure the mounting distance from front mounting flange to the center of the pulley groove farthest from the compressor body.

When installing a new or rebuilt V5 compressor, it is recommended that a molecular-sieve filter/drier be installed in the liquid line of the air conditioning system. The molecular-sieve desiccant is capable of removing fine aluminum particles from the refrigerant to prevent clogging the inlet screen in the metering device. This filter may itself become clogged, in which case it must be removed or replaced.

Hitachi

The Hitachi compressor has been used on Datsun/Nissan car lines since 1980 and on the Honda Civic in 1982 and 1983. There are many different styles and types of the nine models of Hitachi compressor. This compressor is also, on occasion, labeled Nihon Radiator. When it is necessary to replace a Hitachi compressor, a careful match of the original compressor must be made. The primary consideration is its mounting configuration. There are basically three mounting configurations:

❏ Bosses on top with pads on bottom with regular sump and no side cover or deep sump with side cover

❏ Bosses on top and bottom

❏ Female threaded holes with left-side suction fitting (primarily Honda) or right-side suction fitting (Datsun/Nissan pickup trucks)

Keihin

The Keihin compressor was used on Honda car lines from 1983 through 1991. There are seven models, each for a specific application. The best method of identifying a Keihin compressor is by using the OEM type code found on a decal or tag on the rear of the unit. It may be necessary to first remove the heat shield to gain access to the code label.

 WARNING: The Keihin compressor has no oil reserve sump. It is therefore important that the proper amount of oil be added after any service to ensure proper lubrication.

Some compressors have little or no oil reserve. Since oil is often lost with refrigerant, it is unwise to "top off" a system without determining the cause.

Certain models of the Keihin OEM compressor are being phased out. They are being replaced with a Sanden compressor and a conversion package. The conversion package often includes a mount and drive kit that is supplied by Honda.

Matsushita/Panasonic

The Matsushita compressor has been found on Honda car lines since 1985, on Subaru car lines since 1990, and on Mazda car lines since 1992. There are eight models, each of which is designated for a particular application. The best method of identifying a Matsushita compressor is by referring to the OEM code found on a decal or tag on the compressor body. If the decal is located on the rear head, it may be necessary to first remove the heat shield to gain access to it. Matsushita and Panasonic compressors are interchangeable.

Mitsubishi

The Mitsubishi, also known as "Mitsubishi Heavy Industries," compressor is manufactured exclusively for Mitsubishi car and truck lines. The car line, however, does not use this compressor exclusively. The Nippondenso compressor may also be found on these car lines. Mitsubishi also manufactures compressors for manufacturers of other car lines. Their model numbers are generally prefixed with the letters *FX*. The best method of identifying a Mitsubishi compressor for replacement is by the OEM code found on a decal or tag attached to the compressor case.

Nihon Radiator

The Nihon Radiator compressor is generally found on Nissan/Datsun car lines. Other manufacturers, such as Calsonic, Diesel Kiki, Hitachi, Seltec, and Zexel, may produce compressors under the Nihon Radiator name. The best way to obtain a proper replacement compressor is to order it by year, make, and model of the vehicle. It may also be necessary to have other information, such as accessory equipment, body style, and/or VIN, available to ensure compatibility.

Nippondenso

Nippondenso has, perhaps by far, the greatest number of makes and models of compressor available for automotive air conditioning service. They are used on many car lines, such as Acura, BMW, Chevrolet, Chrysler, Corvette, Dodge, Ford, Honda, Toyota, Lexus, Lincoln, Mazda, Mercedes Benz, Mercury, Merkur, Mitsubishi, Plymouth, and Porsche.

Compressors must be identified by model number, not just by appearance.

There are over 25 different models and more than one hundred styles of Nippondenso compressors (Figure 6-19) in current use. Other models, such as Ford's FX6 and FX15 and Chrysler's A590 and C171, are also manufactured by Nippondenso. Other models no longer in current production have been replaced, generally by improved design models. Most compressors, which are designated for particular applications, cannot be interchanged. A replacement compressor, then, is best identified by giving the supplier information such as:

❑ OEM number from identification tag

❑ Make, model, and year of vehicle

❑ Grooves in pulley: 1, 2, 4, 5, or 6

❑ Accessories, such as power steering

❑ Series and/or date of manufacture (VIN)

Figure 6-19 Nippondenso's (A) two-cylinder and (B) six-cylinder compressors shown without clutch assemblies. (Courtesy of BET, Inc.)

Before attempting any installation, it is always wise to make a visual inspection to determine if the supplied replacement compressor is comparable with the defective unit. If the supplied compressor is without a clutch assembly, pull the clutch assembly from the defective compressor to make a comparison. Check the shaft size and length. Also, determine if it is a splined or keyed shaft. Many clutches are not interchangeable.

Servicing the Nippondenso Compressor

The procedures for servicing the Nippondenso compressor are given in three sections: replacing the shaft seal, checking and adding oil, and servicing the clutch. These procedures assume that the compressor has been removed from the vehicle and that the services are to be performed "on the bench."

Replacing the Shaft Seal

Seal replacement for the Nippondenso compressor is somewhat different from most other compressors in that the front head assembly must first be removed. See Photo Sequence 7 for a typical procedure for removing and replacing the Nippondenso shaft seal assembly.

Checking and/or Adding Oil to the Nippondenso Compressor

The Nippondenso compressor is factory charged with 13 oz. (384 mL) of refrigeration oil. It is not recommended that the oil level be checked as a matter of routine unless there is evidence of a severe loss.

Some service procedures can be performed on the vehicle if space permits.

Special Tools

Bearing remover/pulley installer
External snap ring plier
Graduated container
Hub remover
Three-jaw puller
Shaft key remover
Shaft protector
Shaft seal seat installer
Shaft seal seat remover

Classroom Manual
Chapter 4,
page 90

Photo Sequence 7
Typical Procedure for Removing and Replacing the Nippondenso Compressor Shaft Seal Assembly

P7-1 After removing the clutch and coil assemblies, use the shaft key remover to remove the shaft key.

P7-2 Remove the felt oil absorber and retainer from the front head cavity.

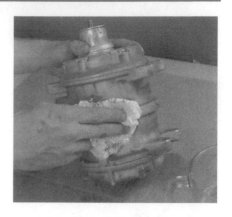

P7-3 Clean the outside of the compressor with pure mineral spirits and air dry. Do not submerge the compressor into mineral spirits. Drain the oil into a graduated measure.

P7-4 Remove the six through bolts from the front head. Use the proper tool; some require a 10-mm socket and others require a 6-mm Allen wrench. Discard the six brass washers (if so equipped) and retain the six bolts.

P7-5 Gently tap the front with a plastic hammer to free it from the compressor housing. Remove and discard the head-to-housing O-ring and the head-to-valve plate gasket.

P7-6 Place the front head on a piece of soft material, such as cardboard, cavity side up. Use the shaft seal seat remover to remove the seal seat.

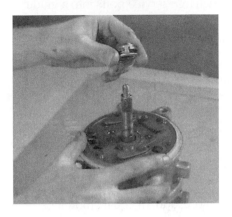

P7-7 Using both hands, remove the shaft seal cartridge.

P7-8 Liberally coat all seal parts, compressor shaft, head cavity, and gaskets with clean refrigeration oil. Carefully install the shaft seal cartridge, making sure to index the shaft seal on the crankshaft slots.

P7-9 Install the seal seat into the front head using the seal seat installer.

P7-10 Install the head-to-valve plate gasket over the alignment pins in the compressor housing. Install the head-to-housing O-ring. Carefully slide the head onto the compressor housing, making sure that the alignment pins engage in the holes in the head.

P7-11 Using six new brass washers (if required), install the six compressor through bolts. Using a 10-mm socket or a 6-mm Allen wrench, as required, tighten the bolts to a 260 lb.-in. (29.4 N·m) torque. **SERVICE TIP:** Use an alternate pattern when torquing the bolts.

P7-12 Replace the oil with clean refrigeration oil. Install the crankshaft key, using a drift. Align the ends of the felt and its retainer and install them into the head cavity. Be sure the felt and retainer are fully seated against the seal plate. Replace the clutch and coil assembly.

Do not attempt to remove the service valves before the refrigerant has been removed from the system.

The following procedure assumes that the suction and discharge service valves have been removed from the compressor.

Draining the Compressor

1. Drain the compressor oil through the suction and discharge service ports into a graduated container.

2. Rotate the crankshaft one revolution to be sure that all oil is drained.

3. Note the quality and quantity of the oil drained. Inspect the drained oil for brass or metallic particles, which indicate a compressor failure. Record in ounces or milliliters the amount of oil removed.

4. Discard the old oil as required by local regulations.

Refilling the Compressor

Be sure to use the proper type and grade of oil to ensure refrigerant compatibility.

1. Add oil, as follows:

   a. If the amount of oil drained was 3 oz. (89 mL) or more, add an equal amount of clean refrigeration oil.

   b. If the amount of oil drained was less than 3 oz. (89 mL), add 5–6 oz. (148–177 mL) of clean refrigeration oil.

2. If the compressor is to be replaced, drain all of the oil from the new or rebuilt compressor and replace the oil as outlined in step 1 (a or b, as applicable).

 SERVICE TIP: Oil is added into the suction and/or discharge port(s). Rotate the compressor crankshaft at least five revolutions by hand after adding oil.

Servicing the Nippondenso Compressor Clutch

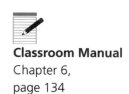

Classroom Manual
Chapter 6,
page 134

The Nippondenso compressor may be equipped with either a Nippondenso or Warner clutch assembly. Though these two clutches are similar in appearance, their parts are not interchangeable. Complete clutch assemblies are, however, interchangeable on this compressor.

The apparent difference in the two clutches is that the Nippondenso pulley (Figure 6-20) has two narrow single-row bearings that are held in place with a wire snap ring. The Warner clutch (Figure 6-21) has a single wide double-row bearing that is staked or crimped in place.

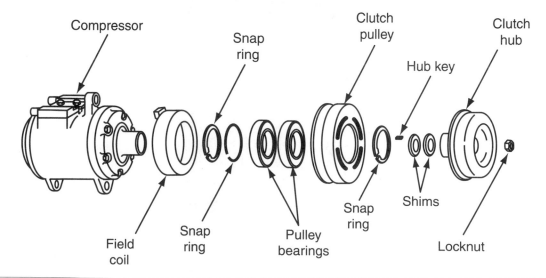

Figure 6-20 Exploded view of a Nippondenso compressor with a Nippondenso clutch. (Courtesy of Ford Motor Company)

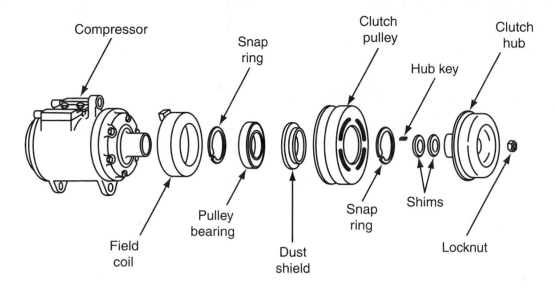

Figure 6-21 Exploded view of a Nippondenso compressor with a Warner clutch.

Removing the Clutch

1. Remove the hub nut.

2. Use the hub remover and remove the clutch hub (Figure 6-22).

   ✔ **SERVICE TIP:** The shaft/hub key need not be removed. Take care not to lose the shim washer(s).

3. Use the snap ring plier to remove the pulley retainer snap ring.

4. With the shaft protector in place (Figure 6-23), remove the pulley and bearing assembly with the three-jaw puller.

   ▲ **WARNING:** Make certain that the puller jaws are firmly and securely located behind the pulley to avoid damage.

Tools that are not properly secured and used may cause damage to the part.

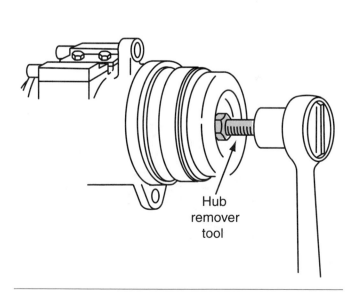

Figure 6-22 Use the hub remover to remove the clutch hub.

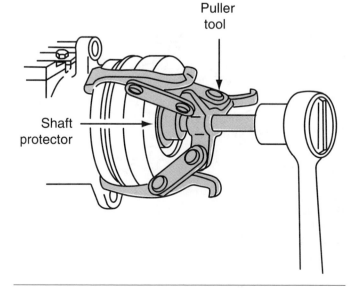

Figure 6-23 With the shaft protector in place, use a three-jaw puller to remove the pulley and bearing assembly.

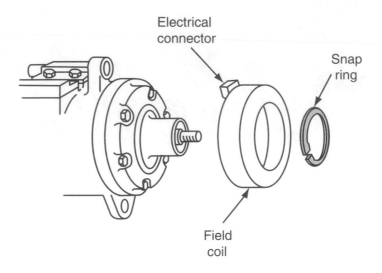

Figure 6-24 Use a snap ring plier to remove the snap ring and remove the clutch coil assembly.

5. Use the snap ring plier to remove the field coil retaining snap ring (Figure 6-24).
6. Note the location of the coil electrical connector and lift the field coil from the compressor.

Replacing the Pulley Bearing

 SERVICE TIP: If the compressor is equipped with a Nippondenso clutch, use a small screwdriver and remove the bearing retaining snap ring before proceeding.

1. Support the pulley with the proper clutch pulley support.
2. Drive out the bearing(s) using a hammer and bearing remover (Figure 6-25).

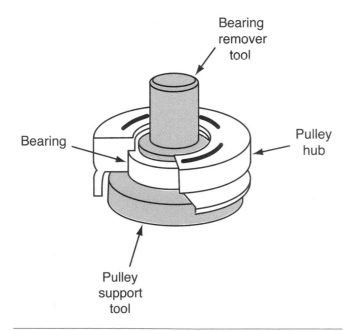

Figure 6-25 Use a hammer (not shown) and bearing remover tool to drive out the bearing *after* placing the hub on a pulley support.

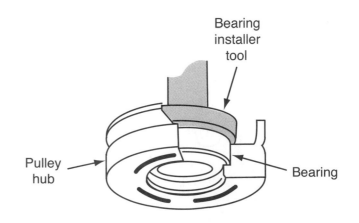

Figure 6-26 Use a bearing installer and hammer to drive the new bearing(s) into the pulley hub.

3. Lift out the dust shield and retainer or leave them in place. Make sure that the dust shield is in place *before* installing the bearing(s).

4. Install the new bearing(s) using the bearing installer and the hammer (Figure 6-26). The bearing(s) must be fully seated in the rotor.

5. Replace the wire snap ring if the clutch is a Nippondenso. If it is a Warner clutch, stake the bearing in place using the prick punch and the hammer.

Installing the Clutch

✔ **SERVICE TIP:** Before reassembly, use pure mineral spirits to clean all parts, including the pulley bearing surface and the compressor front head.

1. Install the field coil. Be sure the locator pin on the compressor engages with the hole in the clutch coil.

2. Install the snap ring. Be sure the bevel edge of the snap ring faces out.

3. Slip the rotor/bearing assembly squarely on the head. Using the bearing remover/pulley installer tool, *gently tap* the pulley onto the head (Figure 6-27).

4. Install the rotor/bearing snap ring. The bevel edge of the snap ring must face out.

5. Install shim washers and/or be sure they are in place. Check the shaft/hub key to ensure proper seating.

6. Align the hub keyway with the key in the shaft. Press the hub onto the compressor shaft using the hub replacer tool (Figure 6-28).

▲ **WARNING:** Do not drive (hammer) the hub on; to do so will damage the compressor.

7. Using a nonmagnetic feeler gauge, check the air gap between the hub and rotor (Figure 6-29). The air gap should be 0.021–0.036 in. (0.53–0.91 mm).

8. Turn the shaft (hub) one-half turn and recheck the air gap. Change the shim(s) as necessary to correct the air gap.

9. Install the locknut and tighten to 10–14 ft.-lb. (13.6–19.0 N·m).

10. Recheck the air gap. See steps 7 and 8.

If index pins are not engaged properly, misalignment occurs.

A snap ring installed backward will not hold properly.

Hub damage may occur if the key is not properly aligned.

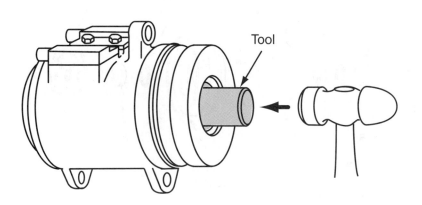

Figure 6-27 Gently tap the pulley assembly onto the compressor head.

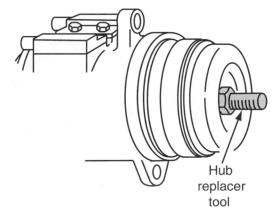

Figure 6-28 Press the hub onto the shaft using the hub replacer tool.

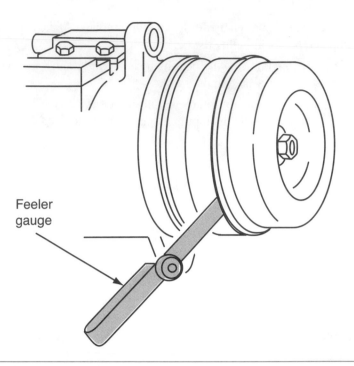

Figure 6-29 Check the air gap between the hub and rotor.

Panasonic (Matsushita)

Classroom Manual
Chapter 6,
pages 139 and 142

The Panasonic rotary vane type compressor (Figure 6-30) was first introduced in 1993 by Ford Motor Company on the Probe. The compressor, which is belt driven off the engine, uses HFC-134a as a refrigerant.

 WARNING: The use of proper refrigerant is important. Do not mix refrigerants.

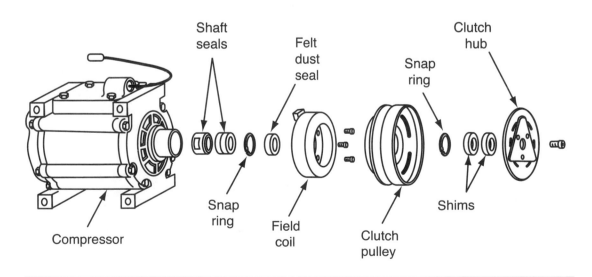

Figure 6-30 Panasonic's rotary vane compressor.

Servicing the Panasonic Vane Type Compressors

The main components of the Panasonic vane type compressor are the rotor with three vanes, a sludge control valve, a discharge valve, and a thermal protector. The only service that may be accomplished is checking and adjusting oil, servicing the clutch, replacing the shaft seal, and servicing the thermal protector, sludge control, and discharge valve.

All service procedures assume that the compressor has been removed from the vehicle and is being worked on on the bench.

Checking and Adjusting Compressor Oil Level

A new Panasonic Rotary compressor contains 6.78 oz. (200 mL) of a special paraffin-base refrigeration oil, designated as YN-9.

 WARNING: Use the proper oil and the correct amount of oil. Too much oil will reduce the system capacity and too little oil will result in insufficient lubrication.

It is necessary to adjust the oil any time the compressor is serviced or when being replaced, as outlined in the following procedure:

Procedure

1. Drain the oil from the defective compressor into a calibrated container and note the amount removed. Allow the compressor to drain thoroughly.
2. Drain the oil from the replacement compressor into a second calibrated container. Allow the compressor to drain thoroughly.
3. Add the same amount of clean refrigeration oil to the replacement compressor that was removed from the defective compressor.
4. Add an additional 0.68 oz. (20 mL) of oil.

Servicing the Clutch Assembly

Removing the Clutch and Coil

1. Using an Allen wrench, remove the clutch armature Allen bolt (Figure 6-31).
2. Remove the clutch armature (Figure 6-32).
3. Remove the shim(s) and set aside.
4. Using **internal snap ring** pliers, remove the clutch rotor/pulley snap ring (Figure 6-33).
5. Remove the clutch rotor/pulley.
6. Using a screwdriver, remove the three clutch field coil screws. Remove the clutch field coil (Figure 6-34).

Oil removed from an air conditioning system or component must be discarded in accordance with EPA regulations.

Special Tools
External snap ring pliers
Internal snap ring pliers

Classroom Manual
Chapter 4, page 90

Figure 6-31 Remove/replace the Allen holding bolt. (Courtesy of BET, Inc.)

Figure 6-32 Remove the clutch armature. (Courtesy of BET, Inc.)

Replacing the Coil and Clutch

1. Replace the clutch field coil and secure with three screws (see Figure 6-34).
2. Replace the clutch rotor/pulley and secure with the internal snap ring (see Figure 6-33).
3. Replace the shim(s).
4. Replace the clutch armature and secure with the Allen bolt (see Figure 6-31).

Figure 6-33 Remove/replace the rotor snap ring. (Courtesy of BET, Inc.)

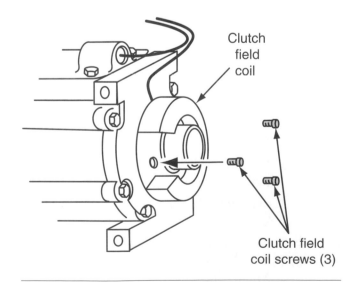

Figure 6-34 Using a screwdriver, remove the three clutch field coil screws.

Servicing the Compressor Shaft Seal

To service the compressor shaft seal, proceed as follows:

Removing the Shaft Seal

1. Remove the clutch. It is not necessary to remove the clutch coil for seal service.
2. Remove the felt dust seal from the seal cavity (Figure 6-35).
3. Using internal snap ring pliers, remove the shaft seal snap ring (Figure 6-36).
4. Using the seal remover, remove the seal seat (Figure 6-37).
5. Using the seal remover/installer, remove the shaft seal (Figure 6-38).

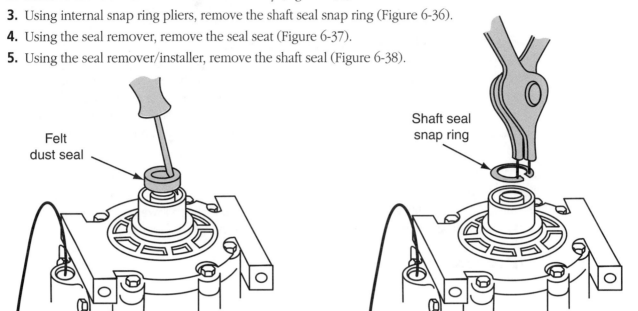

Figure 6-35 Remove/replace the felt dust seal.

Figure 6-36 Remove/replace the shaft seal snap ring.

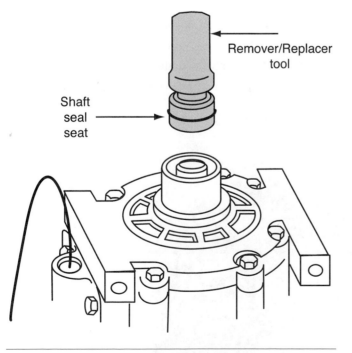

Figure 6-37 Remove/install the shaft seal seat.

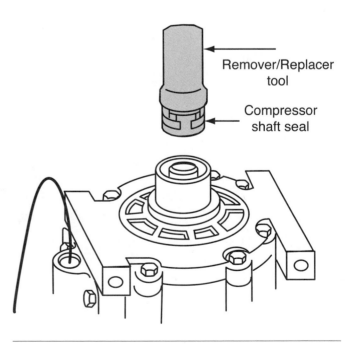

Figure 6-38 Remove/install the shaft seal.

Installing the Shaft Seal

1. Coat all seal parts with clean refrigeration oil.
2. Install the shaft seal, using the remover/installer tool (see Figure 6-38).
3. Install the shaft seal, using the seal remover tool (see Figure 6-37).
4. Replace the shaft seal snap ring (see Figure 6-36).
5. Replace the felt dust seal (see Figure 6-35).
6. Replace the clutch assembly.

A seal, if installed backward, is almost impossible to remove.

Servicing the Compressor

The thermal protector, sludge control, and discharge valve are the only components that may be serviced in the Panasonic rotary compressor. It is not necessary to remove the clutch assembly or the shaft seal assembly for this service. See Photo Sequence 8 for a typical procedure for disassembling the Panasonic vane-type compressor to service the sludge control, discharge valve, and thermal protector.

Reassembly

For the disassembly of the Panasonic vane rotary compressor to service the internal parts, follow the illustrated procedure as outlined in Photo Sequence 8. The following, with reference to the photo sequence illustrations, may then be followed in the reverse order for the reassembly procedure.

Procedure

1. Replace the discharge valve and stopper, P8-11.
2. Secure the discharge valve and stopper with the two bolts removed in disassembly, P8-10.
3. Install the thermal protector and secure it with the snap ring, P8-8.
4. Replace the thermal protector housing with a new gasket and secure it with four cap-screws (Figure 6-39).
5. Secure the thermal protector hold-down bracket with the screw previously removed, P8-6.
6. Replace the two compression springs and spring stoppers, P8-5.
7. With a new gasket in place, install the oil control valve with three bolts, P8-3.
8. Install the housing cover. Secure the housing cover with two Allen bolts and six hex nuts, P8-2 and P8-1.
9. Replace the refrigeration oil.

Before reassembly, liberally coat all parts with clean refrigeration oil that is compatible with the refrigerant requirements of the system.

Figure 6-39 Replace the thermal protector housing. (Courtesy of BET, Inc.)

Photo Sequence 8
Typical Procedure for Disassembling the Panasonic Vane-Type Compressor to Service the Sludge Control, Discharge Valve, and Thermal Protector

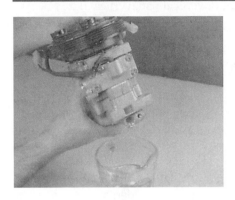

P8-1 Drain the oil from the compressor as outlined in Part I. Remove the six housing cover hex nuts at the rear of the compressor.

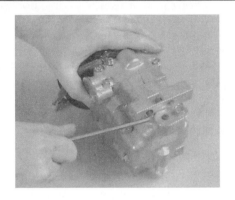

P8-2 Remove the two Allen bolts from the rear of the compressor. Lift off the housing cover.

P8-3 Use a screwdriver to remove the thermal protector hold-down bracket retaining screw.

P8-4 Using a socket wrench, remove the four thermal protector housing bolts.

P8-5 Remove the thermal protector housing. Remove and discard the gasket.

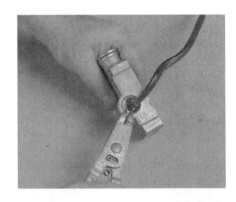

P8-6 Use internal snap-ring pliers and remove the thermal protector snap ring retainer.

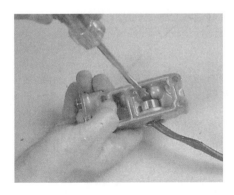

P8-7 Push the thermal protector out of the housing.

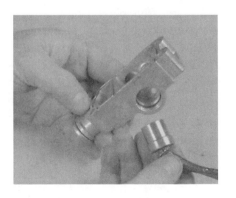

P8-8 Remove the thermal protector.

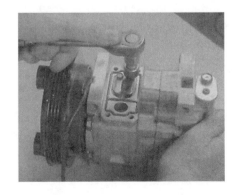

P8-9 Using a socket wrench, remove the two discharge valve stopper bolts.

Sanden

There are about 20 models of the Sanden compressor, formally known as Sankyo. These compressors are used by Chevrolet, Chrysler, Dodge, Fiat, Ford, Honda, Jeep, Mazda, Peugeot, Renault, Subaru, and Volkswagen. The considerations for selecting a replacement Sanden compressor are:

❏ Head style: horizontal or vertical O-ring; horizontal or vertical pad; vertical flare

❏ Clutch diameter: from 3.8 in. (96.5 mm) to 5.6 in. (142 mm)

❏ Number of grooves in clutch rotor: either 1 or 2 V-groove or 4, 5, 6, 7, or 10 poly groove

❏ Mounting boss measurement, front to rear, outside to outside: from 2.85 in. (72.4 mm) to 4.41 in. (112 mm)

❏ Type of refrigerant: CFC-12 or HFC-134a

▲ **WARNING:** Do not mix refrigerants.

 SERVICE TIP: Sanden's compressors are equipped with Buna-N O-rings for CFC-12 service and neoprene O-rings for HFC-134a service. Sanden's Buna-N O-rings are black while their neoprene O-rings are blue.

Servicing the Sanden (Sankyo) Compressor

Servicing the Sanden/Sankyo compressor is limited in this manual to: replacing the shaft oil seal, checking and adjusting the proper oil level, and servicing the clutch. This procedure assumes that the compressor has been removed from the vehicle and is being serviced on the bench.

Replacing the Compressor Shaft Oil Seal

The following procedures may be followed when replacing the compressor shaft seal.

Removing the Shaft Seal

Special Tools

Air gap gauge set
Front plate installer
Face puller
Hammer, small soft
O-ring remover
Seal protector
Seal remover and
   installer
Seal seat remover
Seal seat retainer
Spanner wrench

1. Using a 3/4-in. hex socket and spanner wrench (Figure 6-40), remove the crankshaft hex nut.
2. Remove the clutch front plate (Figure 6-41) using the clutch front plate puller.
3. Remove the shaft key and spacer shims and set them aside.
4. Using the snap ring pliers (Figure 6-42), remove the seal retaining snap ring.
5. Remove the seal seat using the seal seat remover and installer (Figure 6-43).
6. Remove the seal (Figure 6-44) using the seal remover tool.
7. Remove the shaft seal seat O-ring (Figure 6-45) using the O-ring remover.
8. Discard all parts removed in steps 5, 6, and 7.

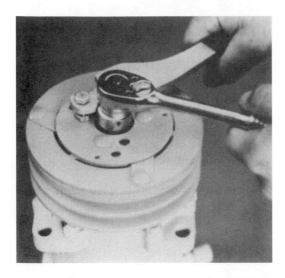

Figure 6-40 Remove the crankshaft hex nut. (Courtesy of Sankyo)

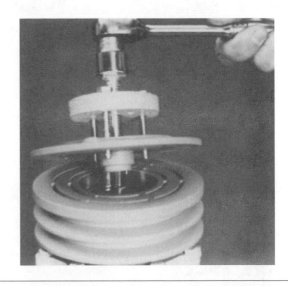

Figure 6-41 Remove the clutch front plate. (Courtesy of Sankyo)

Figure 6-42 Remove the seal seat snap ring. (Courtesy of Sankyo)

Figure 6-43 Remove the seal seat. (Courtesy of Sankyo)

Installing the Shaft Seal

1. Clean the inner bore of the seal cavity by flushing it with clean refrigeration oil.
2. Coat the new seal parts with clean refrigeration oil.

   ⚠ **WARNING:** *Do not touch* the carbon ring face with the fingers. Normal body acids will etch the seal and cause early failure.

3. Install the new shaft seal seat O-ring. Make sure it is properly seated in the internal groove. Use the remover tool to position the O-ring properly.
4. Install the seal protector on the compressor crankshaft. Liberally lubricate the part with clean refrigeration oil.
5. Place the new shaft seal in the seal installer tool and carefully slide the shaft seal into place in the inner bore. Rotate the shaft seal clockwise (cw) until it seats on the compressor shaft flats.

Do not discard parts that may be reused, such as shaft keys, spacers, nuts, bolts, clamps, snap rings, and so on.

Some seal faces are made of a ceramic material. These seals may be damaged if care is not taken during handling.

Figure 6-44 Remove the seal. (Courtesy of Sankyo)

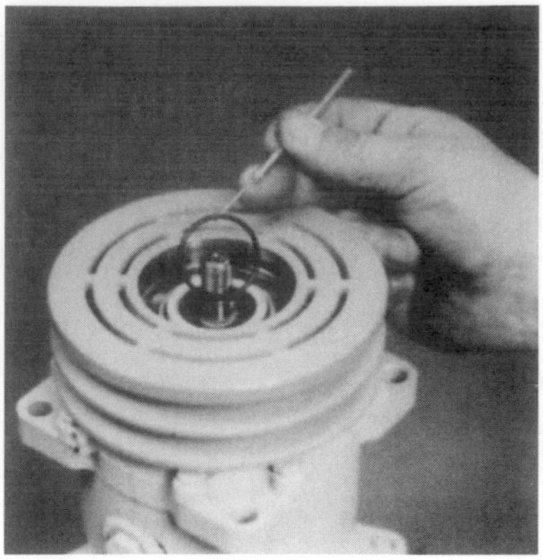

Figure 6-45 Remove the O-ring. (Courtesy of Sankyo)

6. Rotate the tool counterclockwise (ccw) to remove the seal installer tool.

7. Remove the shaft seal protector.

8. Place the shaft seal seat on the remover/installer tool and carefully reinstall the shaft seal in the compressor seal cavity.

9. Replace the seal seat retainer.

10. Reinstall the spacer shims and shaft key.

11. Position the clutch front plate on the compressor crankshaft.

Use nonmetallic feeler gauges to determine clutch air gap.

12. Using the clutch front plate installer tool, a small hammer, and an air gap gauge, reinstall the front plate (Figure 6-46).

13. Draw down the front plate with the shaft nut. Use the air gap gauge for go at 0.016 in. (0.4 mm) and no-go at 0.031 in. (0.79 mm).

14. Using the torque wrench, tighten the shaft nut to a torque of 25–30 ft.-lb. (33.0–40.7 N·m).

Figure 6-46 Use the air gap gauge to check the rotor-to-hub clearance. (Courtesy of Sankyo)

Checking Compressor Oil Level

The compressor oil level should be checked at the time of installation and after repairs are made when it is evident that there has been a loss of oil. The Sankyo compressor is factory charged with 7 fluid oz. (207 mL) of oil. A special angle gauge and dipstick are used to check the oil level. The oil chart (Figure 6-47) compares the oil level with the inclination angle of the compressor.

This procedure may also be followed with the compressor in the vehicle after first ensuring that the refrigerant has been recovered.

Preparing the Compressor

1. Position the angle gauge tool across the top flat surfaces of the two mounting ears.
2. Center the bubble and read the inclination angle.
3. Remove the oil filler plug. Rotate the clutch front plate to position the rotor at the top dead center (Figure 6-48).
4. Face the front of the compressor. If the compressor angle is to the right, rotate the clutch front plate counterclockwise (ccw) by 110°. If the compressor angle is to the left, rotate the plate clockwise (cw) by 110° (Figure 6-49).

Checking the Oil Level

 SERVICE TIP: The dipstick tool for this procedure is marked in eight increments. Each increment represents 1 oz. (29.57 mL) of oil.

Inclination Angle In Degrees	Acceptable Oil Level In Increments
0	6–10
10	7–11
20	8–12
30	9–13
40	10–14
50	11–16
60	12–17

Figure 6-47 Dipstick reading vs. inclination angle. (Courtesy of Sankyo)

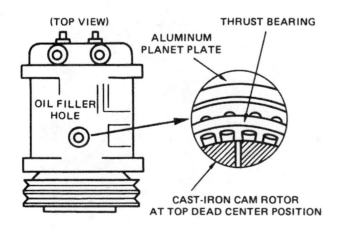

Figure 6-48 Position the rotor to top dead center (TDC). (Courtesy of Sankyo)

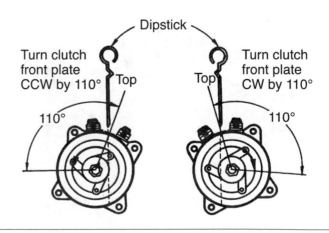

Figure 6-49 Rotate the clutch front plate. (Courtesy of Sankyo)

Classroom Manual
Chapter 4,
pages 86 and 90

Do not attempt to remove the oil plug until after ensuring that the refrigerant has been removed from the system.

TDC is an acronym for "top dead center."

An overcharge of oil has basically the same effect as an overcharge of refrigerant.

Classroom Manual
Chapter 6,
page 134

1. Insert the dipstick until it reaches the stop position marked on the dipstick.
2. Remove the dipstick and count the number of increments of oil.
3. Compare the compressor angle and the number of increments with the table (see Figure 6-47).
4. If necessary, add oil to bring the oil to the proper level. *Do not overfill.* Use only clean refrigeration oil of the proper grade.

Servicing the Clutch

Although this procedure presumes that the compressor is removed from the vehicle, if ample clearance is provided in front of the compressor for clutch service, it need not be removed for this service.

Removing the Clutch

1. Use a 3/4-in. hex socket and spanner wrench to remove the crankshaft hex nut as shown in Figure 6-40.
2. Remove the clutch front plate, using the clutch front plate puller as shown in Figure 6-41.
3. Using the snap ring pliers, remove the internal and external snap rings (Figure 6-50 and Figure 6-51).
4. Using the pulley puller (Figure 6-52), remove the rotor assembly.
5. If the clutch coil is to be replaced, remove the three retaining screws and the clutch field coil. Omit this step if the coil is not to be replaced.

Replacing the Rotor Bearing

1. Using the snap ring pliers, remove the bearing retainer snap ring.
2. From the back (compressor) side of the rotor, knock out the bearing using the bearing remover tool and a soft hammer.
3. From the front (clutch face) side of the rotor, install the new bearing using the bearing installer tool and a soft hammer. Take care not to damage the bearing with hard blows of the hammer (Figure 6-53).
4. Reinstall the bearing retainer snap ring.

Figure 6-50 Remove the internal snap ring. (Courtesy of Sankyo)

Figure 6-51 Remove the external snap ring. (Courtesy of Sankyo)

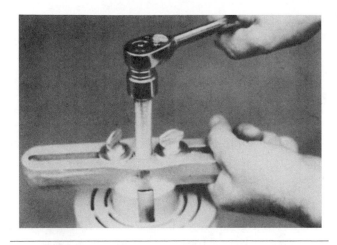

Figure 6-52 Remove the rotor assembly. (Courtesy of Sankyo)

Figure 6-53 Install the rotor bearing. (Courtesy of Sankyo)

Replacing the Clutch

1. Reinstall the field coil (or install a new field coil, if necessary) using the three retaining screws.

2. Align the rotor assembly squarely with the front compressor housing.

3. Using the rotor two-piece installer tools and a soft hammer, carefully drive the rotor into position until it seats on the bottom of the housing.

4. Reinstall the internal and external snap rings using the snap ring pliers.

5. Align the slot in the hub of the front plate squarely with the shaft key.

 WARNING: Alignment is essential in order to prevent damage to mating surfaces.

6. Drive the front plate on the shaft using the installer tool and a soft hammer (Figure 6-54). *Do not use unnecessary hard blows.*

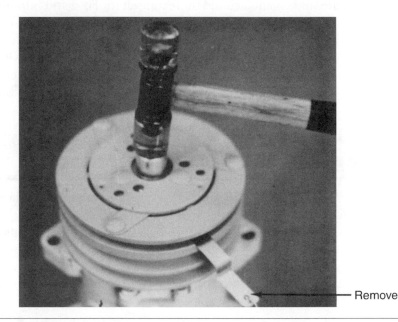

Remove

Figure 6-54 Drive the front plate onto the shaft. (Courtesy of Sankyo)

7. Check the air gap with go and no-go gauges.

8. Replace the shaft nut and tighten it to a torque of 25–30 ft.-lb. (33.9–40.7 N·m) using the torque wrench.

Sankyo

The Sankyo compressor is also known as a Sanden compressor and is covered in this chapter under the heading "SANDEN."

Seiko-Seiki

The Seiko Seiki compressor is found as original equipment on 1986 through 1989 Saab car lines. It may also be found on some other car lines. The best way to identify this compressor is by the OEM type code tag or label found on the compressor body. It is also necessary to identify the clutch type, diameter, and number of grooves. Service port O-rings are refrigerant specific: black Buna-N for CFC-12 and blue neoprene for HFC-134a.

Tecumseh

The HA and HG series compressors are popular for "do-it-yourself" refrigerant recovery systems.

Tecumseh was one of the first manufacturers of automotive air conditioning compressors. A large, heavy, cast iron flywheel-pulley compressor was used from the late 1940s through the late 1950s. Known as Tecumseh's Model HH, it was discontinued in 1958 in favor of a smaller, somewhat lighter model LB. The model LB, which was also made of cast iron, was soon discontinued, however and was replaced by the popular HA and HG series, which are still found on limited applications.

Model HR980

Spring-lock and screw-on hose fittings and connectors are not interchangeable.

The HR980 compressor is manufactured by Tecumseh for Ford and is found on Ford car lines. There are only two versions of the HR980 compressor: spring-lock or screw-on hose fittings. Either version has two-loop/one-bracket mounting provisions. Another version, which is available only as a factory conversion kit, is for spring-lock application and has three loops with no bracket.

This kit comes complete with accumulator, compressor and clutch, crankshaft pulley, discharge line, suction line, and mounting hardware. Either version is available with a single-, double-, or six-groove pulley assembly, each with a standard diameter.

Servicing the HR-980 Compressor

The Tecumseh HR-980 compressor (Figure 6-55) is used on some Ford car lines. Since the internal assembly is not accessible, service is limited to shaft seal and clutch repairs.

This compressor is factory-charged with 8 oz. (236.6 mL) of refrigeration oil. In a balanced system, approximately 4 oz. (118.3 mL) of oil will be found in the compressor, 3 oz. (88.7 mL) in the evaporator, 1 oz. (29.6 mL) in the condenser, and 1 oz. (29.6 mL) in the accumulator.

This service procedure is in three parts: checking and adding oil, servicing the clutch, and replacing the seal.

Figure 6-55 The Tecumseh HR-980 compressor.

Checking and Adding Oil

Special Tools
Graduated container

1. If checking the oil in a serviceable compressor: drain and measure the oil.

   a. If 4 oz. (118.3 mL) or more, replace with the same amount as drained.

   b. If less than 4 oz. (118.3 mL), replace with 4 oz. (118.3 mL) of clean refrigeration oil.

2. If replacing the compressor, drain the new or rebuilt compressor and add 4 oz. (118.3 mL) of clean new refrigeration oil.

Classroom Manual
Chapter 4,
page 90

Servicing the Clutch

The following procedure may be followed for servicing the clutch assembly.

Removing the Clutch Assembly (Figure 6-56)

1. Remove the retaining nut.
2. Using the hub remover tool, remove the clutch hub from the compressor shaft (Figure 6-57). Remove and retain the shim(s).
3. Using a spanner wrench, remove the clutch pulley retaining nut.
4. Remove the pulley and bearing assembly from the compressor by hand. If the assembly cannot be removed by hand, use the shaft protector and pulley remover (Figure 6-58).
5. Remove the field coil from the compressor.
6. Clean the front of the compressor to remove any dirt and/or corrosion.

Classroom Manual
Chapter 6,
page 134

Replacing the Clutch Bearing

1. Place the pulley on the clutch pulley support (Figure 6-59).
2. Use the bearing replacing tool to drive out the bearing.

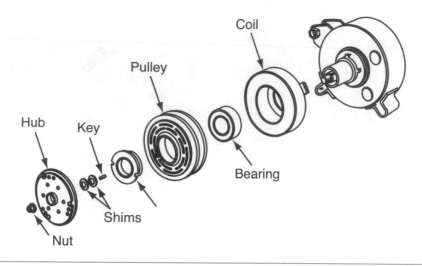

Figure 6-56 An exploded view of the clutch assembly. (Courtesy of Tecumseh)

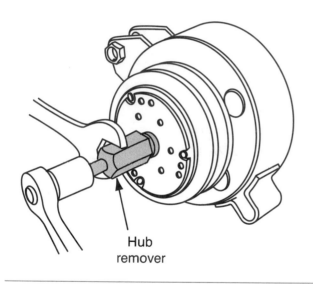

Figure 6-57 Remove the clutch hub. (Courtesy of Tecumseh)

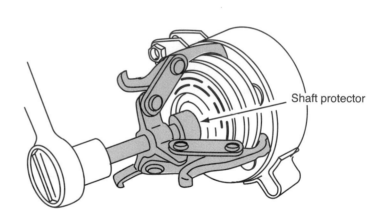

Figure 6-58 Use the shaft protector and three-jaw puller to remove pulley assembly if it cannot be removed by hand. (Courtesy of Tecumseh)

3. Turn the pulley over, flat side atop a clean board.

4. Position the new bearing in the bearing bore of the pulley and use the pulley bearing replacer to seat the bearing (Figure 6-60).

5. Stake the new bearing. Use a blunt drift or punch at three equally spaced places inside the bore. Do not use the same places that were used to retain the old bearing.

Replacing the Pulley Assembly

1. Install the field coil. The slots of the coil should fit over the housing lugs. The electrical connector should be toward the top of the compressor.

▲ **WARNING:** If the coil is not in alignment, the clutch rotor will not fit properly and may "drag" or rub.

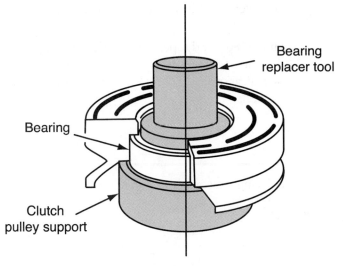

Figure 6-59 Pulley placed atop the clutch pulley support for service. (Courtesy of Tecumseh)

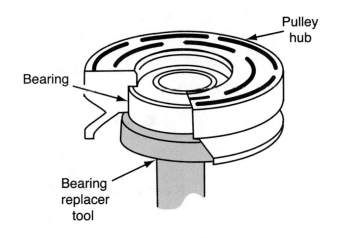

Figure 6-60 Use a bearing replacer to seat the new bearing. (Courtesy of Tecumseh)

2. Install the pulley and bearing assembly on the front of the compressor. If they are properly aligned, the assembly should slide on. If it is difficult to slide the assembly on, use the pulley replacer and tap lightly with a plastic hammer. *Do not use unnecessary force.*

3. Apply a drop of thread lock to the threads of the pulley retainer nut.

4. Install the pulley retainer nut and tighten to 65–70 ft.-lb. (88–94 N·m) using a spanner and torque wrench.

5. Make certain that the key is aligned with the keyway of the clutch hub, then install the hub and shim(s) onto the compressor shaft. Use the hub replacer (Figure 6-61).

⚠️ **WARNING:** *Do not* drive the hub onto the shaft. Compressor damage will result.

6. Install the nut and tighten to 10–14 ft.-lb. (14–18 N·m).

"Greater than" is often shown by the symbol ">" and "less than" by the symbol "<."

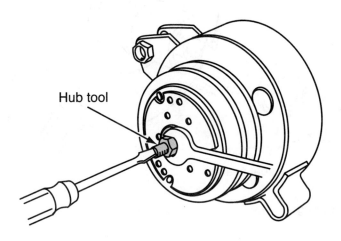

Hub tool

Figure 6-61 Replace the hub. (Courtesy of Tecumseh)

7. Check the air gap at three equally spaced intervals (120°) around the pulley. Record the measurements.

8. Rotate the compressor pulley one-half turn (180°) and repeat step 7. The smallest air gap permitted is between 0.021 in. (0.53 mm) and 0.036 in. (0.91 mm). If the air gap is greater than (>) or less than (<) these specifications, add or remove shims, as required (step 5) to bring the air gap into specifications.

Replacing the Shaft Seal

The following procedure may be followed for replacing the compressor shaft seal.

Removing the Seal (Figure 6-62)

1. Remove the clutch and coil.
2. Remove the key from the compressor shaft.
3. Carefully pry the dust shield from the compressor using a small screwdriver (Figure 6-63). Take care not to damage the end of the compressor housing.
4. Remove the seal snap ring retainer using the internal snap ring pliers.
5. Clean the inside of the seal cavity to prevent entry of foreign material when the seal is removed.
6. Insert the shaft seal seat tool and engage the seal. Tighten the outer sleeve to expand the tool in the seal seat (Figure 6-64).
7. Pull on the tool while rotating it clockwise (cw) to remove the seal seat.
8. Use the O-ring remover and remove the O-ring (Figure 6-65).

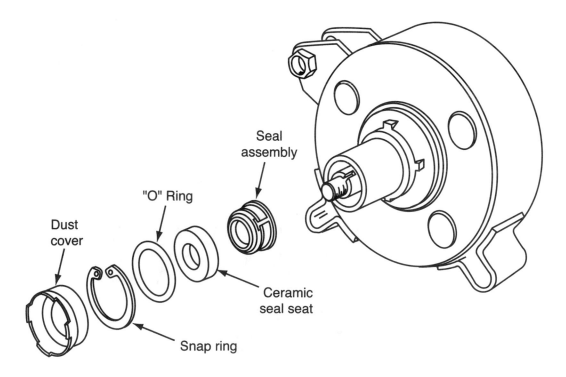

Figure 6-62 Exploded view of shaft seal assembly. (Courtesy of Tecumseh)

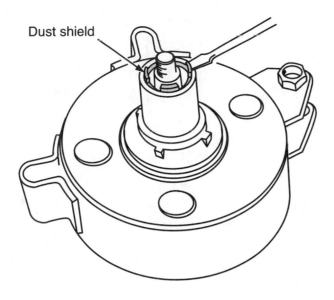

Dust shield

Figure 6-63 Pry the dust shield from the seal cavity. (Courtesy of Tecumseh)

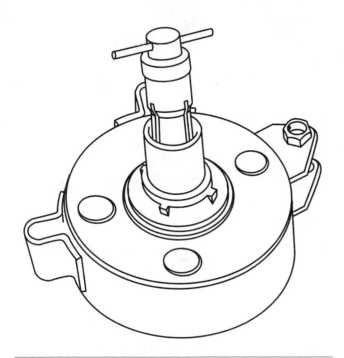

Figure 6-64 Insert seal seat tool into the seal cavity. (Courtesy of Tecumseh)

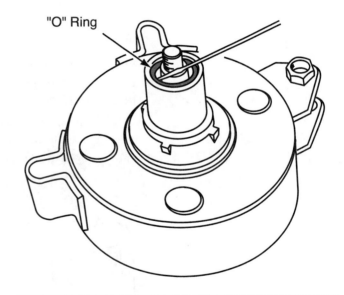

"O" Ring

Figure 6-65 Remove the O-ring. (Courtesy of Tecumseh)

9. Insert the seal assembly tool into the compressor. While forcing the tool downward, rotate it counterclockwise (ccw) to engage the tangs of the seal (Figure 6-66).

10. Pull the seal from the compressor and remove the seal from the tool.

11. Check the inside of the compressor to ensure that all surfaces are free of nicks and burrs.

Figure 6-66 Insert the seal on the remover/replacer tool. (Courtesy of Tecumseh)

Installing the Seal

1. Liberally coat the O-ring with clean refrigeration oil and insert it into the cavity using the O-ring installer, O-ring sleeve, and O-ring guide.

2. With the O-ring in place, remove the tools from the cavity.

Use the same type and grade oil that is used in the system to ensure refrigerant compatibility.

3. Liberally coat the shaft seal with clean refrigeration oil and carefully engage the seal with the seal remover/replacer tool (Figure 6-67A).

4. Carefully place the seal over the shaft and, while rotating it, slide the seal down the shaft until the assembly engages the flats and is in place.

5. Rotate the tool to disengage it from the seal. Remove the tool from the cavity.

6. Liberally coat the seal seat with clean refrigeration oil and engage the seal seat with the remover/replacer tool (Figure 6-67B).

7. Carefully insert the seal seat onto the compressor shaft with a clockwise (cw) rotation. Take care not to disturb the O-ring installed in step 1.

8. Disengage the tool from the seal seat and remove the tool.

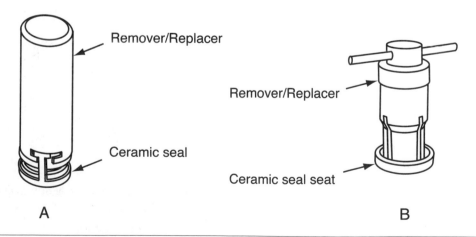

Figure 6-67 (A) Seal remover/replacer tool; (B) seal seat remover/replacer tool. (Courtesy of Tecumseh)

9. Using a snap ring pliers, install the snap ring. The flat side of the snap ring must be against the seal seat.

   ⚠️ **WARNING:** Do not bump or tap the snap ring into place; to do so may damage the seal seat which is made of ceramic.

10. Install the dust shield.

11. Replace the clutch and coil.

A snap ring installed backward will not remain secure.

York

Full-size, two-cylinder York compressors (Figure 6-68) were found on some Ford, Lincoln, and Mercury car lines through 1982. They were also used on some imports, trucks, and aftermarket applications. These compressors are available in 6-, 8-, and 10-in.[3] (0.098-, 0.131-, and 0.164-L) displacements.

The mini-York, one-cylinder, is 6-in.[3] (0.098-L) displacement and is available only with a right-side suction fitting. It is available with either rotolock, flangetop, or tube-O fittings. The full-size compressor is available with a left or right suction fitting for either flangetop or rotolock valves. The tube-O valve provision is only available in 10-in.[3] (0.164-L) models with a left-side suction fitting location.

Rotary

Only 50,000 York rotary compressors (Figure 6-69) were manufactured before production was discontinued. New York rotary compressors are very scarce. Remanufactured compressors are available for service replacement, however. These compressors are found mainly on import and aftermarket applications. Since only four models were produced, they are relatively simple to identify. The best method is by clutch diameter: 5 in. (127 mm), 5.25 in. (133.4 mm), 5.34 in. (135.6 mm), or 6 in. (152.4 mm). This compressor is not interchangeable with any other York compressor.

The rotary compressor is not sensitive to an overcharge of oil. An overcharge, however, may reduce overall system capacity and efficiency.

Figure 6-68 York's two-cylinder compressor. (Courtesy of BET, Inc.)

Figure 6-69 York's rotary compressor. (Courtesy of BET, Inc.)

Checking and Adding Oil to the York Vane Rotary Compressor

An overcharge of oil does not harm the compressor; an undercharge may cause a slight vane chatter, though be sufficient for lubrication. Normal oil charge is 6–9 oz. (177–266 mL) depending upon system refrigerant capacity (Figure 6-70). The normal oil charge is 2–4 oz. (59–118 mL) regardless of the refrigerant charge.

Preparing the Compressor

1. Start and run the engine at idle speed for 10 minutes with the air conditioning controls set for maximum cooling and medium fan speed.

2. Stop the engine.

3. Recover the refrigerant.

4. Loosen the mounting hardware and belt(s) to adjust the attitude of the compressor so that the service valves are vertical.

5. Remove the suction and discharge fittings. Discard the O-rings.

 WARNING: Note the color of the O-rings to ensure refrigerant compatibility.

6. Rotate the compressor shaft by hand counterclockwise (ccw) five to ten revolutions.

System R-12 Charge		Oil Change	
Pounds	Liters	oz.	mL
2	0.946	6	177.4
3	1.419	7	207.0
4	1.892	8	236.6
5	2.365	9	266.2

Figure 6-70 Oil requirements are based on the refrigerant capacity of the system.

Figure 6-71 Check the oil level. (Courtesy of BET, Inc.)

Checking the Oil Level

1. Use the proper dipstick (Figure 6-71) to measure the oil level. Oil level should be 2–4 oz. (59–118 mL).

2. If the oil level is less than 2 oz. (59 mL), add oil to the correct level; if it is more than 2 oz. (59 mL), the oil level is considered adequate.

 SERVICE TIP: Oil may be added through the discharge port (stamped "D" on the compressor sump).

Add only the proper type and grade of oil. Refrigeration oil types should not be mixed.

Returning the Compressor to Service

1. Install suction and discharge fittings with new O-rings.

2. Replace the belt(s) and reposition the compressor to tension belts. If using a belt tension gauge, proper tension is 80 lb./strand for a dual belt or 100–120 lb./strand for a single belt.

3. Evacuate and charge the system.

Pounds/strand is the same as ft.-lb. (N·m).

Zexel

The Zexel, a rotary compressor manufactured by the Zexel Corporation, is found on some Nissan, Toyota, and General Motors car lines.

 CASE STUDY

A customer brings her vehicle into the shop because the air conditioner does not work. Initial inspection reveals that the clutch does not energize when the air conditioner is turned on, but the blower motor is operative.

The technician checks all of the fuses and circuit breakers that may affect the clutch circuit and finds that all are good. Next, the technician uses a voltmeter to check the available voltage at the clutch coil connector. The test indicates that 12.6 volts are available.

The technician then performs a voltage drop test across the ground side of the clutch coil. The voltmeter indicates 12.6 volts. The conclusion is that the ground provision of the clutch coil is defective and must be repaired.

After the ground wire, which is bonded to body metal, is cleaned and reconnected, the clutch functions properly and the air conditioner is fully operational.

Terms to Know

Belt	Internal snap ring	Seal
Carbon seal face	Mounting boss	Seal seat
Decal	Mounting	Shaft
Etching	OEM	Shaft seal
Flange	Off-road	Snap ring
Graduated container	Overcharge	Spade-type
High-torque clutch	Pin-type	Superheat switch
Hub	Running design change	Troubleshoot

ASE-Style Review Questions

1. All of the following statements are correct, EXCEPT:
   A. A slight overcharge of refrigerant in Ford's FS6 compressor will result in certain failure.
   B. The DA6, HR6, and HR6HE are basically the same design of a six-cylinder compressor.
   C. A compressor changes refrigerant from a low-pressure vapor to a high-pressure liquid.
   D. A liquid line molecular-sieve drier should be installed when replacing a V5 compressor.

2. Compressors are being discussed:
   Technician A says that the C171 compressor used by Chrysler is manufactured by Sankyo.
   Technician B says that the A590 compressor used by Chrysler is manufactured by Nippondenso.
   Who is right?
   A. A only
   B. B only
   C. Both A and B
   D. Neither A nor B

3. Compressor application is being discussed:
   Technician A says an important consideration is the clutch and pulley type.
   Technician B says the mounting bracket dimensions and location are important.
   Who is right?
   A. A only
   B. B only
   C. Both A and B
   D. Neither A nor B

4. *Technician A* says that it is virtually impossible to insert the seal seat backward when using the appropriate tool.
   Technician B says that it is almost impossible to remove a seal seat that has been installed backward.
   Who is right?

   A. A only
   B. B only
   C. Both A and B
   D. Neither A nor B

5. Most air conditioning systems that have HFC-134a as the refrigerant also use ___ as a lubricant.
   A. Polyalkaline glycol (PAG)
   C. Polyol ester (POE)
   B. Mineral oil
   D. Alkylbenzene oil

6. *Technician A* says that a prick punch is often used to secure a bearing into the rotor.
   Technician B says that an external snap ring is often used to secure a bearing into the rotor.
   Who is right?
   A. A only
   B. B only
   C. Both A and B
   D. Neither A nor B

7. O-rings are being discussed:
   Technician A says that HFC-134a O-rings are usually black.
   Technician B says that CFC-12 O-rings are blue or green.
   Who is right?
   A. A only
   B. B only
   C. Both A and B
   D. Neither A nor B

8. *Technician A* says that the seal cavity may be flushed with clean refrigeration oil.
   Technician B says that the seal cavity may be flushed with clean mineral spirits.
   Both agree that the residue must be disposed of in a proper manner.
   Who is right?
   A. A only
   B. B only
   C. Both A and B
   D. Neither A nor B

9. *Technician A* says that oil level in some compressors may be checked using a dipstick.
Technician B says some compressors require that the oil be drained and measured to determine the level.
Who is right?
   A. A only
   B. B only
   C. Both A and B
   D. Neither A nor B

10. *Technician A* says that the air gap is not easily adjusted with the use of a feeler gauge due to the magnetic influence of the clutch coil.
Technician B says that the air gap is automatically adjusted if the armature shaft nut (or bolt) is torqued to proper specifications.
Who is right?
   A. A only
   B. B only
   C. Both A and B
   D. Neither A nor B

ASE Challenge

1. During the compression stroke of an air conditioner compressor, the suction valve is closed by the:
   A. Discharge valve
   B. Discharge pressure
   C. Valve spring
   D. Suction pressure

2. If the lubricant drained from a defective compressor contains metallic particles, all of the following should be done, EXCEPT:
   A. Install a filter in the liquid line
   B. Replace the accumulator or receiver-drier
   C. Clean the inlet screen of the metering device
   D. Install a new or rebuilt compressor

3. The angle of the swash plate in a variable displacement compressor is controlled by:
   A. Suction pressure differential in the crankcase
   B. Pressure differential between low and high side
   C. Both A and B
   D. Neither A nor B

4. The shims shown in the illustration below are used to:
   A. Align the clutch pulley
   B. Secure the hub key
   C. Backup the lock nut
   D. Adjust the air gap

5. The purpose of the compressor is to:
   A. Pump low-pressure vapor to a high-pressure vapor
   B. Pump low-pressure vapor to a low-pressure liquid
   C. Pump low-pressure liquid to a high-pressure liquid
   D. None of the above

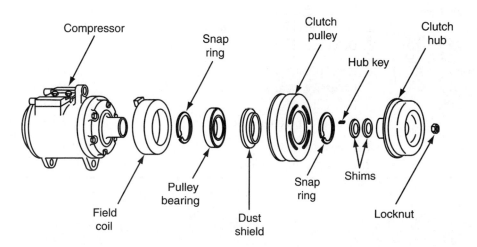

Compressor   Snap ring   Clutch pulley   Clutch hub   Hub key   Field coil   Pulley bearing   Dust shield   Snap ring   Shims   Locknut

Table 6-1 ASE TASK

Identify required lubricant type; inspect and correct oil level in A/C compressor.

Problem Area	Symptoms	Possible Causes	Classroom Manual	Shop Manual
LOSS OF OIL	Low oil level readings	Refrigerant leak	152	249
UNKNOWN OIL TYPE	Oil not miscus with refrigerant	Improper oil used in system	101–105	215, 218, 228 213, 221, 225 224, 232

Table 6-2 ASE TASK

Inspect, adjust, and replace A/C compressor drive belts and pulleys.

Problem Area	Symptoms	Possible Causes	Classroom Manual	Shop Manual
NOISE	Rattles	1. Loose pulley	136–137	410
		2. Defective bearing	136–137	205, 208 222, 225 231
	Squeaks	1. Worn belt(s)	136–137	410
		2. Misaligned belt(s)	136–137	410
		3. Misaligned pulley(s)	136–137	410

Table 6-3 ASE TASK

Inspect, test, service, and replace A/C compressor clutch components or assembly.

Problem Area	Symptoms	Possible Causes	Classroom Manual	Shop Manual
NOISE	Rattles or grinds	1. Clutch slipping	165	352, 354
		2. Defective clutch bearings	165	352, 354
		3. Defective clutch armature or rotor	165	352, 354
POOR COOLING	Clutch slipping	1. Improper air gap	165	352, 354
		2. Excessive wear	165	352, 354
		3. Low voltage	222–225	351–352
		4. Poor ground connection	222–225	351–352
	Clutch seized	1. Excessive heat	165	352, 354
		2. Improper application	165	352, 354

Table 6-4 ASE TASK

Inspect, test, service, or replace A/C compressor.
Inspect, repair, or replace A/C compressor mountings.

Problem Area	Symptoms	Possible Causes	Classroom Manual	Shop Manual
NOISE	Rattles when A/C is turned on	1. Defective bearing(s) in compressor	145–146	107
		2. Worn internal compressor parts	145–146	107
		3. Overcharge of refrigerant	86–89	93
		4. Loose or defective compressor mounts	145–146	187
POOR OR NO COOLING	Compressor inoperative	1. Low or no voltage to clutch coil	222–225	351–352
		2. Belt(s) broken	164	410
		3. Clutch seized	165	352, 354
		4. Compressor seized	145–146	107

Table 6-5 ASE TASK

Identify, inspect, and replace A/C system service valves (gauge connections).

Problem Area	Symptoms	Possible Causes	Classroom Manual	Shop Manual
LOSS OF REFRIGERANT	Poor to no cooling	1. Leaking service valve core (Schrader type)	168–169	249
		2. Leaking service valve seat (shut-off type)	168–169	249
		3. Leaking service valve flange gasket(s)	168–169	249

Name _____ Date _____

Compressor Identification

Upon completion of this job sheet you should be able to identify the different types of compressors used in automotive air conditioning systems.

Tools and Materials

Several vehicles equipped with air conditioning systems.

Describe the vehicle being worked on:

Vehicle 1

Year _____ Make _____ Model _____

VIN _____ Engine type and size _____

Vehicle 2

Year _____ Make _____ Model _____

VIN _____ Engine type and size _____

Vehicle 3

Year _____ Make _____ Model _____

VIN _____ Engine type and size _____

Vehicle 4

Year _____ Make _____ Model _____

VIN _____ Engine type and size _____

Procedure

1. Visually inspect the air conditioning system and describe its overall condition.

   _____

   _____

Vehicle	One	Two	Three	Four
Compressor:				
Type				
Cylinders				
Refrigerant:				
Type				
Charge:				
Lb./oz	__/__	__/__	__/__	__/__
mL	_____	_____	_____	_____

Lubricant:

Type   _____

Charge:

Oz.   __/__   __/__   __/__   __/__

mL   _____   _____   _____   _____

Belt Type   _____

☑ **Instructor's Check** _____

JOB SHEET 20

20

Name _____ Date _____

Check and Correct Compressor Oil Level

Upon completion of this job sheet you should be able to check and correct compressor lubricant levels.

ASE Correlation

This job sheet is related to the the ASE Heating and Air Conditioning Systems Test's content area: *B. Refrigeration System Component Diagnosis and Repair, 1. Compressor and Clutch* ,Task: *Identify required lubricant type; inspect and correct level in A/C compressor.*

Tools and Materials

An air conditioning system compressor
Service manual
Selected air conditioning tools
Lubricant, if required

Describe the vehicle being worked on:

Year _____ Make _____ Model _____

VIN _____ Engine type and size _____

Procedure

1. What type of refrigerant is the compressor designed for?

   _____

2. What type of lubricant is the compressor designed for?

   _____

3. What is the lubricant capacity?
   Oz. _____ mL _____

4. Following procedures outlined in the Service manual drain the lubricant from the compressor.

5. How much lubricant was drained from the compressor?
   Oz. _____ mL _____

6. When refilling, how much clean fresh lubricant should be added to compressor?
   Oz. _____ mL _____

7. Should the lubricant removed in step 4 be reused?

   _____

   Why?

   _____

☑ **Instructor's Check** _____

JOB SHEET 21

21

Name _____ Date _____

Removing and Replacing a Compressor Clutch

Upon completion of this job sheet you should be able to remove and replace a typical compressor clutch.

ASE Correlation

This job sheet is related to the ASE Heating and Air Conditioning Systems Test's content area: *B. Refrigeration System Component Diagnosis and Repair, 1. Compressor and Clutch,* Task: *Inspect, test, service, and replace A/C compressor clutch components or assembly.*

Tools and Materials

Compressor with clutch
Service manual
Selected air conditioning tools
Clutch components, as required

Describe the vehicle being worked on:

Year _____ Make _____ Model _____

VIN _____ Engine type and size _____

Procedure

Following procedures outlined in the service manual, perform the following task and write a short summary of each step in the space provided:

1. Remove the clutch hub and plate assembly.

   _____
   _____

2. Visually inspect the hub and plate assembly. Note any problems.

   _____
   _____

3. Remove the pulley and bearing assembly.

   _____
   _____

4. Carefully inspect the bearing for signs of wear or roughness. Note any problems.

   _____
   _____

5. Make an electrical resistance check of the coil.
   Coil resistance should be: _____ ohms _____   Coil resistance is: _____ ohms

   _____
   _____

6. Replace parts, as needed.

_____

_____

7. Reassemble the clutch.

_____

_____

☑ **Instructor's Check** _____

System Servicing and Testing

Upon completion and review of this chapter you should be able to:

❑ Leak-test an air conditioning system using soap solution.

❑ Leak-test an air conditioning system using a halogen leak detector.

❑ Leak-test an automotive air conditioning system using dye solution.

❑ Remove CFC-12 refrigerant from the system using approved refrigerant recovery equipment.

❑ Remove HFC-134a refrigerant from the system using approved recovery equipment.

❑ Evacuate an air conditioning system using the single evacuation method.

❑ Evacuate an air conditioning system using the triple evacuation method.

❑ Charge the system with refrigerant CFC-12.

❑ Charge the system with refrigerant HFC-134a.

Introduction

Generally, the first piece of equipment that a service technician reaches for is the manifold and gauge set (Figure 7-1). The manifold and gauge set, to the technician, is much the same as a blood pressure test kit is to a physician. It provides a means to "see" what is happening inside the system. Pressures inside an air conditioning system are as important to the system as pressures inside the body are to the human.

Procedures for the proper use of the manifold and gauge set are given in Chapter 3 of this manual. Review of Chapter 3 would be a good idea at this time.

CAUTION: Safety glasses must be worn while working with refrigerants. Remember, **liquid** refrigerant splashed in the eyes can cause blindness.

Figure 7-1 Manifold and gauge set.

Basic Tools

Manifold and gauge set (CFC-12 or HFC-134a, as applicable)

Refrigerant recovery system (CFC-12 or HFC-134a, as applicable)

Vacuum pump (oilless, CFC-12 or HFC-134a, as applicable)

Charging cylinder (CFC-12 or HFC-134a, as applicable)

Scales (if using refrigerant cylinders)

Can tap (if using small cans of refrigerant)

Dye injector

Halogen leak detector

Safety glasses

Fender cover

Small brush

Some systems also have a sight glass.

Refrigeration Contamination

Before servicing an air conditioning system, it should first be determined what type of refrigerant is in the system and, perhaps more important, if the refrigerant is contaminated. Recent studies indicate that, on average, 23 out of every 1,000 motor vehicles tested contained some form of **contaminated refrigerant**. That amounts to over 450,000 contaminated systems out of the 20 million vehicles serviced each year. A system is considered to be contaminated if it contains more than 2 percent of a "foreign" substance. Since the average system contains less than 3 lb. (1.42 L) of refrigerant, contaminants may not exceed 0.96 oz. (28.4 mL). There are actually several types of contamination to be found in the automotive air conditioning system. In order of occurance, these contaminants include air, moisture, mixed refrigerant types, and illegal refrigerants.

A refrigerant identifier should be used to determine the purity of the refrigerant before servicing an air conditioning system. Failure to do so may result in personal injury if it contains a flammable substance, or in a contaminated recovery cylinder if it contains other types of refrigerant.

Air

Any time that a system is open for repair service there is the possibility that air may enter the system. Also, air will enter a closed system if the ambient air pressure is greater than the system refrigerant pressure. For example, assume that an air conditioning system is low of refrigerant due to a leak in the low side. If the low-pressure cut-off switch is defective, or there is a problem that permits the low-side pressure to fall below zero gauge (ambient) pressure, air will enter the system at the same location that the refrigerant leaked out. To remove air from an air conditioning system, it must be leak free and thoroughly evacuated with a quality vacuum pump. To remove air from recovered refrigerant, follow manufacturers instructions included with the recovery/recycle equipment.

Moisture

Because moisture is in the air—humidity—it enters the system with air during service. The higher the humidity, the greater the moisture for any given quantity of air. Moisture removal, however, is not as simple as air removal. To remove all moisture from a system, one must use a vacuum pump capable of achieving and maintaining a deep vacuum for several hours.

In a deep vacuum, moisture will boil off and be carried out of the system as a vapor. The length of time required depends on three major factors: the amount of moisture in the system; the ambient temperature; and the efficiency of the vacuum pump which relates to the amount of vacuum that will be applied to the air conditioning system. Assuming sea-level atmospheric pressure, an air conditioning system evacuation should be conducted when the ambient temperature is 60°F (15.6°C) or above. Moisture will boil at this temperature in a vacuum of 29.4 (2.3 kPa absolute) or better. A temperature/vacuum pressure chart is given in Chapter 7 (Figure 7-11) of the Classroom Manual for moisture removal at various vacuum levels.

A few drops of water in a bell jar provide an excellent illustration of lowering the boiling point of water in a vacuum. In the average air-conditioned laboratory under a vacuum of about 28.9 in. Hg (3.75 kPa absolute), droplets of water begin to bubble (boil) and are eventually removed from the bell jar. The bell jar experiment, however, does not perfectly illustrate "real world" experience with automotive air conditioning systems. In an automotive air conditioning system, moisture is not as "free." In addition to any free moisture in the system, trapped moisture must also be boiled out of the lubricant and desiccant.

The average service facility may not possess or maintain equipment capable of achieving a sufficient vacuum for complete moisture removal. Because of this, if a high level of moisture is suspected in an automotive air conditioning system, the general remedy is to replace the accumulator-drier or receiver-drier that contains a desiccant and then evacuate the system to remove excess moisture.

Periodic maintenance of the vacuum pump is essential to ensure maximum performance. The vacuum pump lubricant should be changed at intervals recommended by the manufacturer.

Mixed Refrigerant Types

It is not uncommon to find that HFC-134a has been added to a CFC-12 system or vice versa. Sometimes it is found that a blend refrigerant has been added to a CFC-12 or HFC-134a system. If this is the case, the refrigerant must be recovered and the system should be evacuated and charged with the proper refrigerant. It is not possible to separate the refrigerants with currently available equipment. Recovery should be made using equipment dedicated to contaminated refrigerant and stored in a dedicated cylinder also designated for contaminated refrigerant.

In addition to the flammable refrigerants, identified below as hazardous, the following refrigerants are NOT approved for automotive use under the Significant New Alternatives Policy (SNAP) of the Environmental Protection Agency (EPA). R-176 contains CFC-12 and therefore is inappropriate as a CFC-12 substitute, and R-405A contains perfluorocarbons, which have been implicated in global warming.

Hazardous Refrigerant

A refrigerant identifier must be used to ensure the purity of the refrigerant before service is performed on the vehicle. Hazardous refrigerants are generally those that contain an excessive amount of a flammable substance and are therefore not approved by the SNAP program. If a system is known or suspected to contain a flammable substance, it must be recovered in a manner consistent with the specific instructions provided with the particular recovery equipment used.

At the present time, flammable blend refrigerants that are NOT approved for automotive (or any other) use include OZ-12®, HC-12a®, and Duracool 12a.

Leak Testing the System

Before undertaking any leak-detection procedures, perform a visual inspection of all system components, fittings, and hoses for signs of lubricant leakage, damage, wear, or corrosion.

Note that polyalkaline glycol (PAG) lubricant may evaporate and therefore not be visible. Mineral oil, on the other hand, does not evaporate and will leave a visible stain. To prevent an inaccurate or false reading with an electronic leak detector, make sure there is no refrigerant vapor or tobacco smoke in the vicinity of the vehicle being checked. Also, because there could be more than one leak in the air conditioning system, when a leak has been found, continue to check for additional leaks. Perform the leak test in a relatively calm area so the leaking refrigerant is not dispersed in air movement. With the engine not running:

1. Connect a proper manifold and gauge set (CFC-12 or HFC-134a) to the air conditioning system's low- and high-side service ports.
2. Ensure that the refrigerant pressure in the air conditioning system is at least 50 psig (345 kPa) and that the ambient temperature is 60°F (15.6°C) or above.

   NOTE: If less than specified, recover, evacuate and recharge the system with enough refrigerant to perform the leak test.

   NOTE: At temperatures below 60°F (15.6°C), leaks may not be detected since the system pressure may not reach 50 psig (345 kPa).
3. Conduct the leak test from the high side to the low side at points shown in Figure 7-2, as follows.

Compressor

Check the high- and low-side hose fittings, relief valve, gaskets, and shaft seal. Check the service valves, if compressor-mounted, with the protective covers removed.

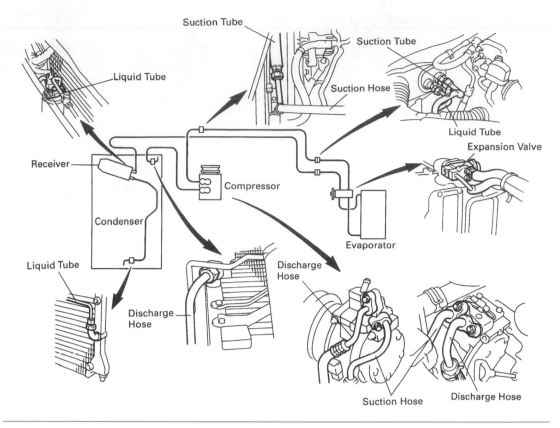

Figure 7-2 Test points for leak detection.

Accumulator

Check the inlet and outlet fittings, pressure switch, weld seams and the fusible plug, and low-side service fitting (with cap removed), if equipped.

Receiver

Check the inlet and outlet hose fittings, pressure switch, weld seams and the fusible plug, and sight glass, if equipped.

Service Valves

Check all around the low- and high-side service valves with the caps removed. Ensure that the service valve caps are secured on the service valves after testing to prevent future leaks.

NOTE: After removing the manifold gauge hoses from the service valves, wipe away residue to prevent any false readings by the leak detector. Blowing low pressure air across the service valve should clear any refrigerant residue vapor. If a service valve port is found to be leaking, repair it as required before replacing the valve cap.

Evaporator

In some systems the blower motor resistor block may be removed to gain access to the evaporator core for testing. Since refrigerant is heavier than air, however, the leak may be best detected at the evaporator condensate drain hose. If using an electronic leak detector, place the probe near the drain hose for 10–15 seconds immediately after stopping the engine. Take care not to contaminate the end of the test probe. If it is a dual air conditioning system, do not forget to check both evaporators, front and rear.

Condenser

Check all around the discharge (inlet) line and liquid (outlet) line connections. Check the front (face) of the condenser for any leaks that may be due to road damage. If the air conditioning system has an auxiliary condenser check it for leaks as well.

Metering Device

Carefully check all connections, inlet and outlet, of the metering device if a thermostatic expansion valve (TXV) or the spring lock coupling if an orifice tube. In a dual air conditioning system, check both front and rear metering devices.

Hoses

Although barrier-type refrigeration hoses, now required by the EPA, are relatively leak proof, they may on occasion develop a pinhole leak. Visually inspect all hoses carefully for telltale traces of lubricant and/or dye that indicate a leak. Leaks are not so easily detected visually since PAG lubricant may evaporate from the surface of the hose. Any of the leak test methods may be employed if a leak in a hose is suspected and is not visually detected. If a dual air conditioning system, check all hoses, front to back, of the vehicle.

Pressure Controls

Check all around a pressure control. Though rare, a pressure control has been known to leak past the seal of the electrical connector. Remove the connection and thoroughly check the control.

Methods

The three most popular methods of leak detection today use soap bubbles, halogen, and ultraviolet dye. They are covered in the following sections. The once popular halide (gas) leak detector is now seldom used. It is only effective for detecting CFC and HCFC refrigerants and is considered hazardous in the view of the danger of encountering flammable refrigerants.

Soap Solution

The soap solution is used as a method of leak detection when it is impractical or impossible to pinpoint the exact location of a leak by using the halide or halogen leak detector methods. A commercial soap solution is available (Figure 7-3) that is generally more effective than a homemade solution. A good grade of sudsing liquid dishwashing detergent may be used, however, if a prepared commercial solution is not available.

Preparing the System

1. Connect the manifold and gauge set to the system.
2. Make certain that the high- and low-side **manifold hand valves** are in the closed (cw) position.
3. If the system is equipped with manual service valves, place the high- and low-side valves in the cracked position.
4. Open the low- and high-side hose **shut-off valves**.
5. Determine the presence of refrigerant in the system. A minimum value of 50 psig (348 kPa) is needed.
6. If there is an insufficient **charge** of refrigerant in the system, continue with the next step, "Adding Refrigerant for Leak Testing." If the charge is sufficient, omit the next step and proceed with leak testing.

The least expensive method is by soap bubbles.

Classroom Manual
Chapter 7,
page 153

Some detergents may be used undiluted.

Figure 7-3 Leak test (soap) solution. (Courtesy of BET, Inc.)

Figure 7-4 R-12 and R-134a cans. (Courtesy of BET, Inc.)

Adding Refrigerant for Leak Testing

> ⚠️ **WARNING:** Use a refrigerant identifier to determine that the source refrigerant is the same type—CFC-12 or HFC-134a (Figure 7-4)—that is used in the air conditioning system.

1. Attach the center manifold service hose to a source of refrigerant.
2. Open the refrigerant container service valve.
3. Open the high-side manifold hand valve until a pressure of 50 psig (348 kPa) is reached on the low-side gauge. Then close the high-side hand valve.
4. Close the refrigerant container service valve.
5. Close the service hose shut-off valve.
6. Remove the hose from the refrigerant container.

Procedure

1. Apply soap solution to all joints and/or suspected areas by using the dauber supplied with the commercial solution or by using a small brush with household solution.
2. Leaks are exposed when a bubble forms (Figure 7-5).
3. Repair any leaks found.

Halide (Gas) Leak Detection

The halide (gas) leak detector is essentially little more than a propane torch. The halide leak detector was once the most popular leak detector because of its low initial cost and relatively low upkeep. Other than propellant replacement, the only maintenance required is an occasional reactor plate replacement.

When leak testing, all joints and fittings should be free of oil. This precaution eliminates the possibility of a false reading caused by refrigerant absorbed in the oil. Cigarette smoke, purging of another unit nearby, and refrigerant vapors in the surrounding air can also give false readings on the detector.

> ⬛ **CAUTION:** A halide leak detector must only be used in a well-ventilated area. It must never be used in spaces where explosive gases are present. When refrigerant comes into contact with an open flame, toxic gases are formed. Never inhale the vapors or fumes from the halide leak detector—they can be poisonous.

Classroom Manual
Chapter 4,
page 96

A minimum of 50 psig (348 kPa) is required for leak testing.

The hose shut-off valve is provided to reduce emissions.

Shop Manual
Chapter 4,
page 110

Figure 7-5 Leaks are detected when a bubble forms.

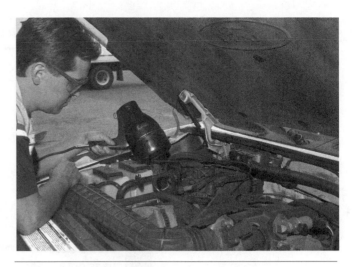

Figure 7-6 Ultraviolet (UV) lamp used for leak testing.

In the presence of CFC and HCFC refrigerant, a color change will occur in the flame above the reactor plate:

❑ Pale blue: no refrigerant loss

❑ Yellow: small amount of refrigerant loss

❑ Purplish-blue: large amount of refrigerant loss

❑ Violet: heavy amount of refrigerant loss; the volume may be great enough to extinguish the flame

Classroom Manual
Chapter 7,
page 152

Tracer Dye Leak Detection

As a rule of thumb, the presence of oil at a fitting or connection generally indicates a refrigerant leak. This is not always the case, however, because oil is used on fittings as an aid in assembly procedures. If a leak is suspected, the area should be wiped clean and the leak verified. This may be accomplished by either of the several methods discussed in this chapter.

A popular method of refrigerant leak detection is with the use of a fluorescent tracer dye that is easily detected with an ultraviolet (UV) lamp (Figure 7-6). The fluorescent tracer dye lasts for about 500 hours of air conditioning system use. When expended, another injection of tracer dye is required. To inject dye into the system, the system pressure must be above 80 psig (551.6 kPa). For proper use and accurate results always follow the instructions included with the tracer dye.

To pinpoint the leak, scan all the air conditioning system components, fittings, and hoses with the ultraviolet lamp. The exact location of a leak will be revealed by a bright yellow -green glow of the tracer dye. Leaks are best detected in low ambient-light conditions. In areas where the lamp cannot be used, such as where the ambient-light is high, a mechanic's mirror may be used. The technician may also wipe the suspected area with a disposable, nonfluorescent towel, which is then examined with the lamp for traces of the dye.

After the leak has been repaired, the dye can be removed from the exterior of the leaking area by using an oil solvent. To verify that the repair has been made, operate the air conditioning system for about five minutes and reinspect the area with the UV lamp. Since more than one leak may occur, it is wise to check the entire system.

Many technicians prefer to add a dye trace solution to the refrigerant any time the air conditioning system is opened for service. With the high cost of refrigerant and technician labor, this practice is well worth the additional cost at the time of repairs.

Halogen (Electronic) Leak Detection

Classroom Manual
Chapter 7,
pages 152

The halogen (electronic) leak detector (Figure 7-7) is the most sensitive of all types of leak detectors. These leak detectors, to comply with SAE standard J-1627, must sense a refrigerant leak of as little as 0.5 oz. (14.78 mL) per year. This type of leak detector can be of great value in detecting the "impossible" leak.

When using an electronic leak detector, ensure that the instrument, if required, is calibrated, and set properly according to the operating instructions provided by the **manufacturer**. In order to use the leak detector properly, read the operating instructions and perform any specified maintenance.

Other vapors in the service area or any substances on the components—such as antifreeze, windshield washing fluid, or solvents and cleaners—may falsely trigger the leak detector. Make sure that all surfaces to be checked are clean. Do not permit the detector sensor tip to come into contact with any substance; a false reading can result and the leak detector can be damaged.

CAUTION: A halogen electronic leak detector must be used in well-ventilated areas only. It must never be used in spaces where explosive gases may be present.

The following list and Photo Sequence 9 illustrate typical procedures for using an electronic leak detector:

1. Hold the probe in position about 3/16 in. (5 mm) from the area to be checked (Figure 7-8).
2. When testing, circle each fitting with the probe (Figure 7-9).
3. Move the probe along the component about 1–2 in. (25–50 mm) per second (Figure 7-10).
4. If a leak is detected, verify it by fanning or blowing compressed air into the area of the suspected leak, then repeat the leak check.

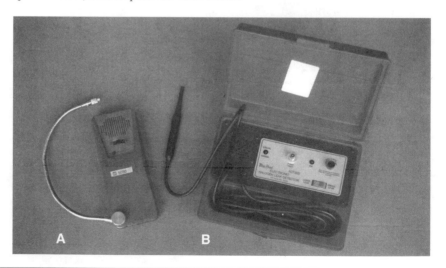

Figure 7-7 Electronic leak detectors: (A) cordless; (B) corded.

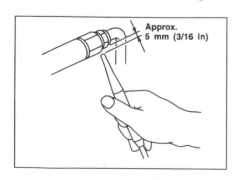

Figure 7-8 Hold probe about 3/16 in. (5 mm) away.

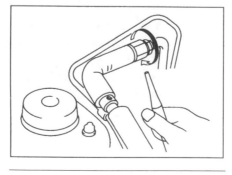

Figure 7-9 Circle each fitting.

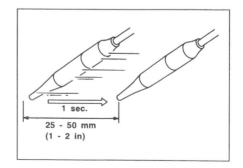

Figure 7-10 Move probe 1–2 in. (25–50 mm) per second.

Photo Sequence 9
Typical Procedure for Checking for Leaks

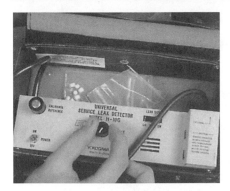

P9-1 Turn the control and/or sensitivity knobs to OFF or ZERO. If the leak detector is corded, connect it to an approved voltage source. Skip this step if it is cordless.

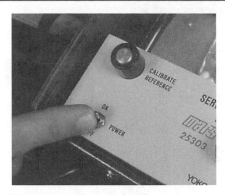

P9-2 Turn on the switch. Allow a warmup period of about five minutes. There is usually no warmup period required for cordless models.

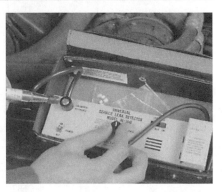

P9-3 Place the probe at the reference leak. Adjust the control and/or sensitivity knobs until the detector reacts.

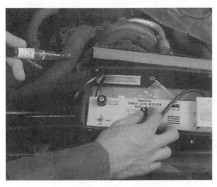

P9-4 Remove the probe. The reaction should stop.

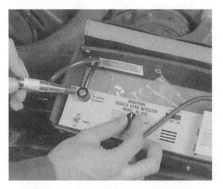

P9-5 If the reaction continues, the sensitivity control is adjusted too high. Repeat the procedure of step P4-3. If the reaction stops, the sensitivity adjustment is adequate.

P9-6 Slowly move the search hose under and around all of the joints and connections.

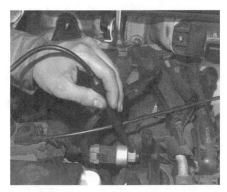

P9-7 Check all seals and screw-in pressure control devices.

P9-8 Check the service fittings. It will be necessary to remove the service cap for this test.

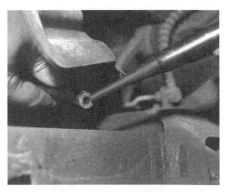

P9-9 Check the evaporator at the outlet ducts.

WARNING: Do not keep the probe in contact with refrigerant any longer than is necessary to locate the leak. Do not deliberately place the probe in a stream of raw refrigerant or in an area where a severe leak is known to exist. The sensitive components of the leak detector can be severely damaged.

SERVICE TIP: If a leak is located, the electronic leak detector reacts in the same manner as it does when placed by the reference leak.

Repair System

1. After the leak is located, recover the system refrigerant as outlined in this chapter.
2. Repair the leak as indicated.
3. Check the compressor oil.
4. Add or change oil, if required.
5. Recheck for leaks.
6. **Evacuate** the system.
7. Charge the system with refrigerant.
8. Perform other service procedures as necessary.

Be sure to use the proper refrigerant, only R-12 or R-134a, as appropriate.

Classroom Manual
Chapter 7,
page 156

Evacuating the System

An important step in the repair of an automotive air conditioning system is proper **evacuation** charging with refrigerant. The air conditioning system must be evacuated whenever it is serviced to the extent that the refrigerant was recovered. Proper evacuation rids the system of all air and most moisture that may have entered during repair service. Photo Sequence 10 illustrates the typical procedure for evacuating the system. The following procedure assumes that the system has been serviced and does not contain refrigerant.

Checking for Leaks

A **standing vacuum test** may be made to leak test the system. Proceed as follows:

Only enough refrigerant to increase the system pressure to 50 psi (345 kPa) is required.

1. Evacuate the systems outlined in Photo Sequence 10.
2. Close the manifold hand valves and turn off the vacuum pump.
3. Note the low-side guage reading; it should be 29 in. Hg. (3.4 kPa absolute) or lower.
4. Allow the system to "rest" for five-minutes, then again note the low-side gauge reading.
   NOTE: The low-side gauge needle should not raise faster than 1 in. Hg. (3.4 kPa absolute) in five minutes.

Classroom Manual
Chapter 7,
page 160

If the system does not meet the requirement of step 4, a leak is indicated. A partial charge of refrigerant must be installed and the system must be leak checked. After the leak is detected the refrigerant must be recovered. After the leak is repaired, again perform the standing vacuum test, starting with step 1 above.

Triple Evacuation Method

The basic steps in the **triple evacuation** method are given here. The procedures assume that the system is sound after the refrigerant has been removed and repairs, if any, have been made.

Photo Sequence 10
Typical Procedure for Evacuating the System

P10-1 Connect the manifold and gauge set low- and high-side service hoses to the system.

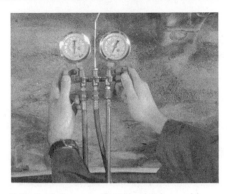

P10-2 Make sure that the high- and low-side manifold hand valves are in the closed position.

P10-3 If the system is equipped with shut-off type service valves, place them in the cracked position.

P10-4 Remove the protective caps from the inlet and exhaust of the vacuum pump.

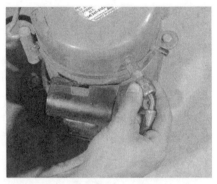

P10-5 Connect the center manifold service hose to the inlet of the vacuum pump.

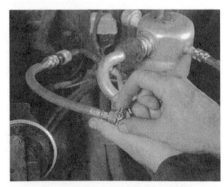

P10-6 Open the shut-off valve of the three service hoses.

P10-7 Connect the power cord of the vacuum pump to an approved power source.

P10-8 Turn on the vacuum pump.

P10-9 Open the low-side manifold hand valve and observe the low-side gauge needle. The needle should be immediately pulled down to indicate a slight vacuum.

P10-10 After about five minutes, the low-side gauge should indicate 20 in. Hg (33.8 kPa absolute) or less. The high-side gauge needle should be slightly below the zero index of the gauge.

P10-11 If the high-side needle does not drop below zero, unless restricted by a stop, a blockage in the system is indicated. If the system is blocked, discontinue the evacuation. Repair or remove the obstruction. If the system is clear, continue the evacuation.

P10-12 Operate the pump for another 15 minutes and observe the gauges. The system should be at a vacuum of 24–26 in. Hg (20.3–13.5 kPa absolute). If it is not, close the low-side hand valve.

P10-13 Observe the compound (low-side) gauge. If the needle rises, indicating a loss of vacuum, there is a leak that must be repaired before the evacuation is continued. If no leak is evident, continue the evacuation.

P10-14 Reopen the low-side manifold hand valve.

P10-15 Open the high-side manifold hand valve.

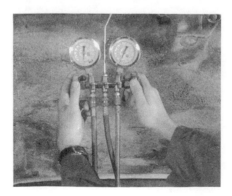

P10-16 Allow the vacuum pump to operate for a minimum of 30 minutes, longer if time permits. After pump-down, close the high- and low-side manifold hand valves. Turn off the vacuum pump and close the service hose shut-off valves. Turn off the vacuum pump valve, if equipped. Then, disconnect the manifold service hose from the vacuum pump. Replace the protective caps, if any.

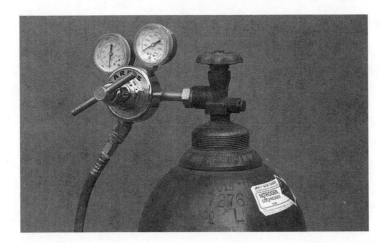

Figure 7-11 Typical dry nitrogen setup.

Procedure

As previously detailed, connect the manifold and gauge set into the system. Be sure that all hoses and connections are tight and sound, and that all appropriate valves are in the closed position.

First Stage

1. Pump a vacuum to the highest efficiency for 25–30 minutes.
2. Close the manifold hand valves.
3. Close the service hose shut-off valves.
4. Close the vacuum pump shut-off valve, if equipped. If the system is not equipped with a shut-off valve, turn off the pump.
5. Disconnect the service hose from the vacuum pump.
6. Connect the service hose to a **dry nitrogen** (Figure 7-11) source.

   ■ **CAUTION:** Make sure that the nitrogen supply has proper regulators and that supply pressure does not exceed 75 psig (517 kPa).

7. Open the service hose shut-off valve.
8. Open the nitrogen supply valve.
9. Open the low- and high-side service hose shut-off valves.
10. Open the low-side manifold hand valve to break the vacuum:
    a. Slowly increase the pressure to 1–2 psig (6.8–13.7 kPa).
    b. Close all valves: the manifold **low-side hand valve**, service hose shut-off valve, low- and high-side hose shut-off valves, and the nitrogen supply valve.
11. Disconnect the service hose from the nitrogen supply.

Second Stage

1. Allow one-half hour, which is sufficient time for the dry nitrogen to "**stratify**" the system.
2. Reconnect the service hose to the vacuum pump.
3. Open the vacuum pump shut-off, if equipped. If not equipped, turn on the pump.
4. Open the service hose shut off valve.
5. Open the manifold low- and **high-side hand valves**.
6. Pump a vacuum to the highest efficiency for 25–30 minutes.
7. Repeat steps 2 through 11 as outlined in "First Stage" procedures.

Several different types of service hose shut-off valves are available.

Be sure the nitrogen pressure is regulated BEFORE opening the valve.

Third Stage

1. Follow steps 1 through 6 as outlined in "Second Stage" procedures.
2. Close all valves: the service hose shut-off valve, vacuum pump shut-off valve, if equipped (if not equipped, turn off the pump), low- and high-side service hose shut-off valves, and manifold low- and high-side hand valves.
3. Turn off the vacuum pump (if not previously done).
4. Remove the service hose from the vacuum pump.

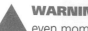 **WARNING:** The system is now under a vacuum. Opening any valves and/or fittings, even momentarily, will allow moisture-laden air to enter the system.

5. The system is now ready for charging. Follow the appropriate procedure outlined in this chapter.

Refrigerant Recovery

To **purge** an air-conditioning system, in general terms, is to remove all of its contents, primarily refrigerant. While the term is generally understood to refer to refrigerant, it may also include air and moisture. Purging a system of refrigerant is usually necessary when a component is to be serviced or replaced. Until recently, to "purge" was to vent refrigerant into the atmosphere. The **Federal Clean Air Act** Amendments of 1990, however, required that after July 1, 1992, no refrigerants could be intentionally vented.

It is now necessary that a refrigerant **recovery system** (Figure 7-12) be used to purge an air conditioning system of refrigerant. Manufacturers specifications and procedures should be followed

Air contains moisture in the form of humidity.

Classroom Manual
Chapter 7,
page 162

To recover refrigerant is to remove it, in any condition, from the system.

Figure 7-12 Typical refrigerant recovery system (Courtesy Robinair Division—SPX Corp.)

to ensure safe and adequate performance. The following procedure is typical and should be used only as a guide.

 CAUTION: Adequate ventilation must be maintained during this procedure.

Procedure

1. Connect the manifold and gauge set into the system.
2. Start the engine and adjust the speed to 1,250–1,500 rpm.

NOTE: Some system malfunctions, such as a defective compressor, may make the next four steps impossible to perform.

3. To stabilize the system, set all air conditioning controls to MAX cold position with the blower on HI.
4. Reduce the engine speed to 1,000–1,200 rpm and operate for 10-15 minutes.
5. Return the engine speed to normal idle to prevent dieseling.
6. Turn off the air conditioner controls.
7. Shut off the engine.

See Photo Sequence 11 for a typical procedure for recovering (purging) refrigerant from the system.

Charging the System

Three typical methods of charging an automotive air conditioning system refrigerant are given in this service procedure: from **pound** cans with the system off, from pound cans with the system operating, and from a bulk source.

 **CAUTION:** Above 130°F (54.4°C), liquid refrigerant completely fills a container. Hydrostatic pressure builds up rapidly with each degree of temperature added. Never heat a refrigerant container above 125°F (51.7°C).

Refrigerant is generally less expensive per unit in larger packages.

Classroom Manual
Chapter 7,
pages 163

Photo Sequence 11
Typical Procedure for Recovering (Purging) Refrigerant from the System

P11-1 Attach the manifold and gauge set service hose to the refrigerant recovery system.

P11-2 Open all hose shut-off valves.

P11-3 Open both the high- and low-side manifold hand valves.

Photo Sequence 11
Typical Procedure for Recovering (Purging) Refrigerant from the System (continued)

P11-4 Connect the refrigerant recovery system to an approved electrical power supply.

P11-5 Turn on the main switch.

P11-6 Turn on the recovery (compressor) switch.

P11-7 Operate until a vacuum pressure is indicated. The recovery system will automatically shut off. If it is not equipped with an automatic shut-off, turn the compressor switch to OFF after achieving a vacuum pressure.

P11-8 Observe the gauges for at least five minutes. If the vacuum does not rise, complete refrigerant recovery. If the vacuum rises but remains at 0 psig (0 kPa) or below, a leaking system is indicated. Complete refrigerant recovery and repair the system.

P11-9 If the vacuum rises to a positive pressure, above 0 psig (0 kPa), the refrigerant was not completely removed from the system. Repeat the recovery procedure, starting with step P6-5 of this procedure.

P11-10 Repeat step P6-7 until the system holds a stable vacuum for at least two minutes.

P11-11 After all of the refrigerant has been recovered from the system, close all valves. Close the service hose shut-off valves, the low- and high-side manifold valves, and the recovery system inlet valve.

P11-12 Disconnect all the hoses from the system service valves or fittings. Cap all fittings and hoses to prevent dirt, foreign matter, or moisture from entering the system. This is most important for an HFC-134a system. The lubricant used in this system is very hygroscopic.

The following additional safety precautions must also be followed when handling refrigerant.

- ❏ Do not deliberately inhale refrigerant.
- ❏ Do not apply a direct flame to a refrigerant container.
- ❏ Do not place an electric resistance heater close to a refrigerant container.
- ❏ Do not abuse a refrigerant container.
- ❏ Do not use pliers or vice grips to open and close refrigerant valves. Use only approved wrenches.
- ❏ Do not handle refrigerant without suitable eye protection.
- ❏ Do not discharge refrigerant into an enclosed area.
- ❏ Do not expose refrigerant to an open flame.
- ❏ Do not lay cylinders flat. Store containers in an upright position only. Secure large cylinders with a chain to prevent them from tipping over.

Eye protection should include side shields.

Preparing the System

This procedure assumes that the system has been evacuated. If it has not been evacuated, refer to the proper procedure for evacuation *before* attempting to charge the system.

If charging from pound cans, set up the refrigerant recovery system according to the equipment manufacturer's instructions to be used to recover any residual refrigerant that remains in a can while charging the system.

Install the Can Tap Valve on a "Pound" Container

This procedure may be followed prior to servicing a system using pound cans.

1. Be sure that the **can tap valve** stem is in the fully counterclockwise (ccw) position.
2. Attach the valve to the can (Figure 7-13). Secure the valve with the locking nut, if so equipped. Secure the valve with the clamping lever, if so equipped.
3. Make sure that the manifold service hose shut-off valve is closed.
4. Connect the manifold service hose to the can tap port.
5. Pierce the can by turning the valve stem all the way in the clockwise (cw) direction.
6. Back the can tap valve out, turning in a counterclockwise (ccw) direction.
7. The center (service) hose is charged with refrigerant to the shut-off valve, and is under a vacuum between the manifold and shut-off valve.

There are two basic types of can tap valve: locking nut and clamping ring.

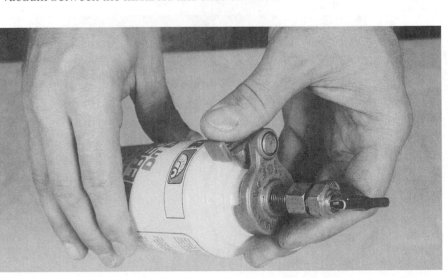

Figure 7-13 Attach the can tap valve.

WARNING: Do not open the high- or low-side hand valves at this time.

Checking the System for Blockage

This procedure may be followed when charging the system from pound cans.

1. Open the service hose shut-off valve.
2. Open the low- and high-side service hose shut-off valves.
3. Open the high-side gauge manifold hand valve. Observe the low-side gauge pressure. Close the high-side hand valve.
4. Close the hose shut-off valves. Close the service hose and the low- and high-side hoses.

 **SERVICE TIP:** If the low-side gauge does not move from the vacuum range into the pressure range, a system blockage is indicated. Correct the blockage, then evacuate and continue with the appropriate service procedure.

If the discharge valve and plate are in good condition, refrigerant must circulate through the system to impress pressure on the low-side gauge.

Using Pound Cans (System Off)

1. Open the service hose shut-off valve.
2. If not previously done, fully open the can tap valve.
3. Open the high-side gauge manifold hand valve.
4. Observe the low-side pressure gauge. If the gauge indication does not move from the vacuum range to the pressure range, a system blockage is indicated. If the system is not blocked, proceed with step 5. If the system is blocked, correct the condition and evacuate the system before continuing with step 5.
5. Invert the container (Figure 7-14) and allow the liquid refrigerant to enter the system.
6. Tap the refrigerant container on the bottom. An empty can produces a hollow ringing sound.

 WARNING: Charging liquid refrigerant into a compressor while it is running may cause damage to the compressor and injury to the technician.

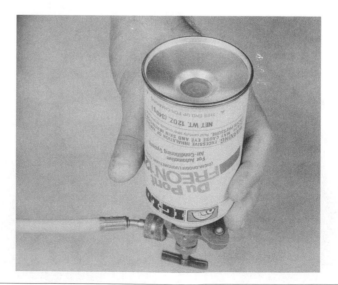

Figure 7-14 Invert the container to allow liquid refrigerant to flow.

7. Use the recovery system to remove any residual refrigerant from the empty can.

8. Repeat steps 5, 6, and 7 with additional cans of refrigerant as required to charge the air conditioner. Refer to the manufacturer's specifications for system capacity.

Photo Sequence 12 illustrates a typical procedure for completing the system charge.

The system generally requires 10 percent less R-134a than was required for R-12.

Using Pound Cans (System Operational)

1. Start the engine and adjust the speed to about 1,250 rpm by turning the idle screw or the setting on the high cam.

2. Make sure that both the manifold hand valves are closed.

3. Adjust the air conditioning controls for MAX cooling with the blower on HI.

4. Open the service hose shut-off valve.

5. Open the low- and high-side hose shut-off valves.

Photo Sequence 12
Typical Procedure for Completing the System Charge

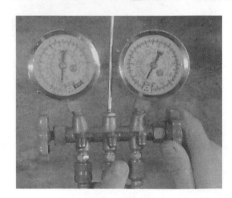

P12-1 Close the high-side manifold hand valve.

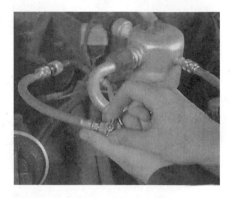

P12-2 Close the service hose shut-off valve.

P12-3 Rotate the compressor clutch by hand through two or three revolutions to ensure that liquid refrigerant has not entered the low side of the compressor.

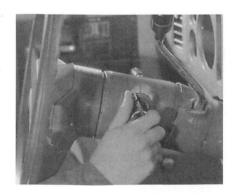

P12-4 Start the engine and set it to fast idle.

P12-5 Engage the clutch to start the compressor.

P12-6 Set all controls to MAX cooling. Open manifold valves and hose shut-off valves so refrigerant charging can take place.

Photo Sequence 12
Typical Procedure for Completing the System Charge (continued)

P12-7 Conduct a performance test. Add refrigerant. Tap another can of refrigerant, if necessary.

P12-8 Turn off all air conditioning controls.

P12-9 Reduce the engine to regular idle speed. Stop the engine.

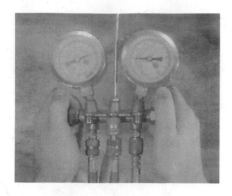

P12-10 Close all valves, the service hose shut-off valve, low- and high-side service hose valves, low- and high-side manifold hand valves (if open), can tap valve (if refrigerant remains in the can), and back seat service valves, if equipped.

P12-11 Remove the manifold and gauge set from the system.

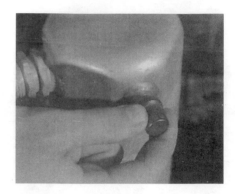

P12-12 Replace all protective caps and covers.

The 1/4-in. (6.4-mm) service hose acts like a capillary tube. At 40 psig (276 kPa), the refrigerant will vaporize by the time it travels the 6 ft. (1.8 m) of hose.

6. If not previously done, open the can tap valve.
7. With the can in an upright position, open the low-side manifold hand valve.
8. After the pressure on the low side drops below 40 psig (377 kPa absolute), the can may be inverted to allow more rapid removal of the refrigerant.

⚠️ **WARNING:** Regulate the low-side manifold hand valve so that low-side pressure remains under 40 psig (276 kPa) to ensure that liquid refrigerant does not enter the system.

9. Tap the can on the bottom to determine if it is empty. An empty refrigerant can will give a hollow ringing sound. One may also shake the can to determine if refrigerant is "sloshing around" inside.
10. Repeat steps 7, 8, and 9 with additional cans of refrigerant as required to charge the system completely. Refer to the manufacturer's specifications regarding system capacity.

11. Close all valves: the can tap valve (if refrigerant remains in the can), service hose shut-off valve, low- and high-side service hose shut-off valves, low- and high-side manifold hand valves, and **back seat** system service valves, if equipped.

12. Remove the refrigerant container. If refrigerant remains in the can, close the charging hose shut-off valve and remove the can tap from the center service hose. If the can is empty, use the recovery system to remove any residual refrigerant from the manifold, hoses, and can.

13. Remove the manifold and gauge set.

14. Replace all protective caps and covers.

Take care not to overcharge the air conditioning system.

Charging from a Bulk Source

Shops that perform a large volume of air conditioning service can obtain bulk refrigerant in 10-, 15-, 25-, 30-, 50-, and 145-pound (4.5-, 6.8-, 11.3-, 13.6-, 22.7-, and 65.8-**kilogram**) cylinders. The use of bulk containers requires a set of scales or other approved measuring device to determine when the proper system charge is obtained.

The high cost of refrigerant, particularly CFC-12, has prompted many service facilities to charge for it by the ounce. An ounce (29.6 mL) of CFC-12 now generally costs more than a pound (473 mL) did before the federal restrictions on refrigerants.

Regulations by the Bureau of Weights and Measures, in most states, require that certified scales be used to ensure that the costs of refrigerant to the customer are proper. Two types of scale easily meet this requirement: electronic and beam.

An electronic scale (Figure 7-15) is more properly referred to as an electronic charging meter. Some charging meters display refrigerant weight as it is being charged from the tank; others may be programmed to automatically charge a selected amount of refrigerant and will shut off when the programmed amount has been reached.

Some charging meters measure refrigerant in increments of 0.5 oz. (14.8 mL), while others are accurate to and measured in 0.25 oz. (7.4 mL) increments. Charging meters can generally accept cylinders having a gross weight of 100 lb. (45.4 kg) or more. For heavier cylinders, a beam scale, such as that shown in Figure 7-16 must be used. The platform of a scale or charging meter should be corner balanced so that no matter where the cylinder is placed on the platform surface its weighing accuracy and function will not be affected. When buying refrigerant in any size container, it is suggested that the cylinder be checked for weight and a sample of its contents be tested for purity. At today's prices, one cannot be too careful.

Classroom Manual
Chapter 4,
page 97

Hanging scales are often used to weigh refrigerant.

Figure 7-15 Typical electronic charging meter (scale).

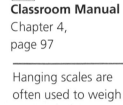
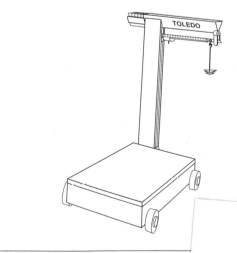

Figure 7-16 Typical certified scale Toledo Scale Co.)

Most electronic charging meters are compatible with all refrigerants and have a tare function that may be zeroed (0.00 oz.). Some meters have a digital display in either pounds (lb.) or kilograms (kg). Battery-powered charging meters generally include an ac adaptor and have a "sleep mode" to ensure extended battery life.

The following procedure assumes that the system has been properly prepared for charging procedures. If not, consult the appropriate heading for the proper procedure *before* continuing.

Connecting the Refrigerant Container

Be sure to use the correct refrigerant. R-12 and R-134a are not compatible.

1. Make sure that all service valves are in the off or closed position. Check the service hose shut-off valves, manifold hand valves, compressor service valves, if equipped, and refrigerant source valve.
2. Connect the center manifold service hose to the supply refrigerant cylinder.
3. Open the refrigerant cylinder hand valve.
4. The system is now under a vacuum from the manifold to the service hose shut-off valve, and under a refrigerant charge from the cylinder to the hose shut-off valve.

Charging the System

1. Open the service hose shut-off valve.
2. Open the low- and high-side manifold hose shut-off valves.
3. Briefly open the high-side manifold hand valve. Observe the low-side gauge.
4. System blockage is indicated if the low-side gauge needle does not move from the vacuum range into the pressure range. If system is not blocked, proceed with step 5. If the system is blocked, correct the blockage and reevaluate the system before continuing with step 5.
5. Start the engine and adjust the speed to about 1,250 rpm.
6. Adjust the air conditioning controls for MAX cooling with the blower on HI.

CAUTION: Keep the refrigerant cylinder in an upright position at all times. Liquid refrigerant must not be allowed to enter the compressor. This can cause serious damage and possible injury.

7. Place the refrigerant cylinder on an approved scale and note the **gross weight**.
8. Open the low-side manifold hand valve to allow refrigerant to enter the system.
9. When the system is fully charged, close the low-side manifold hand valve.
10. Close the refrigerant cylinder service valve.
11. Close the service hose shut-off valve.
12. Remove the service hose from the refrigerant cylinder.
13. Note the gross weight now shown on the scale. Refrigerant used, by weight, is the difference between what the cylinder weighed in step 7 and what it now weighs.
14. Conduct performance tests or other tests as required.
15. Return the engine to its normal idle speed.
16. Turn off all air conditioning system controls.
17. Shut off the engine.
18. Close all valves. Close the low- and high-side service hose shut-off valves. Back seat the compressor service valves, if equipped.
19. Use the recovery system to remove any residual refrigerant from the manifold and hoses.
20. Remove the manifold and gauge set.
21. Replace all protective caps and covers.

Adding Dye or Trace Solution

A dye or trace solution can be added to an air conditioning system as an aid to pinpoint a small leak. The dye shows the exact location of a leak by depositing a visible colored film around the leak. A trace solution produces a latent ultraviolet fluorescent color to show the exact location of a leak. Depending on the dye used, the film may be orange-red or yellow. Refrigerant manufacturers produce refrigerants that includes a dye solution (Figure 7-17).

Preparing the System

It will be assumed that the system is partially charged with refrigerant or that it has been properly prepared to receive a partial charge. It is never wise to operate a system that is contaminated or moisture-laden, even for testing purposes.

1. Make sure that all valves are in the closed position.
2. Connect the manifold and gauge set to the system.
3. Connect the charging cylinder to the center service hose (Figure 7-18).
4. Connect a short hose to the inlet of the charging cylinder (Figure 7-19).
5. Attach the other end of the hose to a refrigerant source (Figure 7-20).
6. Remove the plug and add the dye solution to the charging cylinder. Replace the plug.
7. Start the engine and operate it at idle speed.
8. Set the controls for MAX cooling.
9. Open the service hose shut-off valve.
10. Open the low-side hose shut-off valve.
11. Open the refrigerant source valve.
12. Slowly open the low-side manifold hand valve to allow the dye or trace solution to enter the system.
13. Charge the system to at least half capacity. Allow the system to operate for 15 minutes. Then turn off the controls and shut off the engine.

An ultraviolet (UV) lamp used to check for leaks gets very hot.

Operating a system that is contaminated may cause serious problems in a short period of time.

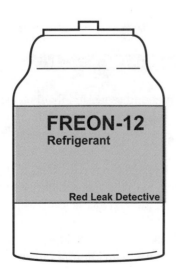

Figure 7-17 Typical "pound" can of Refrigerant-12 with a red dye additive. Freon and Dytel are trademarks of E. I. DuPont.

Figure 7-18 Connect the charging cylinder to the center service hose.

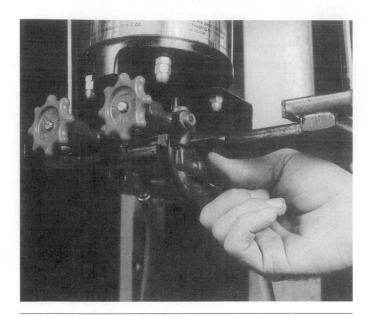

Figure 7-19 Connect a short hose to the inlet of the charging cylinder.

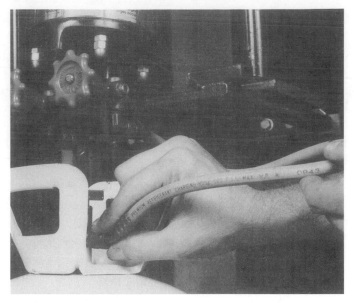

Figure 7-20 Attach the other end of the hose to a refrigerant source.

14. Observe the hoses and fittings for signs of the dye solution. If no signs of a leak are evident at this time, arrange to have the car available the following day for diagnosis and repair. If leak(s) are detected, make repairs as required. (The dye solution will remain in the system without causing harm or reduced performance.)

15. Close all valves: the refrigerant source valve, service hose shut-off valve, low-side manifold gauge valve, and compressor service valve(s), if equipped.

16. Remove the manifold and gauge set.

17. Remove the charging cylinder. Disconnect the hose from the refrigerant source. Remove the charging cylinder from the service hose. Remove the short hose from the charging cylinder.

18. Replace all protective caps.

Protective caps help guard against leaks and help to prevent the entrance of dirt and debris.

CASE STUDY

A customer complained of an inoperative air conditioner and requested that Freon be added. The technician advised the customer that if refrigerant were needed, there must be a leak in the system. The customer responded "I have to add Freon every couple of months—just put it in."

The technician attempted to explain the problems with just adding refrigerant, such as loss of oil, harm to the environment, and possible damage to the air conditioner system components, such as the compressor. The customer still insisted that he only wanted Freon. Politely and tactfully, the technician refused the service. "You have come to the wrong place," she told the customer. "This service facility employs only ASE-certified technicians who are dedicated to their profession. To perform a service in an improper manner violates the essence of ASE certification."

The somewhat surprised, but impressed, customer left the facility without further ado. He returned the next day and had the air conditioner properly repaired.

Terms to Know

Back seat	Flammable refrigerant	Pound
Can tap valve	Fluorescent tracer dye	Purge
Charge	Gross weight	Recovery system
Contaminated refrigerant	High-side hand valve	Shut-off valve
Dry nitrogen	Kilogram	SNAP
Electronic charging meter	Liquid	Standing vacuum test
Evacuate	Low-side hand valve	Stratify
Evacuation	Manifold hand valve	Triple evacuation
Federal Clean Air Act	Manufacturer	

ASE Style Review Questions

1. *Technician A* says that a system is contaminated if it contains more than 2 percent of a foreign substance.
   Technician B says that air is considered a contaminant if it exceeds 2 percent of the system capacity.
   Who is right?
   - **A.** A only
   - **B.** B only
   - **C.** Both A and B
   - **D.** Neither A nor B

2. *Technician A* says that tobacco smoke will not affect refrigerant leak detection.
   Technician B says halide is the best and an inexpensive method of leak detection.
   Who is right?
   - **A.** A only
   - **B.** B only
   - **C.** Both A and B
   - **D.** Neither A nor B

3. *Technician A* says that special electronic leak detectors are available that are used for HFCs.
   Technician B says that there are electronic leak detectors available that will detect CFCs as well as HFCs.
   Who is right?
   - **A.** A only
   - **B.** B only
   - **C.** Both A and B
   - **D.** Neither A nor B

4. *Technician A* says that the system is purged of refrigerant if the manifold gauges read a slight vacuum.
   Technician B says that the system may be purged of refrigerant even if the manifold gauges read a slight pressure.
   Who is right?
   - **A.** A only
   - **B.** B only
   - **C.** Both A and B
   - **D.** Neither A nor B

5. *Technician A* says that one need not evacuate the system if the "sweep and purge" method is used.
   Technician B says that one need not evacuate the system if it has been "opened" for less than five minutes.
   Who is right?
   - **A.** A only
   - **B.** B only
   - **C.** Both A and B
   - **D.** Neither A nor B

6. *Technician A* says that "purging" is the same as "evacuation."
   Technician B says that "pumping down" is the same as "evacuation."
   Who is right?
   - **A.** A only
   - **B.** B only
   - **C.** Both A and B
   - **D.** Neither A nor B

7. *Technician A* says when it is used improperly, refrigerant can cause blindness.
   Technician B says when it is used improperly, refrigerant can create a harmful vapor.
   Who is right?
   - **A.** A only
   - **B.** B only
   - **C.** Both A and B
   - **D.** Neither A nor B

8. *Technician A* says that the minimum pressure recommended for leak testing a CFC-12 system is 60 psig (414 kPa).
   Technician B says the minimum recommended pressure for leak testing an HFC-134a system is 60 psig (414 kPa).
   Who is right?
   - **A.** A only
   - **B.** B only
   - **C.** Both A and B
   - **D.** Neither A nor B

9. *Technician A* says fluorescent tracer dye is easily detected using an infrared lamp.
Technician B says the dye should be removed from the system after leak detection.
Who is right?
   A. A only
   B. B only
   C. Both A and B
   D. Neither A nor B

10. *Technician A* says a standing vacuum test may be held to check an air conditioning system for leaks.
Technician B says a standing vacuum test does not reveal how many leaks there are in the system..
Who is right?
   A. A only
   B. B only
   C. Both A and B
   D. Neither A nor B

ASE Challenge

1. When an air conditioning system is opened for repairs or service there is a possibility that _____ will enter system.
   A. Air
   B. Moisture
   C. Both A and B
   D. Neither A nor B

2. Which of the following is LEAST likely to cause a higher than normal high-side pressure?
   A. Overcharge of refrigerant
   B. Overcharge of lubricant
   C. Refrigerant contamination
   D. Air

3. Which of the following is MOST likely to cause a lower than normal low-side pressure?
   A. Undercharge of refrigerant
   B. Undercharge of lubricant
   C. Refrigerant contamination
   D. Air

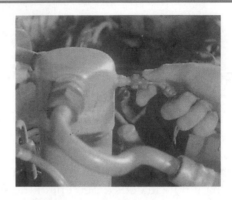

4. The illustration above shows:
   A. High-side hose being connected to the receiver-drier
   B. High-side hose being connected to the accumulator
   C. Low-side hose being connected to the receiver-drier
   D. Low-side hose being connected to the accumulator.

5. All of the following are lubricants that may be used with selected refrigerants, EXCEPT:
   A. PING
   B. PAG
   C. POG
   D. Mineral oil

Table 7-1 ASE TASK

Leak-test A/C system; determine needed repairs.

Problem Area	Symptoms	Possible Causes	Classroom Manual	Shop Manual
SYSTEM LEAKS	Poor to no cooling	1. Hose, fitting, or O-ring leak	152–156	151
		2. Compressor seal or gasket leak	152–156	187
		3. Component weldment leak	152–156	164, 166
		4. Evaporator core leak	152–156	166

Table 7-2 ASE TASK

Identify and recover A/C system refrigerant.

Problem Area	Symptoms	Possible Causes	Classroom Manual	Shop Manual
AVOID CROSS-CONTAMINA-TION	Identify refrigerant type	1. CFC-12 system charged with HFC-134a	195–199	147
		2. HFC-134a system charged with CFC-12	195–199	147
		3. CFC-12 or HFC-134a system charged with an unknown refrigerant	195–199	147
EPA RULES	Avoid venting	1. Improper recovery of CFC-12 refrigerant	199, 202	318, 321
		2. Improper recovery of HFC-134a refrigerant	199, 202	318, 321

Table 7-3 ASE TASK

Evacuate A/C system.

Problem Area	Symptoms	Possible Causes	Classroom Manual	Shop Manual
EVACUATE A/C SYSTEM	System does not hold a vacuum	1. Leak in the system	152–156	249
		2. Defective vacuum pump	158–159	254
		3. Leak/defect in manifold gauge set	152–156	67
	System holds vacuum but slight "bleed down" after five minutes	1. Insufficient time for proper evacuation	159–161	254
		2. Vacuum not "deep" enough	158–159	254
		3. Excessive moisture in the system	158–159	254

Table 7-4 ASE TASK

Clean A/C system components and hoses.

Problem Area	Symptoms	Possible Causes	Classroom Manual	Shop Manual
HOUSEKEEPING	Untidy and unclean	1. Dirt and debris in fins of condenser	116	166
		2. Dirt and debris in fins of evaporator	120	166
		3. Dirt and debris allowed to enter system during service	158	149, 156
		4. Engine/road soils on external components	158	149, 156

Table 7-5 ASE TASK

Charge A/C system with refrigerant (liquid or vapor).

Problem Area	Symptoms	Possible Causes	Classroom Manual	Shop Manual
REFRIGERANT IDENTIFICATION	Excessive pressure	1. Improper refrigerant	101	147
		2. (Cross) contaminated refrigerant	101	147
CHARGING WITH VAPOR REFRIGERANT	Slow charging through the low side	1. Low-pressure switch open	225–226	96
		2. Service valve not fully open	67–70	150
		3. System malfunction	84–89	110
CHARGING WITH LIQUID REFRIGERANT	Noise, with system running	Liquid slugging of the compressor	89	77
	Slow charging with system not running	Service valve not fully open	67–70	150

JOB SHEET 22

22

Name _____ Date _____

Recover and Recycle Refrigerant

Upon completion of this job sheet you should be able to recover and recycle refrigerant.

ASE Correlation

This job sheet is related to the ASE Heating and Air Conditioning Systems Test's content area: *A/C System Diagnosis and Repair*, Task: *Identify and recover A/C system refrigerant*; and *Refrigerant Recovery, Recycling, Handling, and Retrofit*, Task: *Identify and recover A/C system refrigerant*.

Tools and Materials

Vehicle with air conditioning system in need of service
Selected air conditioning system service tools
Manifold and gauge set with hoses
Refrigerant recovery/recycle machine
Recovery tank

Describe the vehicle being worked on:

Year _____ Make _____ Model _____
VIN _____ Engine type and size _____

Procedure

After each of the following procedures, briefly explain how you performed the task:

1. Connect the gauge manifold hoses to the air conditioning system service ports.

_____

_____

2. What type connectors are found:

On the low-side service port? _____

On the high-side service port? _____

3. Observe the gauges.

The low-side gauge reads _____

The high-side gauge reads _____

NOTE: If zero (0 psig or 0 kPa) or below pressure is observed in step 2 it may be assumed that there is no refrigerant in the system. If it is presumed that there are residual traces of refrigerant in the lubricant, proceed with step 4. If not, one may proceed with Job Sheet 23.

_____

_____

4. Connect the manifold gauge hose to the recovery unit.

_____

_____

5. What type connector is found on the recovery unit?

_____

_____

6. Start the recovery unit, open the appropriate hand valves, and recover air conditioning system refrigerant following instructions provided by the recovery equipment manufacturer.

_____

_____

7. Close the hand valves (opened in step 6) and turn off the recovery unit. Observe the gauges. What is the reading in psig or kPa?

	Now	After 5 Min.	After 10 Min.	After 15 Min.
Low Side	_____	_____	_____	_____
High Side	_____	_____	_____	_____

8. Explain the conclusion of the results of Step 7.

_____

_____

9. Carefully remove the manifold and gauge set hoses ensuring that no ambient air enters the system or proceed with Job Sheet 23 or 24.

_____

_____

☑ **Instructor's Check** _____

JOB SHEET 23

23

Name _____ Date _____

System Evacuation

Upon completion of this job sheet you should be able to evacuate an air conditioning system.

ASE Correlation

This job sheet is related to the ASE Heating and Air Conditioning Systems Test's content area: *A/C System Diagnosis and Repair*, Task: *Evacuate A/C system*.

Tools and Materials

Vehicle with air conditioning system in need of evacuation
Gauge and manifold set
Vacuum pump

Describe the vehicle being worked on:

Year _____ Make _____ Model _____

VIN _____ Engine type and size _____

Procedure

After each of the following procedures, briefly explain how you performed the task:

1. Connect the gauge manifold hoses to the air conditioning system service ports.

_____

_____

2. What type connectors are found:

On the low-side service port? _____

On the high-side service port? _____

3. Observe the gauges.

The low-side gauge reads _____

The high-side gauge reads _____

NOTE: If pressure is observed in step 2, perform Job Sheet 22 before proceeding with this job sheet.

4. Connect the manifold gauge hose to the vacuum pump.

_____

_____

5. What type connector is found on the vacuum pump?

_____

_____

6. Turn on the vacuum pump, open the appropriate hand valves, and evacuate the air conditioning system for the length of time suggested in the service manual, _____hr/min.

_____

_____

7. Close the hand valves (opened in step 6) and turn off the vacuum pump. Observe the gauges. What is the reading in psig or kPa?

	Now	After 5 Min.	After 10 Min.	After 15 Min.
Low Side	_____	_____	_____	_____
High Side	_____	_____	_____	_____

8. Explain the conclusion of the results of step 7.

_____

_____

9. Carefully remove the manifold and gauge set hoses ensuring that no ambient air enters the system or proceed with Job Sheet 24 .

_____

_____

☑ **Instructor's Check** _____

JOB SHEET 24

24

Name _____ Date _____

Charge Air Conditioning System with Refrigerant

Upon completion of this job sheet you should be able to charge or recharge an air conditioning system.

ASE Correlation

This job sheet is related to the ASE Heating and Air Conditioning Systems Test's content area: *A/C System Diagnosis and Repair.* Task: *Charge A/C system with refrigerant (liquid or vapor).*

Tools and Materials

Manifold and gauge set with hoses
Source of refrigerant
    Small "pound" cans
    Bulk source
    Recovery system

Describe the vehicle being worked on:

Year _____ Make _____ Model _____

VIN _____ Engine type and size _____

Procedure

After each of the following procedures, briefly explain how you performed the task:

    1. Connect the gauge manifold hoses to the air conditioning system service ports.

    _____

    _____

    2. What type connectors are found:

    On the low-side service port? _____

    On the high-side service port? _____

    3. Observe the gauges.

    The low-side gauge reads _____

    The high-side gauge reads _____

  NOTE: If pressure is observed in step 2 there is refrigerant in the system. Either perform Job Sheet 22 before proceeding or "top off" the refrigerant as instructed by your instructor following this job sheet.

    4. Connect the manifold gauge hose to the refrigerant source.

    _____

    _____

5. What type connector is found on the refrigerant source?

_____

_____

6. Open the appropriate hand valves and charge the air conditioning system following equipment manufacturers instructions or those outlined in the service manual, as applicable.

_____

_____

7. After charging, close the hand valves (opened in step 6). Observe the gauges. What is the reading in psig and kPa?

	psig	kPa
Low Side	_____	_____
High Side	_____	_____

8. Explain the resultant low- and high-side pressures noted in step 7. (Are they normal? etc.)

_____

_____

9. Carefully remove the manifold and gauge set hoses ensuring that no refrigerant is allowed to escape.

_____

_____

☑ **Instructor's Check** _____

JOB SHEET 25

25

Name _____ Date _____

Air Conditioning System Diagnosis

Upon completion of this job sheet you should be able to make basic air conditioning system diagnostic checks.

ASE Correlation

This job sheet is related to the ASE Heating and Air Conditioning Systems Test's content area: *A/C System Diagnosis and Repair*, Task: *Diagnose A/C system problems indicated by sight, sound, smell, and touch procedures; determine needed repairs.*

Tools and Materials

A vehicle with an air conditioning system
Selected air conditioning service tools

Describe the vehicle being worked on:

Year _____ Make _____ Model _____
VIN _____ Engine type and size _____

Procedure

The following observations are to be made with the engine OFF. Record your procedure and findings in the space provided.

1. Inspect the compressor drive belt for condition and tightness.

Type belt _____ Number of belts _____

Procedure _____

Findings _____

2. Inspect water pump direct drive fan (if applicable).

Procedure _____

Findings _____

3. Inspect all coolant carrying hoses, fittings, and shut-off valve.

Procedure _____

Findings _____

4. Inspect all refrigerant carrying hoses and fittings.

Procedure _____

Findings _____

5. Inspect all vacuum hoses, fittings, and components.

Procedure _____

Findings _____

Make the following checks with the engine running and air conditioning system controls to MAX COOL. Give your procedure and findings in the space provided.

6. Check the operation of the compressor clutch.

Procedure _____

Findings _____

7. Check the operation of the electric cooling fan, if applicable.

Procedure _____

Findings _____

8. Check blower motor operation; LO, LO-MED, HI-MED, and HI.

Procedure _____

Findings _____

9. Check for proper air flow from outlets.

Procedure _____

Findings _____

10. Check refrigerant flow in sight glass, if applicable.

Procedure _____

Findings _____

11. Carefully check suction and liquid line temperature.

Procedure _____

Findings _____

12. Check temperature of air flow from dash outlets.

Procedure _____

Findings	OUTLET	TEMP°F	TEMP°C
	Right	_____	_____
	Center	_____	_____
	Left	_____	_____

On conclusion of testing, turn the air conditioning system OFF and stop the engine. Remove all tools, such as the thermometer used in step 12.

☑ **Instructor's Check** _____

Case and Duct Systems

Upon completion and review of this chapter you should be able to:

- ❏ Remove and replace blowers.
- ❏ Remove and replace blower motors.
- ❏ Remove and replace linkage.
- ❏ Remove and replace controls.

- ❏ Adjust linkage and controls.
- ❏ Troubleshoot, service, and adjust the operation of in-vehicle mode circuits such as vent, hi-lo, MAX (cool/heat), and defrost.

Introduction

A maze of ducts, vents, motors, wiring, and vacuum hoses makes up the typical automotive air conditioning case and duct system in today's modern vehicle (Figure 8-1). The somewhat inaccessibility of most of its components adds to the mystique of this often neglected component of the air conditioning system.

    While it is true that there are literally hundreds of variations, troubleshooting and servicing are not difficult if one is familiar with the system.

Fresh Air Inlet

Most of the heater and air conditioning mode functions are performed with some outside air except when MAX is selected. Though not generally noticeable, the quality of the air conditioning system can be affected if the fresh air inlet screen is blocked with leaves or other **debris**. In time, if neglected, this debris can deteriorate and be pulled into the heater/evaporator case where it can cause serious air flow blockage through the evaporator and/or heater core. The fresh air inlet (Figure 8-2) is often concealed by the hood and is therefore overlooked during preventative maintenance. Cleaning this area should be a part of a periodic preventive maintenance schedule.

Component Replacement

The greatest problem arises from the lack of data necessary for properly servicing any particular unit or system. It is necessary to have the manufacturer's service manuals for specific step-by-step

Basic Tools

Basic mechanic's tool set

Fender cover

Safety glasses

Manifold and gauge set (for CFC-12 or HFC-134a, as applicable)

Calibrated cup

Classroom Manual
Chapter 8,
page 174

HI/LO is also referred to as BILEVEL.

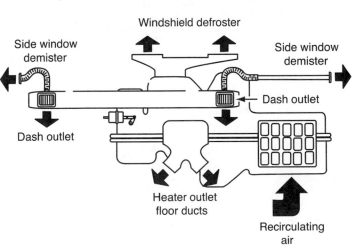

Figure 8-1 A typical case/duct system.

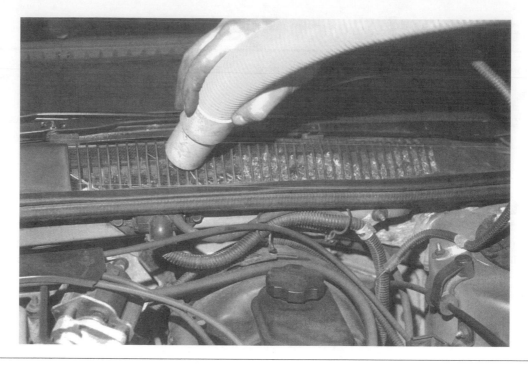

Figure 8-2 Clean debris from fresh air inlet.

procedures. For example, consider the replacement of a hi-lo **actuator** motor. The "big three" automakers differ for one year-model, as follows:

Chrysler

1. Remove the left and right underpanel silencer ducts.
2. Remove the floor console.
3. Remove the center floor heat adaptor duct.
4. Remove the rear seat heat forward adaptor duct.
5. Loosen the center support bracket; pry rearward to gain access to the actuator.
6. Remove the actuator retaining screws.
7. Remove the actuator (Figure 8-3).
8. Remove the electrical connections from the actuator.
9. To reinstall, reverse order of removal.

Ford

1. Disconnect the vacuum hose.
2. Remove the retaining screws.
3. Remove the actuator from the linkage.
4. Remove the actuator (Figure 8-4).
5. For installation, reverse the removal procedures.

General Motors

General Motors calls their air bag deployment system a "Supplemental Inflatable Restraint (SIR)."

1. Disable the air bag deployment system.
2. Remove the battery negative (-) cable and fuse.
3. Remove the instrument panel.
4. Remove the floor outlet assembly.
5. Remove the windshield defroster vacuum hoses.

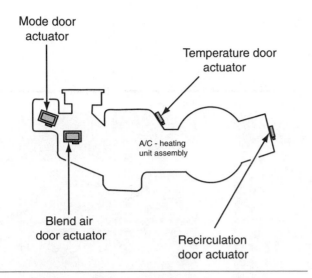

Figure 8-3 Typical Chrysler mode door actuator location.

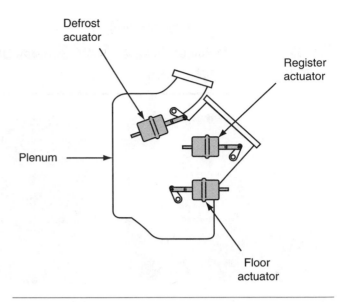

Figure 8-4 Typical Ford mode door actuator details.

6. Remove the windshield defroster air distribution assembly.
7. Remove the vacuum hose from the upper/lower mode valve actuator.
8. Remove the retaining nuts or screws.
9. Remove the actuator (Figure 8-5).
10. To install the new actuator, reverse the preceding procedure.

This comparison is not to suggest that Ford's procedure is the simplest or that General Motors' procedure is the most difficult. The procedures vary considerably for all year-model applications. The example, which was randomly selected, is intended to provide a general comparison of what may be expected in the day-to-day service of air conditioning systems and to express the importance of having at hand an appropriate service manual.

Blower Motor

Blower motor replacement is generally a little more straightforward than some of the other case/duct components (Figure 8-6). See Photo Sequence 13. This procedure, however, should be considered typical for any type vehicle.

Classroom Manual
Chapter 10,
pages 221–222

Silicone rubber, available in tube form, is ideal for sealing mating surfaces.

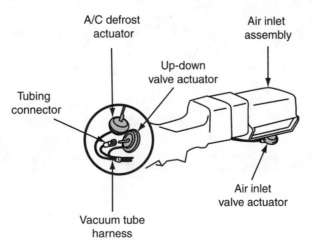

Figure 8-5 Typical General Motors mode door actuator details.

Photo Sequence 13
Typical Procedure for Removing a Blower Motor

P13-1 Disconnect the battery ground (–) cable.

P13-2 Disconnect or remove any wiring, brackets, or braces, hampering blower motor service.

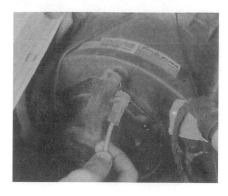

P13-3 Disconnect the blower motor lead(s).

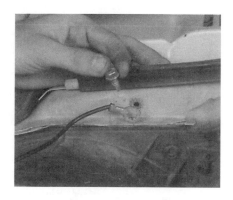

P13-4 Disconnect the blower motor ground (–) wire.

P13-5 Disconnect the blower motor cooling tube (if applicable).

P13-6 Remove the attaching screws.

P13-7 Remove the blower and motor assembly. The sealing gasket often acts as an adhesive. If this is the case, carefully pry the blower flange away from the case.

P13-8 Remove the shaft nut or clip, if applicable.

P13-9 Remove the blower wheel. Do not lose spacer. Use it on replacement motor.

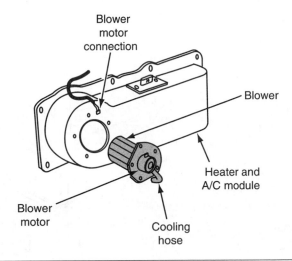

Figure 8-6 Blower motor and plenum details.

For reassembly or the installation of a new **blower** motor assembly, reverse P13-9 through P13-1. If the gasket, P13-7, was damaged or destroyed, replace it with a new gasket or seal the mating surfaces with a suitable caulking material.

Replacing the Power Module or Resistor

Special Tools

Coolant recovery system, if applicable

Hose clamp pliers, if applicable

CAUTION: This component may be very hot. Take care before touching it with the bare hands.

1. Remove the brace(s) or cover(s) that may restrict access to the **power module** or **resistor** (Figure 8-7).
2. Remove the electrical connector(s).
3. Remove the retainer(s), if equipped.
4. Remove the retaining screws or nuts.
5. Remove the power module or resistor.
6. For replacement, reverse the procedure.

Classroom Manual
Chapter 8, page 175

Replacing the Heater Core

Access to the heater core is gained by following directions given in specific service manuals. This procedure is typical and assumes the procedure for access to the heater core is available.

1. Drain the cooling system into a clean container. The coolant may be reused, reclaimed, or discarded in a manner consistent with **Environmental Protection Agency (EPA)** guidelines.
2. Disconnect the battery ground (–) cable.
3. Disconnect the heater hoses at the bulkhead. This is a good opportunity to inspect the heater hoses and replace any that show signs of deterioration.

WARNING: Do not use undue force when connecting the heater hoses. Damage to the new heater core may occur if care is not taken.

Classroom Manual
Chapter 10, page 239

Classroom Manual
Chapter 11, page 267

Special Tools

Coolant recovery system

Refrigerant recovery system

Hose clamp pliers

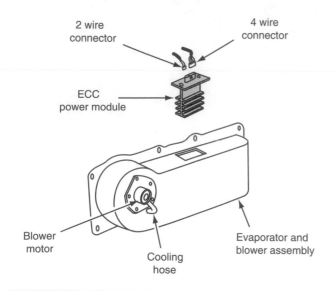

2 wire connector   4 wire connector

ECC power module

Blower motor

Cooling hose

Evaporator and blower assembly

Figure 8-7 Typical power module.

Heater core

Heater unit

Figure 8-8 Removing the heater core from the case.

4. Gain access to the heater core as outlined in the appropriate service manual.
5. Remove the retaining screws, brackets, or straps.
6. Remove the core from the case (Figure 8-8).

Replacing the Evaporator Core

To illustrate the importance of proper service manuals for this service, it may be noted that the 1994 Ford Taurus/Mercury Sable manual instructs the technician, "Using a **hot knife**, cut the top of the air conditioning evaporator housing between the raised outline." An illustration is included in the manual to show the area to be cut (Figure 8-9). The 1993 Ford Probe manual simply says to remove the blower motor and assembly screws before separating the case halves and removing the evaporator (Figure 8-10).

The following procedure assumes that access to the evaporator has been determined.

1. Recover the refrigerant.
2. Drain the radiator if the heater hose(s) has to be removed to gain access.

Drain coolant into a clean container so it may be reused.

Hacksaw

Internal hinge line

A/C evaporator housing

Figure 8-9 Use a hacksaw or hot knife to cut the case.

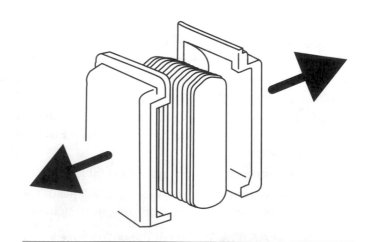

Figure 8-10 Split case halves to remove evaporator.

3. Remove the heater hose(s), if necessary.
4. Remove any **wiring harness**, heat shields, brackets, covers, and braces that may restrict access to the evaporator core.
5. Remove the liquid line at the thermostatic expansion valve (TXV) or fixed orifice tube (FOT).
6. Remove the suction line at the evaporator or accumulator outlet.
7. Gain access to the evaporator core as outlined in the service manual.
8. Lift the evaporator from the vehicle.
9. Drain the oil from the evaporator into a calibrated cup.
10. For replacement, reverse the preceding procedure. First, replace the oil with the same amount and type as drained in step 9.

Replace any heater hoses found to be brittle or damaged.

Discard all O-rings. They should be replaced with new O-rings on reassembly.

Dispose of used oil in accordance with local ordinances.

Odor Problems

An odor emitting from the air conditioning system ducts may be caused by by a leaking heater core or hose inside the heater/evaporator case. An odor may also be caused by refrigeration oil leaking into the heater/evaporator case due to a leaking evaporator. The remedy is to repair or replace the leaking parts.

A musty odor is usually due to water leaks, a clogged evaporator drain tube, or mold and mildew on the evaporator core. Mold and mildew, which are fungi, are most common in air conditioning systems in vehicles operated in hot and humid climates. The odor, generally noted during startup, may be caused by debris in the heater/evaporator case or by microbial fungi growth on the evaporator core.

This is generally a temporary condition: as climate conditions change, the odor will disappear. The problem can also often be eliminated by clearing the evaporator condensate drain tube. If the odor persists, however, it will be necessary to take corrective action by removing debris from the heater/evaporator case and using a disinfectant to clean the evaporator core of any mold or mildew. This is typically accomplished by the following procedure.

Clean the Evaporator Core

CAUTION: This procedure should only be performed on a cold vehicle to prevent the disinfectant from coming in contact with hot engine components. Take extreme care not to get disinfectant in eyes or on hands or clothing. Wash thoroughly with soap and water immediately after handling.

CAUTION: If disinfectant gets into the eyes, hold the eyelids open and flush with a steady, gentle stream of water for 15 minutes. Immediately seek professional medical attention.

This procedure should be considered as only typical. For specific procedures, always follow the instructions included with the disinfectant kit.

1. Put on protective clothing, including rubber gloves and safety goggles.
2. Mix disinfectant as directed.
3. Check under the vehicle to be certain that the evaporator drain outlet is not restricted.
4. If vehicle battery is used to power the cleaning gun, connect a battery charger to the vehicle to avoid draining the battery during the cleaning procedure.
5. Remove the blower resistor. Leave electrical connectors attached.

CAUTION: Do not permit the coils of the blower resistor to become grounded to any metal surface.

NOTE: If there is no blower resistor, a hole must be drilled in the ductwork. Follow the manufacturer's instructions for hole size and placement. Failure to do so may result in serious damage.

6. Remove inlet filter, if equipped. Inspect case for debris and remove any that is found through the blower resistor opening or hole drilled in step 5.

NOTE: If debris cannot be removed through the opening, remove the core from the vehicle and clean it. At this time, inspect the air inlet screen and repair or replace it as necessary.

7. Open all windows and doors.
8. Position a large fan, turned on high, to provide adequate ventilation.
9. Place the ignition switch in the ON position. Do not start the engine.
10. Set air conditioning system mode selector to VENT, blower speed to LO, and temperature to full COLD.
11. Place a 2-qt. (1.9-L) drain pan under the evaporator drain tube to collect the disinfectant and rinse water.
12. Insert the nozzle of the spray gun through the blower resistor opening or drilled hole and insert the siphon hose into container of disinfectant.

NOTE: Ensure adequate coverage of the evaporator. The core should be saturated completely.

13. Turn the ignition switch to the OFF position.
14. While allowing the core to "soak" for five minutes or so, check underneath the vehicle to verify proper evaporator drain operation. If improper, correct as necessary by unclogging or increasing drain slits.
15. Again, turn the ignition switch to the ON position. Do not start the engine.
16. Rinse the evaporator core with clean water. Use the spray gun to remove all disinfectant residue.
17. Return the air conditioning system mode selector, blower speed, and temperature selector to their original positions, as in step 10.
18. Again, turn the ignition switch to the OFF position.
19. Reinstall the blower resistor or place a cover over the hole (see step 5).
20. Perform housekeeping:
    a. Properly dispose of used disinfectant and rinse water solution.
    b. Carefully, disconnect the battery charger.
    c. Install new inlet filter (if removed in step 6).
    d. Turn off and remove fan.
    e. Close vehicle doors and windows.
    f. Perform personal hygiene: Wash with soap and water. Carefully remove protective clothing, gloves, and goggles.

Delayed Blower Control

An aftermarket delayed blower control may be installed in many systems to reduce the probability of a recurrence of odors caused by mold and mildew. It is installed following the instructions included with the package or as given in a manufacturer's **technical service bulletin** (TSB). The delayed blower control is used to dry out the evaporator and air distribution system. After air conditioning compressor has been in operation for four or five minutes, the control will cause the blower motor to run for about 45 minutes after the ignition switch is placed in the OFF position. The delayed blower control will operate the blower motor at high speed for five minutes to clear the evaporator core of accumulated condensate, thereby reducing the recurrence of odors caused by mold and mildew.

The Air Door Control System

The air door control system uses vacuum-operated or electrically powered motors to position the air doors, also referred to as mode doors, to provide the desired in-vehicle air delivery conditions. These generally include OFF, MAX, VENT, BILEVEL, HTR, BLEND, and DEF. The air flow pattern for each of these conditions are given in Chapter 8 of the Classroom Manual.

There are basically three types of control systems: vacuum, rotary vacuum, and vacuum solenoid and motor.

Vacuum Control

In the vacuum control system, vacuum actuators, also called vacuum motors, are used to position the air doors and valves. This system relies on a vacuum signal from either the engine manifold or an on-board vacuum pump. Vacuum is applied to the selected actuator by a rotary vacuum valve (RVV) or a vacuum solenoid in the main control panel.

Rotary Vacuum Valve

When a rotary vacuum valve system (Figure 8-11) is used to control air doors and valves, the master control head selector rotates a vacuum switch that aligns the vacuum passages in the valve to direct a vacuum signal to the appropriate vacuum actuator(s) for the mode selected.

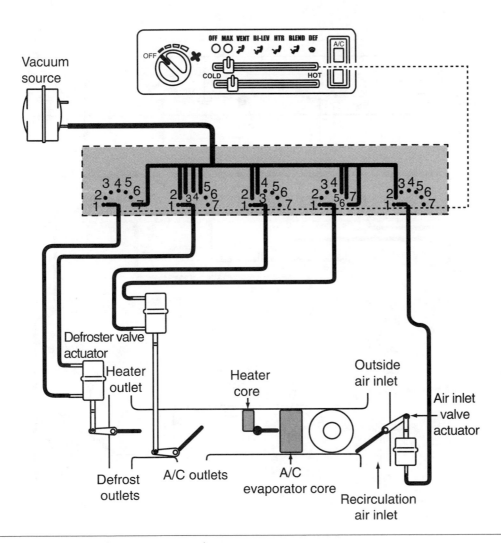

Figure 8-11 Typical rotary vacuum valve system.

Vacuum Solenoids

When vacuum solenoids (Figure 8-12) are used to control doors and valves, the master control head selector contains electrical circuits that provide a ground path for the selected vacuum solenoid. The selected (energized) solenoid allows vacuum to be applied to the selected actuator.

An automatic air distribution system often uses vacuum solenoids that are located inside the programmer to control the position of the mode doors. The programmer controls the electrical ground side of the solenoids to establish a ground path to the selected vacuum solenoid. When the ground path is removed, the vacuum actuator is allowed to vent.The temperature door is controlled by a five-wire motor that is controlled by the programmer. The programmer provides power and ground to the selected motor circuits. The direction of motor rotation depends on which motor circuit the programmer provides power to. At the same time, the other motor circuit is provided ground. To stop the motor, the programmer provides a ground path to both circuits.

A potentiometer (pot) inside the motor, mechanically linked to the air door, provides feedback to the programmer as to the actual position of the door. This feedback is an essential programmer input that allows the motor to operate until the desired air door position is reached, then turns the motor off.

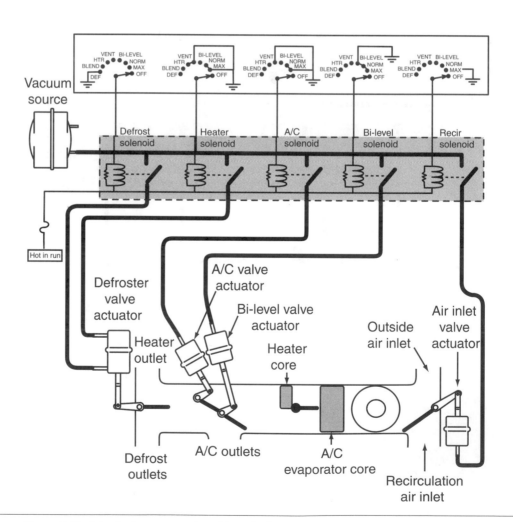

Figure 8-12 A typical vacuum solenoid control.

Figure 8-13 A typical vacuum motor (pot).

Testing the Vacuum System

An inoperative **vacuum motor** could be due to a loss of **vacuum signal** at the appropriate time. This could be the fault of the vacuum source, vacuum switch, **check valve**, **reserve tank**, hose, **restrictor**, or vacuum motor. To determine the cause, disconnect the suspected vacuum motor (Figure 8-13) and substitute another vacuum source, such as the vacuum pump. If the motor is inoperative, it should be replaced. If it is operative, check further for the source of the problem.

If an inoperative fresh air door or mode door is the problem, a fault in the vacuum control system is again indicated. Most older systems use vacuum motors with a vacuum selector valve at the control head to control the operation of these doors. Some vehicles have vacuum motors controlled by electric solenoids, while others use electric motors at the doors. There are also systems in which all the mode doors have electric motor control (Figure 8-14).

Generally, a vacuum system problem can be traced to a cut, kinked, crimped, or disconnected vacuum hose. A faulty selector valve, vacuum actuator, storage tank, or check valve may also be the problem.

In an electric solenoid or electric motor system, an electrical system defect may be responsible for improper mode door operation. Since these systems function electrically, it is wise to consult the appropriate manufacturer's service manual for specific troubleshooting procedures. One must be extremely careful when troubleshooting electrical systems under the dash. Just one improper test point could cause serious damage to one of the on-board computers.

The area under the dash is cramped and congested—filled with wires, vacuum hoses, and ducts as well as various electrical and mechanical components and assemblies. It is not, therefore, easy to gain access any of the components, especially those associated with the vacuum control system. The control panel on many vehicles, however, may be pulled out far enough from the dash to gain access. Extreme caution must be exercised when gaining access to any under-dash component. Failure to do so could trigger the air bag restraint system.

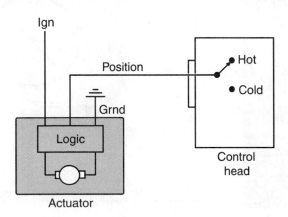

Figure 8-14 Typical mode door motor schematic.

Classroom Manual
Chapter 10,
page 228–233

Special Tools

Vacuum gauge
Short piece of
vacuum hose

A vacuum leak can
often be heard.
Listen carefully for
any change in pitch
of a hissing sound
during service.

Classroom Manual
Chapter 10,
page 216, 218

In some vehicles the vacuum connection will be at the base of the carburetor.

The first step in troubleshooting the vacuum system is to ensure that manifold vacuum is available at the selector switch. Vacuum diagrams are generally provided in the manufacturer's service manual to help identify color coding and connections (Figure 8-15). The following is a typical procedure to quickly check of the vacuum control system for improper or erratic direction of air flow from the outlets .

1. Are other vacuum-operated devices operational?
2. Do other vacuum motors operate properly?
   a. If yes, there is a vacuum source.
   b. If no, proceed with step 3.
3. Disconnect the hose at the manifold inlet.
4. Connect a vacuum gauge, a short hose, and a vacuum gauge in line with the vacuum source and system (Figure 8-16).
5. Is there now a vacuum signal?
   a. If yes, check for a defective check valve or hose(s).
   b. If no, check for blockage or restriction at the manifold fitting and correct it as required.

Vacuum Switch

The vacuum control provides a vacuum passage for selected circuits in the control system. To test for a defective vacuum control:

1. Disconnect the hose from the inoperative vacuum motor at the switch.
2. Connect a vacuum gauge to the vacant port.
3. Move the switch through all of its positions.
4. Is there a vacuum at the port in either position?
   a. If yes, the switch is probably all right.
   b. If no, the switch may be defective; proceed with step 5.
5. Is there a vacuum signal at any of the other ports?
   a. If yes in step 5 but no in step 4, the switch is defective and should be replaced.
   b. If no, the problem may be a defective restrictor, check valve, hose, or reserve tank.

Check valves are used to prevent a loss of vacuum during acceleration and after engine shut-down.

Special Tools

Vacuum pump

Vacuum gauge

Vacuum hose

Check Valve

A check valve (Figure 8-17) allows flow in one direction and blocks (checks) the flow in the other direction. See Photo Sequence 14 for testing a check valve.

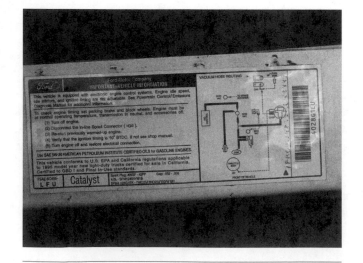

Figure 8-15 An underhood vacuum system decal.

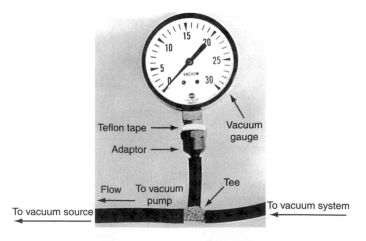

Figure 8-16 Check the vacuum source.

Photo Sequence 14
Typical Procedure for Testing a Check Valve

P14-1 Remove the check valve from the vehicle.

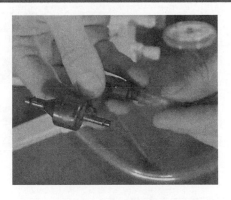

P14-2 Attach a vacuum source, such as a vacuum pump. The direction of flow should be away from the pump.

P14-3 Turn on the pump and observe the gauge. If there is a vacuum, the check valve is good. Proceed with step 4. If there is no vacuum, the check valve is defective and must be replaced.

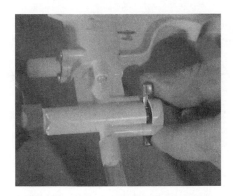

P14-4 Turn off the pump.

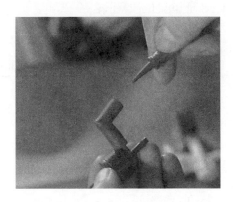

P14-5 Disconnect the check valve.

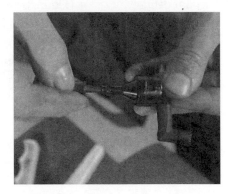

P14-6 Reverse and reconnect the check valve.

P14-7 Turn on the vacuum pump and observe the gauge. If there is a vacuum, the check valve is defective and must be replaced. If there is now no vacuum but the pump held a vacuum for P9-3, the check valve is good and may be returned to the vehicle.

Figure 8-17 A typical vacuum check valve.

Special Tools

Vacuum pump

Vacuum gauge

Vacuum hose

Vacuum reserve tanks may be made of plastic or metal.

Special Tools

Heat gun

Reserve Tank

Using the same setup as for testing the check valve, insert the hose onto the vacuum reserve tank (Figure 8-18) instead of the check valve.

1. Start the vacuum pump.
2. Observe the vacuum gauge.
3. Turn off the vacuum pump. If there is a vacuum and it holds for five minutes, the tank may be considered all right. If there is little or no vacuum and/or it does not hold for five minutes, the tank is defective.

Leaks in vacuum tanks usually may be repaired by using a fiberglass-reinforced resin.

Hose

A vacuum hose is often made of synthetic rubber or nylon. Deterioration, cracking, and splitting are problems found with vacuum hoses. The best way to determine the condition of vacuum hoses is by visual inspection. If a hose shows signs of deterioration, it should be replaced.

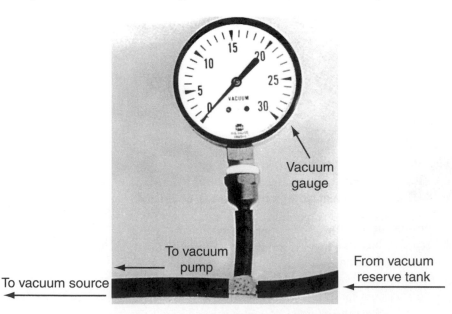

Figure 8-18 Testing the vacuum reserve tank for leaks. (Courtesy of BET, Inc.)

Restrictor

A restrictor is generally a porous bronze filter whose purpose is to prevent minute particles of dust and debris from entering the vacuum system where they could restrict control circuits or cause component damage. It is not practical to clean a restrictor. If in doubt, the simplest remedy is to replace it.

Classroom Manual
Chapter 10, page 232

Temperature Door Cable Adjustment

Remove access panels or components to gain access to the **temperature door** (Figure 8-19) and proceed as follows:

1. Loosen the cable attaching fastener at the heater case assembly.
2. Make sure that the cable is properly installed and routed to ensure no binding and freedom of movement.
3. Place the temperature control lever in the full cold position and hold it in place.
4. Tighten the cable fastener that was loosened in step 1.
5. Move the temperature control lever from full cold to full hot to full cold positions.
6. Repeat step 5 several times and check for freedom of movement.
7. Recheck the position of the door. If it is loose or out of position, repeat steps 2 through 7. If it is still in position and secure, replace the access panels and covers.

Air Filter

The introduction of air filters in the automotive air conditioning systems of domestic vehicles has been slow. The first was found in the 1938 Nash; the next was not until over 35 years later, in Oldsmobile's 1974 Toronado and 88. It would be another 20 years before electrostatically charged air

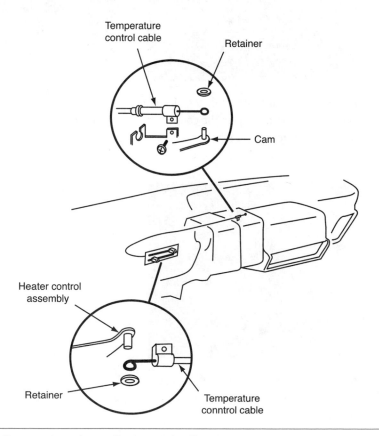

Figure 8-19 Temperature door adjustment details.

Photo Sequence 15
Removing and Replacing the Vacuum Selector Switch

The following is a typical procedure for removing a vacuum selector switch. The switch is replaced by reversing the procedure given. For specific procedures, consult the appropriate manufacturer's service manual.

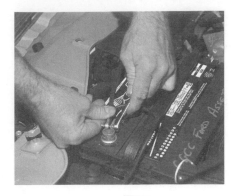

P15-1 Disconnect the battery ground-cable.

P15-2 Remove the instrument panel finish applique.

P15-3 Remove the screws holding the control assembly to the instrument panel.

P15-4 Pull the control assembly from the instrument panel sufficiently to gain access to the rear electrical connector, control cable, and vacuum switch.

P15-5 Depress the latches of the electrical connector to disengage the connector from the control assembly.

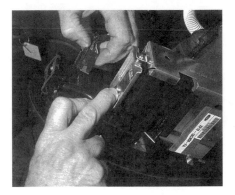

P15-6 Disconnect the temperature control cable from the control assembly.

P15-7 Disconnect the vacuum harness from the control assembly. Discard the used pushnut. Remove the knob from the vacuum selector switch.

P15-8 Remove the vacuum selector switch attaching screw(s) and remove the switch.

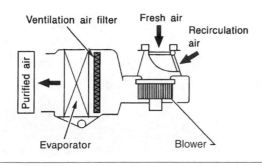

Figure 8-20 Some systems today have an air filter.

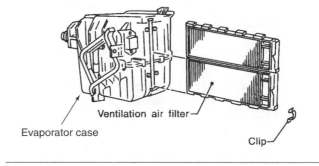

Figure 8-21 Remove air filter from evaporation case.

filters appeared in some Ford Contour and Mercury Mystique car lines. Since 1988, however, many European car lines have included some type of air conditioning system air filter (Figure 8-20).

Referred to by many as a "cabin air filter," the electrostatically charged filters are designed to remove particles as small as 0.02 microns—less than one-millionth of an inch in diameter. Electrostatic filters eliminate such unwanted particles as pollen, mold spores, road dust, bacteria, and tobacco smoke. The filter ensures that the air inside the passenger compartment is clean during either the recirculate or fresh air mode. The filter, usually located in the case/duct system should be serviced annually or every 15,000 miles (24,000 kilometers). A clogged filter can create an air pressure drop placing a greater demand on the blower motor and perhaps leading to an early failure. Because it restricts airflow, a clogged filter will also affect air conditioning, heating, and defroster performance.

The procedure for cleaning or replacing the filter varies from vehicle to vehicle, so it is important to follow manufacturer's recommended procedures. If the vehicle is equipped with an air filter, instructions may be found in the owner's manual or on a label inside the glove box. The following is a typical procedure:

1. Remove dash undercover.
2. Remove glove box.
3. Remove instrument reinforcement from instrument panel.
4. Remove filter retaining clip.
5. Remove air filter from case/duct (Figure 8-21).
6. Clean and/or install new filter.
7. Replace components in reverse order used to remove them.

Summary

Block or ladder diagrams, covering several pages, are often used in manufacturer's service manuals to troubleshoot problems with automotive air conditioning duct systems. It is therefore recommended that the appropriate service manuals be consulted for specific troubleshooting procedures. Table 8-1 lists the problems that typically are encountered in the case/duct system.

TABLE 8-1 TYPICAL CASE/DUCT SYSTEM PROBLEMS

VACUUM SYSTEM	MOTOR SYSTEM
No vacuum to air conditioner master control	No power to air conditioner mode selector
Air conditioner control leaks vacuum	High resistance connection in air conditioner control
Damaged, kinked, or pinched vacuum hose	Broken, loose, or disconnected electrical wiring
Damaged or leaking vacuum motor	Damaged or defective actuator motor
Actuator arm disconnected at door crank	Actuator linkage disconnected at door crank
Damaged or leaking vacuum reserve tank	Defective fuse, circuit breaker, or fusible link

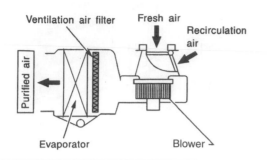

Figure 8-22 Some systems have an air filter.

Figure 8-23 Wear a grounding bracelet when working.around sensitive electronic components.

Damaged or leaking check valve	Defective diode or component in programmer

Always follow a manufacturer's recommended procedures and heed its cautions when troubleshooting any control system. The unintentional grounding of some circuits can cause immediate and permanent damage to delicate electronic components. The use of a test light, powered or nonpowered, is not recommended for under dash service. The battery in a powered test light or the added resistance on a nonpowered test lamp may be sufficient to cause failure to the delicate balance of solid-state electronic circuits.

Be especially cautious when servicing circuits that have a warning symbol such as that shown in Figure 8-22. These circuits are susceptible to damage by electrostatic discharge (ED) merely by touching them. Electrostatic discharge is a result of static electricity, which "charge" a person simply by their sliding across a seat, for example. To provide an extra margin of safety, the technician should wear a grounding bracelet (Figure 8-23), an electrical conducting device that surrounds the wrist and attaches to a known ground source. This device ensures that the body will not store damaging static electricity by providing a path to ground for it to be discharged.

 CASE STUDY

A customer brings a late-model vehicle into the shop with the complaint that air does not come out of the dash outlets regardless of the mode selected. Before attempting to check the underdash air distribution system, it is noted that the vehicle is equipped with an air bag system. The service manual cautions that the air bag system should be disarmed before performing any underdash service. Following service manual procedures, the technician disarms the air bag. In this case, the technician disconnects and tapes the negative (–) battery terminal, removes the fuses, disconnects the wiring harness and, finally, removes the air bag module from the vehicle. By taking time to heed the service manual warnings, possible air bag deployment and injury are avoided.

Terms to Know

Actuator	Debris	Power module
Blower	Disinfectant	Reserve tank
Blower motor	Environmental Protection Agency (EPA)	Resistor
Check valve	Fungi	Restrictor
	Hot knife	Technical service bulletin
	Mildew	Temperature door
	Mold	Vacuum motor

ASE Style Review Questions

Vacuum signal

Wiring harness

1. *Technician A* says that a slight amount of conditioned air is made available at the defroster duct outlet at all times to prevent windshield fogging.
Technician B says that a positive in-vehicle pressure is maintained at all times to prevent exhaust gas infiltration.
Who is right?
   A. A only
   B. B only
   C. Both A and B
   D. Neither A nor B

2. *Technician A* says that a vacuum reserve tank helps maintain a vacuum in the system at all times only when the engine is running.
Technician B says that the check valve prevents vacuum loss when the engine is stopped.
Who is right?
   A. A only
   B. B only
   C. Both A and B
   D. Neither A nor B

3. *Technician A* says that a check valve prevents vacuum flow in either direction.
Technician B says that a check valve permits vacuum flow in either direction.
Who is right?
   A. A only
   B. B only
   C. Both A and B
   D. Neither A nor B

4. *Technician A* says that an inoperative vacuum motor usually means a defective diaphragm in the motor.
Technician B says that a defective diaphragm will not prevent a vacuum motor from operating properly.
Who is right?
   A. A only
   B. B only
   C. Both A and B
   D. Neither A nor B

5. A musty odor from the evaporator may be caused by all of the following, EXCEPT:
   A. Mold
   B. Mildew
   C. Coolant
   D. Refrigerant

6. A vacuum actuator is also called a vacuum:
   A. Rod (arm)
   B. Source
   C. Relay
   D. Motor

7. *Technician A* cautions that one can be burned by touching a power module or resistor with bare hands.
Technician B says that the danger can be eliminated by first disconnecting the battery.
Who is right?
   A. A only
   B. B only
   C. Both A and B
   D. Neither A nor B

8. Servicing an air conditioning system is being discussed:
Technician A says that it is always necessary to disconnect the battery ground (–) cable first.
Technician B says that it is always necessary to disconnect the air bag restraint system first.
Who is right?
   A. A only
   B. B only
   C. Both A and B
   D. Neither A nor B

9. *Technician A* says that most automotive air conditioning repair procedures are covered in detail in independent repair guides.
Technician B says that one should consult specific manufacturers' service manuals.
Who is right?
   A. A only
   B. B only
   C. Both A and B
   D. Neither A nor B

10. Static electricity is being discussed.
Technician A says that static electricity can damage delicate electronic components.
Technician B says that wearing a grounded wrist strap, can prevent static electricity.
Who is right?
   A. A only
   B. B only
   C. Both A and B
   D. Neither A nor B

ASE Challenge

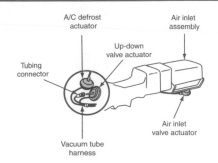

1. All of the following statements about a typical air conditioning system set to MAX cooling are true, EXCEPT:
   - **A.** The heater coolant flow control valve is open
   - **B.** The compressor clutch coil is energized
   - **C.** The blower motor is running
   - **D.** The outside/recirculate door is positioned to recirculate

2. List the steps for removing a blower motor in the proper order:
   - **a.** Remove the attaching screws and hardware
   - **b.** Lift the motor out of the case
   - **c.** Remove the battery ground wire/cable
   - **d.** Disconnect the electrical leads

A. a-b-c-d	**C.** c-d-a-b
B. b-c-d-a	**D.** d-c-a-b

3. The LEAST likely cause of an inoperative vacuum motor is:
   - **A.** A split hose
   - **B.** A defective check valve
   - **C.** A defective vacuum switch
   - **D.** A kinked hose

4. The illustration above is that of:
   - **A.** An electrical wiring harness
   - **B.** An under dash case/duct system
   - **C.** A fresh/recirculate air ventilation system
   - **D.** A vacuum hose harness

5. During normal comfort control operation with the windows closed, harmful gases are not allowed to enter the vehicle because:
   - **A.** The vehicle is air tight when the windows are closed
   - **B.** They are removed by natural convection in the ambient air stream
   - **C.** They are carried away by the force of the ram air
   - **D.** Of a slight in-vehicle positive pressure

Table 8-1 ASE TASK

Diagnose window fogging problems; determine needed repairs.

Problem Area	Symptoms	Possible Causes	Classroom Manual	Shop Manual
POOR VISION	Windshield fogging	1. Heater core leaking	267	285
		2. Heater coolant control valve leaking	267	422–423
		3. Heater coolant control valve out of adjustment	267	422–423
		4. Misdirected air flow	249–250	295

Table 8-2 ASE TASK

Perform cooling system tests; determine needed repairs.

Inspect and replace engine cooling and heater system hoses and belts.

Inspect, test, and replace radiator, pressure cap, coolant recovery system, and water pump.

Inspect, test, and replace thermostat, bypass, and housing.

Problem Area	Symptoms	Possible Causes	Classroom Manual	Shop Manual
ENGINE OVERHEATS	Coolant loss	1. Defective belt(s)	258–259	410
		2. Defective (leaking) radiator hose(s)	264	420
		3. Defective (leaking) heater hose(s)	264	420
		4. Leaking radiator	250–253	405
		5. Defective (clogged) radiator	250–253	405
		6. Defective or incorrect pressure cap	253	408
		7. Defective or incorrect thermostat	256–257	409
		8. Leaking thermostat housing or gasket	256–257	409
		9. Leaking thermostat by-pass hose	256–257	409
		10. Defective water pump	253	407
		11. Leaking coolant recovery system	267	421

Table 8-3 ASE TASK

Identify, inspect, recover coolant; flush and refill system with proper coolant.

Problem Area	Symptoms	Possible Causes	Classroom Manual	Shop Manual
ENGINE OVERHEATS	Higher than normal engine temperature	1. Old (depleted) coolant	268–271	430
		2. Low coolant level	272	430
DRAIN (RECOVER) COOLANT	Poor cooling	1. Depleted coolant	268–271	430
		2. Rust in cooling system	268–271	430
		3. Lack of antifreeze solution	268–271	430
		4. Unknown coolant	268–271	430

Table 8-4 ASE TASK

Inspect, test, and replace heater coolant control valve (manual, vacuum, and electrical types).

Problem Area	Symptoms	Possible Causes	Classroom Manual	Shop Manual
LOSS OF COOLANT	Engine overheating	Leaking heater control valve	267	422–423
POOR HEATER PERFORMANCE	Cool air from heater duct(s)	1. Misadjusted heater control valve cable	228–232	422–423
		2. Loss of vacuum signal to heater control valve	228, 230	422–423
		3. Defective vacuum motor on heater control valve	229	422–423
		4. Defective wiring to motor on heater control valve	216–217	422–423
		5. Defective heater control valve	267	422–423

Table 8-5 ASE TASK

Inspect, flush, and replace heater core.

Problem Area	Symptoms	Possible Causes	Classroom Manual	Shop Manual
COOLANT LOSS	Engine overheating	Leaking heater core	267	285
POOR HEATER PERFORMANCE	Cool air from heater duct(s)	1. Clogged heater core coolant passage	267	285
		2. Clogged heater core air passage	267	285

Table 8-6 ASE TASK

Inspect, test, repair, and replace A/C-heater blower motors, resistors, switches, relay/modules, wiring, and protection devices.

Diagnose the cause of failures in the vacuum and mechanical switches and controls of the heating, ventilating, and A/C systems; determine needed repairs.

Problem Area	Symptoms	Possible Causes	Classroom Manual	Shop Manual
INSUFFICIENT HEATING OR COOLING	Poor air flow from vents	1. Blower motor	221-222	283, 252
		2. Blower motor resistor or speed control	216–217	285
		3. Blower motor control switch	216–217	351–352
		4. Blower motor relay	213	351–352, 360
		5. Blower motor wiring	216–217	351–352, 360
		6. Misadjusted mechanical cable control	216	295
INSUFFICIENT COOLING	Warm air from vent(s)	1. Defective high-pressure control	226	354
		2. Defective low-pressure control	225	354
		3. Loss of vacuum signal to vacuum motor	228	357
		4. Defective vacuum motor	229	281
		5. Defective or misadjusted mechanical control	216	295

Table 8-7 ASE TASK

Inspect, test, service, or replace heating, ventilating, and A/C control panel assemblies.

Inspect, test, adjust, and replace heating, ventilating, and A/C control cables and linkages.

Inspect, test, or replace heating, ventilating, and A/C vacuum actuators (diaphragms/ motors) and hoses.

Identify, inspect, test, and replace heating, ventilating, and A/C vacuum reservoir, check valve, and restrictors.

Problem Area	Symptoms	Possible Causes	Classroom Manual	Shop Manual
INADEQUATE COOLING OR HEATERING	Loss of vacuum signal to control devices	1. Defective control panel assembly	216	348–349
		2. Defective vacuum control switch	228–232	291
		3. Defective or disconnected vacuum hose	232–233	295
		4. Defective check valve	230–231	291
		5. Leaking vacuum reserve tank	230–231	295
		6. Clogged restrictor	230–231	295
	Loose or misadjusted cable or linkage	1. At the control head	228–232	295
		2. At the control	228–232	295

Table 8-8 ASE TASK

Inspect, test, adjust, repair, or replace heating, ventilating, and A/C ducts, doors, and outlets.

Diagnose air distribution system problems; determine needed repairs.

Inspect, test, adjust, and replace temperature blend door/actuators.

Problem Area	Symptoms	Possible Causes	Classroom Manual	Shop Manual
INADEQUATE AIR FLOW	Poor heating or cooling	1. Duct disconnected	171–173	281
		2. Mode door binding or inoperative	175–184	295
		3. Hole or tear in duct hose	175	295, 298
		4. Hose disconnected	267	295, 298
		5. Hose improperly connected	267	295, 298
		6. Outlet blocked or restricted	175–184	295, 298

JOB SHEET 26

26

Name _____ Date _____

Case/Duct System Diagnosis

Upon completion of this job sheet you should be able to make basic checks of the vacuum system of an air conditioning case/duct system.

ASE Correlation

This job sheet is related to the ASE Heating and Air Conditioning Systems Test's content area: *Operating Systems and Related Controls Diagnosis and Repair, 2. Vacuum/Mechanical*, Task: *Inspect, test, adjust, repair, or replace heating, ventilating, and A/C ducts, doors, hoses, and outlets.*

Tools and Materials

Vehicle with manually controlled, factory-installed air conditioning system
Service manual
Chapter 8 of Classroom Manual
Selected air conditioning system tools
Vacuum pump

Describe the vehicle being worked on:

Year _____ Make _____ Model _____

VIN _____ Engine type and size _____

Procedure

Disconnect the vacuum source hose and connect a vacuum pump to the vacuum reserve tank. It will not be necessary to run the engine for a vacuum source. Gain access to the vacuum motors of the mode doors and determine the vacuum signal applied to each for the following air delivery conditions.

Use the abbreviations:
    fv for full vacuum
    pv for partial vacuum
    nv for no vacuum

Air Delivery Condition	Actuator Motor (Pot)			
	Defroster	A/C	Bilevel	Air Inlet
1. MAX	_____	_____	_____	_____
2. NORM	_____	_____	_____	_____
3. Bilevel (B/L)	_____	_____	_____	_____
4. VENT	_____	_____	_____	_____
5. Heat (HTR)	_____	_____	_____	_____
6. BLEND	_____	_____	_____	_____
7. De-Fog	_____	_____	_____	_____
8. OFF	_____	_____	_____	_____

✔ **Instructor's Check** _____

Job Sheet 27

27

Name _____ Date _____

Air Delivery Selection

Upon completion of this job sheet you should be able to trace the air delivery in each of the six basic air delivery modes.

ASE Correlation

This job sheet is related to the ASE Heating and Air Conditioning Systems Test's content area: *Operating Systems and Related Controls Diagnosis and Repair. 3. Automatic and Semi-Automatic Heating, Ventilating, and A/C Systems,* Task: *Diagnose air distribution system problems; determine needed repairs.*

Tools and Materials

Service manual
Blue and red pencil

Procedure

In each of the diagrams below show the position of the mode doors using a red pencil. Show the air flow using a blue pencil.

1. A typical dual-zone duct system with passenger-side full hot selected.

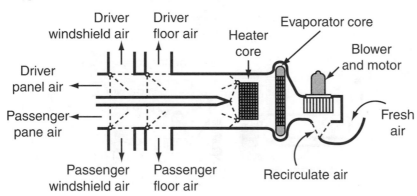

2. A typical dual-zone duct system with passenger-side full cold selected.

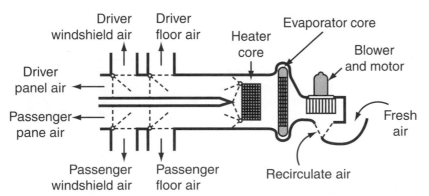

3. Air flow when DEFROST is selected.

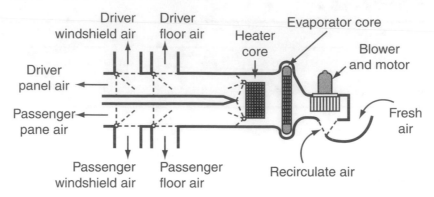

4. Air flow when MIX or HI/LO is selected.

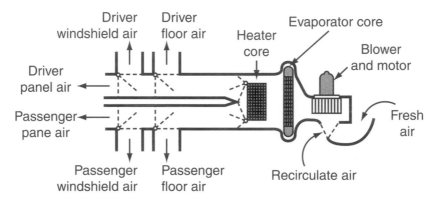

5. Air flow in the cooling mode when BILEVEL is selected.

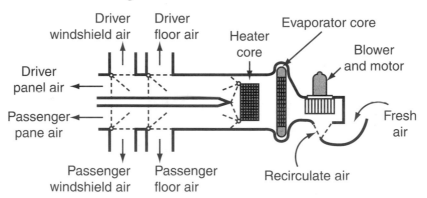

☑ **Instructor's Check** _____

JOB SHEET 28

28

Name _____ Date _____

Replace Case and Duct System Components

Upon completion of this job sheet you should be able to remove and replace case and duct system components.

ASE Correlation

This job sheet is related to the ASE Heating and Air Conditioning Systems Test's content area: *Operating Systems and Related Controls Diagnosis and Repair. 2. Vacuum/Mechanical*, Task: *Inspect, test, adjust, repair, or replace heating, ventilating, and A/C ducts, doors, and outlets.*

Tools and Materials

Vehicle with air conditioning system in need of case/duct service
Service manual
Appropriate tools

Describe the vehicle being worked on:

Year _____ Make _____ Model _____
VIN _____ Engine type and size _____

Procedure

> **CAUTION:** If equipped with air bag(s), follow specific manufacturer's service procedures for replacing defective components.

1. Determine what component part or assembly is in need of replacement. Describe your procedure and how you arrived at your decision.

   Procedure _____

   _____

   Component _____

   _____

2. Look up manufacturer's procedures in the appropriate service manual.

   Service manual (title) _____ Year _____
   Section _____ Page _____ Component _____

3. Following recommended procedures, remove necessary components to gain access and remove the defective part.

   Procedure _____

   _____

4. Obtain the replacement part. Compare it with the part removed. Are they the same or is the new part improved?

   Conclusion _____

   _____

5. Install the new part.

Procedure _____

_____

6. If possible, check the new component for proper operation.

Procedure _____

_____

7. Replace all components removed in step 3 to gain access.

Procedure _____

_____

8. Write a brief summary of any problems encountered with this repair.

_____

_____

☑ **Instructor's Check** _____

JOB SHEET 29

29

Name _____ Date _____

Adjust a Door Cable

Upon completion of this job sheet, you should be able to adjust a case/duct system door cable.

ASE Correlation

This job sheet is related to the ASE Heating and Air Conditioning Systems Test's content area: *Operating Systems and Related Controls Diagnosis and Repair 2. Vacuum/Mechanical*, Task: *3. Inspect, test, adjust, and replace heating, ventilating, and A/C control cables and linkages.*

Tools and Materials

Vehicle with air conditioning system
Service manual
Hand tools, as required

Describe the vehicle being worked on:

Year _____ Make _____ Model _____
VIN _____ Engine type and size _____

Procedure

1. Following procedures outlined in the service manual, gain access to the control cable.

Procedure _____

_____

2. Adjust the cable.

Procedure _____

_____

3. Check cable operation. Readjust, if necessary.

Procedure _____

_____

4. In reverse order, replace components removed in step 1.

Procedure _____

_____

5. What specific problems, if any, were encountered during this procedure?

_____

_____

✓ **Instructor's Check** _____

Retrofit (CFC-12 to HFC-134a)

Upon completion and review of this chapter you should be able to:

❏ Recognize the difference between pure and impure refrigerant by interpreting gauge pressures relating to ambient temperature.

❏ Determine the purity of refrigerant in an air conditioning system or container.

❏ Explain the necessity of using recovery-only equipment for contaminated refrigcrant.

❏ Describe the method of affixing an access saddle valve onto an air conditioning system.

❏ Determine when a system is void of refrigerant and/or air.

❏ Leak test the air conditioning system.

❏ Recover CFC-12 refrigerant from a system.

❏ Diagnose and repair system components.

❏ Evacuate a system prior to charging.

❏ Charge a system with HFC-134a refrigerant.

Basic Tools

Basic mechanic's tool set
Manifold and gauge set
Service hose set
Thermometer
Shop light(s)
Fender cover
Blanket
Vacuum pump
Can tap

Introduction

General information is given in this chapter regarding the proper and safe practices and procedures for retrofitting an automotive air conditioning system. It is most important, however, to follow the manufacturer's instructions when servicing any particular make and model vehicle. This chapter includes, under the appropriate heading, procedures for the following: **purity test**, **access valve** installation, recovery of contaminated refrigerant, and retrofit.

Classroom Manual
Chapter 4, page 89
Chapter 9,
pages 201

A purity test should be held any time there is a concern about the quality of the refrigerant in the system.

Purity Test

A refrigerant identifier, such as the Sentinel by Robinair (Figure 9-1), quickly and safely identifies the purity and type of refrigerant in a vehicle air conditioning system or tank. A display shows if the refrigerant is at least 98 percent CFC-12 or HFC-134a. If the purity or type of refrigerant is not identified, the display will indicate UNKNOWN.

Special Tools

Low-side (compound) gauge with gauge/hose adaptor and service hose
Thermometer

Figure 9-1 Refrigerant indentifier.

The thermometer should be placed in an area where it can "sense" free air.

With the transition to CFC-free air conditioning systems, the likelihood of cross-mixing refrigerants is a growing concern. Different refrigerants, as well as their lubricants, are not compatible and should not be mixed. It is possible, however, for the wrong refrigerant to be mistakenly charged into an air conditioning system, or for refrigerants to be mixed in the same recovery **tank**. Also, since recovery/recycling equipment is generally designed for a particular refrigerant, inadvertent mixing can cause damage to the equipment.

A refrigerant identifier tester is far superior to pressure/temperature comparisons because, at certain temperatures, the pressures of CFC-12 and HFC-134a are too similar to differentiate with a standard gauge. This is easily noted in the chart shown in Figure 9-2. For example, at 90°F (32.2°C), both 95 percent CFC-12 and 95 percent HFC-134a have about the same pressure—111 and 112 psig, respectively. Given that this chart is accurate to plus/minus 2 percent, there is really no way of determining which type refrigerant is in the air conditioning system or tank. Also, because other substitute refrigerants and blends may have been introduced into the automotive air conditioning system, they can contaminate a system or tank and may not be detected by the pressure/temperature method. A refrigerant identifier would conclude the refrigerant in our example to be UNKNOWN.

Use of a refrigerant identifier, often called a purity tester, should be the first step in servicing an automotive air conditioning system. That way, one does not have to be concerned about customer dissatisfaction or damage to the vehicle that could occur if the wrong refrigerant is used. Further, testing refrigerant protects refrigerant supplies and recovery/recycling equipment. At today's prices, preventing just one tank of refrigerant from contamination can save several hundred dollars plus the high cost of disposing of the contaminated refrigerant.

Always follow the manufacturer's instructions for using any type of test equipment. The following procedure for using the Sentinel identifier is typical:

1. Turn on the MAIN POWER switch; the unit automatically clears the last refrigerant sample and is made ready for a new sample.
2. When READY appears on the display, connect a service hose from the tester to the vehicle air conditioning system or tank of refrigerant being tested.
3. The tester automatically pulls in a sample and begins processing it; TESTING shows on the display.

AMB TEMP		R-12/R-134a PERCENT BY WEIGHT										
°F	°C	100/0	98/2	95/5	90/10	75/25	50/50	25/75	10/90	5/95	2/98	0/100
65	18.3	64	67	71	74	83	84	78	73	70	67	64
70	21.1	70	74	79	82	90	92	87	81	77	74	71
75	23.9	77	81	85	91	99	101	96	89	85	83	79
80	26.7	84	88	93	99	107	110	105	98	95	92	87
85	29.4	92	96	101	108	116	120	114	106	103	100	95
90	32.2	100	105	111	116	125	130	125	116	112	109	104
95	35.0	108	114	119	126	135	140	135	126	122	119	114
100	37.8	117	123	127	135	145	151	145	136	133	130	124
105	40.6	127	132	138	146	158	164	159	149	144	141	135
110	43.3	136	142	147	156	170	176	173	164	157	152	146
115	46.1	147	152	159	166	183	192	184	175	168	163	158
120	48.9	158	164	170	177	195	205	196	187	181	176	171

CFC-12/HFC-134a Cross Contamination Chart. All pressures are given in psig. For kPa, multiply psig by 6.895. For example, 100% R-12 at 95°F (35°C) is 108 psig or 744.7 kPa.

Figure 9-2 Temperature/pressure chart of CFC-12 and HFC-134a mixed refrigerants.

Figure 9-3 Attach an appropriate test gauge. (Courtesy of BET, Inc.)

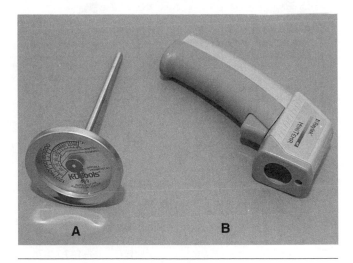

Figure 9-4 Typical (A) dial thermometer and (B) infrared temperature sensor.

4. Within about one minute, the display will show R-12, R-134a, or UNKNOWN. If UNKNOWN is displayed, the refrigerant is a mixture or is some other type refrigerant. In either case, it should not be added to previously recovered refrigerant. Also, it should not be recycled or reused.
5. Turn off the MAIN POWER switch and disconnect the service hose.

If no other method of refrigerant identification is available and there is any doubt as to the condition of the refrigerant in an air conditioning system, the following purity test may be used. It should be noted, however, that for a pressure/temperature test to be valid, there must be some liquid refrigerant in the system. If the refrigerant has leaked to the point that only vapor remains, the pressure will be below that specified at any given temperature. Proceed as follows:

1. Park the vehicle or place the tank inside the shop in an area that is free of drafts and where the ambient temperature is not expected to go below 70°F (21°C).
2. Raise the hood.
3. Determine the type of refrigerant that should be in the system or tank: CFC-12 or HFC-134a.
4. Attach a 0—150 psig (0—1000 kPa) gauge of known accuracy, appropriate for the refrigerant type (Figure 9-3).
5. Place a thermometer of known accuracy (Figure 9-4), in the immediate area of the vehicle or tank to measure the ambient temperature.
6. First thing the following morning:
   a. Note and record the pressure reading shown on the gauge.
   b. Note and record the temperature reading shown on the thermometer.
7. Compare the gauge reading with the appropriate table:
   a. (Figure 9-5): (A) English; (B) metric for CFC-12.
   b. (Figure 9-6): (A) English; (B) metric for HFC-134a.

Access Valves

A saddle clamp access valve (Figure 9-7) may be installed if space does not permit converting the CFC-12 access valve to the HFC-134a valve configuration. Follow this procedure for the typical installation of the **saddle valve**.

Classroom Manual
Chapter 7, page 168
Chapter 3,
pages 67–69

A

Temperature Fahrenheit	Pressure PSIG	kPa	Temperature Fahrenheit	Pressure PSIG	kPa
70	80	551	86	103	710
71	82	565	87	105	724
72	83	572	88	107	738
73	84	579	89	108	745
74	86	593	90	110	758
75	87	600	91	111	765
76	88	607	92	113	779
77	90	621	93	115	793
78	92	634	94	116	800
79	94	648	95	118	814
80	96	662	96	120	827
81	98	676	97	122	841
82	99	683	98	124	855
83	100	690	99	125	862
84	101	696	100	127	876
85	102	703	101	129	889

B

Temperature Celcius	Pressure PSIG	kPa	Temperature Celcius	Pressure PSIG	kPa
21.1	551	80	30.0	710	103
21.7	565	82	30.5	724	105
22.2	572	83	31.1	738	107
22.8	579	84	31.7	745	108
23.3	593	86	32.2	758	110
23.9	600	87	32.8	765	111
24.4	607	88	33.3	779	113
25.0	621	90	33.9	793	115
25.6	634	92	34.4	800	116
26.1	648	94	35.0	814	118
26.7	662	96	35.6	827	120
27.2	676	98	36.1	841	122
27.8	683	99	36.7	855	124
28.3	690	100	37.2	862	125
28.9	696	101	37.8	876	127
29.4	703	102	38.3	889	129

Figure 9-5 Temperature/pressure chart for CFC-12: (A) English and (B) metric.

A

Temperature Fahrenheit	Pressure PSIG	kPa	Temperature Fahrenheit	Pressure PSIG	kPa
70	76	524	86	102	703
71	77	531	87	103	710
72	79	545	88	105	724
73	80	551	89	107	738
74	82	565	90	109	752
75	83	572	91	111	765
76	85	586	92	113	779
77	86	593	93	115	793
78	88	607	94	117	807
79	90	621	95	118	814
80	91	627	96	120	827
81	93	641	97	122	841
82	95	655	98	125	862
83	96	662	99	127	876
84	98	676	100	129	889
85	100	690	101	131	903

B

Temperature Celcius	Pressure PSIG	kPa	Temperature Celcius	Pressure PSIG	kPa
21.1	524	76	30.0	703	102
21.7	531	77	30.5	710	103
22.2	545	79	31.1	724	105
22.8	551	80	31.7	738	107
23.3	565	82	32.2	752	109
23.9	572	83	32.8	765	111
24.4	586	85	33.3	779	113
25.0	593	86	33.9	793	115
25.6	607	88	34.4	807	117
26.1	621	90	35.0	814	118
26.7	627	91	35.6	827	120
27.2	641	93	36.1	841	122
27.8	655	95	36.7	862	125
28.3	662	96	37.2	876	127
28.9	676	98	37.8	889	129
29.4	690	100	38.3	903	131

Figure 9-6 Temperature/pressure chart for HFC-134a: (A) English and (B) metric.

1. Make certain that the system is free of refrigerant. Recover all of the refrigerant as outlined in this chapter if retrofitting the system.
2. Select the proper location for the valve.
   a. Will there be clearance for the hose access adaptor?
   b. Will there be adequate clearance to close the hood and/or replace protective covers?
   c. Will access to other critical components be **restricted** or blocked?
   d. Is the tubing straight, clean, and sound?
3. Select the proper valve for the application.
   a. For low- or high-side use (the low-side valve is larger).
   b. The size of the tube the valve is to be installed on.
4. Position both halves of the saddle valve on the tube (Figure 9-8).

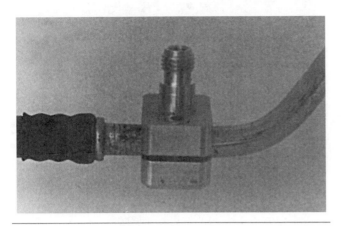

Figure 9-7 A saddle clamp access valve. (Courtesy of BET, Inc.)

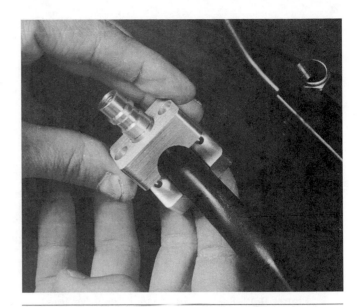

Figure 9-8 Position both halves of the valve on the tube.

> ⚠️ **WARNING:** Make sure that the O-ring is in position.

5. Place the screws (usually socket head) and tighten them evenly. Do not overtighten them; 20–30 in.-lb. (2–3 N•m) is usually recommended.

> ⚠️ **WARNING:** A method other than that outlined in steps 5, 6, and 7 might be recommended. Follow the recommendations provided by the manufacturer of the saddle valve when they differ from those given here.

6. Insert the **piercing pin** in the head of the access port fitting (Figure 9-9).
7. Tighten the pin until the head touches the top of the access port (Figure 9-10).
8. Remove the piercing pin and replace it with the valve core (Figure 9-11).
9. Securely tighten the valve core (Figure 9-12).
10. Install the cap (or **pressure switch**) on the installed fitting.

To ensure compatibility, use only the O-ring included with the saddle valve kit.

Special Tools

In.-lb. torque wrench with socket (to match saddle valve screws)

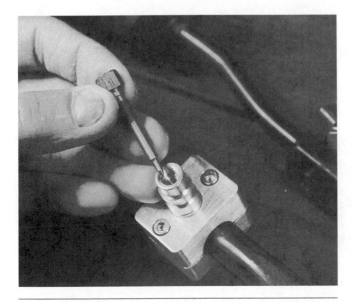

Figure 9-9 Insert the piercing pin.

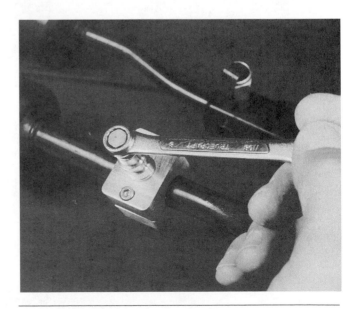

Figure 9-10 Tighten the pin.

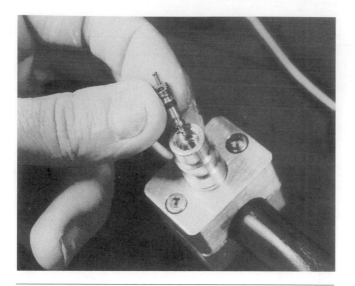

Figure 9-11 Replace the pin with the valve core.

Figure 9-12 Tighten the valve core.

Classroom Manual
Chapter 7,
pages 161
Chapter 9,
pages 262

There are chemicals
available that will
lower the ice bath
temperature to as
much as −15°F
(−26°C).

Recover Only—An Alternate Method

This method of recovery is presented for information only. It should only be accomplished by, or under the direct supervision of, an experienced technician. The most important consideration is that the recovery cylinder will not have been filled to more than 80 percent of capacity (Figure 9-13) when the temperature is increased to ambient.

Refer to the illustration (Figure 9-14) and follow these instructions:

1. Place an identified recovery cylinder into a tub of ice on the floor beside the vehicle.

✓ **SERVICE TIP:** The recovery cylinder should be below the level of the air conditioning system.

2. Add water and ice cream salt. This will lower the temperature to about 0°F (−17.7°C)
3. Connect a service hose from the high-side fitting of the system to the gas valve of the recovery cylinder.

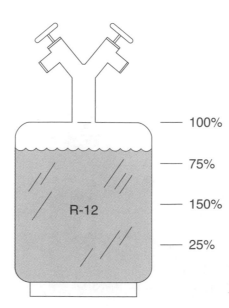

Figure 9-13 Recovery cylinders must not be filled more than 80% capacity.

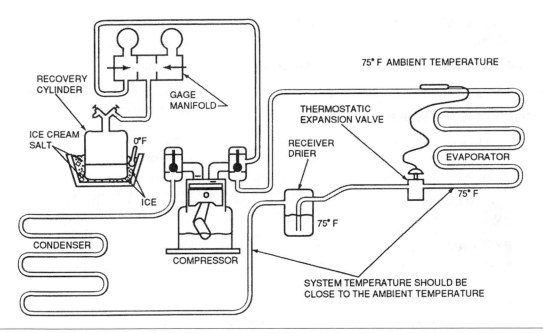

Figure 9-14 Setup for recovering refrigerant. (From Whitman/Johnson, *Refrigeration and Air Conditioning Technology*, 2E © by Delmar Publishers)

4. Open all valves.
5. Cover the recovery cylinder and tub with a blanket to insulate them from the ambient air.
6. Place the shop light(s) or other heat source near the accumulator or receiver.
7. Allow one to two hours for recovery. The actual time that is required will depend upon the ambient temperature and the amount of refrigerant to be recovered.

Retrofit

Specific procedures to retrofit any particular make or model vehicle are provided by the respective vehicle manufacturers. Several aftermarket manufacturers also offer retrofit kits for more generic applications. For example, one such manufacturer claims that three kits are all that are required to retrofit all car lines. According to early information released by automotive manufacturers, however, the procedure, methods, and materials vary considerably from car line to car line.

For example, some require draining mineral oil, while others do not; some require flushing the system, others do not. Also, some require replacing components, such as the accumulator or receiver-drier and/or the condenser, and others do not.

The **Society of Automotive Engineers** (SAE), in mid June 1993, issued their standard J1661 "Procedure for Retrofitting CFC-12 (R-12) Mobile Air Conditioning Systems to HFC-134a (R-134a)." The following service procedure, which is considered typical, is based on SAE's J1661.

Before attempting this procedure, be sure to review Chapter 9 of the Classroom Manual. This contains some very important information that must be understood to successfully retrofit a vehicle air conditioning system.

Procedure

The following step-by-step procedures are to be considered typical for retrofitting any vehicle from refrigerant CFC-12 (R-12) to refrigerant HFC-134a (R-134a). For specific procedures, however, follow the manufacturer's instructions.

Special Tools
Recovery system
Tub (for ice bath)

Classroom Manual
Chapter 9,
pages 193–194

Use only HFC-134a
to retrofit an
automobile air
conditioning system.

WARNING: Do not attempt to use any other type refrigerants.

Connect the Manifold and Gauge Set

Follow this procedure when connecting the CFC-12 manifold and gauge set into the system for service.

Prepare the System

1. Place fender covers on the car to avoid damage to the finish.
2. Remove the protective caps from the service valves (Figure 9-15). Some caps are made of light metal and can be removed by hand; others may require a wrench or pliers.

WARNING: Remove the caps slowly to ensure that refrigerant does not leak past the service valve.

CAUTION: Before beginning the retrofit procedure, perform a purity test to determine the type and quality of the refrigerant in the air conditioning system.

Connect the Manifold Service Hoses

WARNING: The service hoses must be equipped with a Schrader valve **depressing pin** (Figure 9-16). If the hoses are not so equipped, a suitable **adapter** (Figure 9-17) must be used.

Most hand valves are
closed by turning in
the clockwise (cw)
direction.

1. Make sure that the manifold hand shut-off valves (Figure 9-18) are closed.
2. Make sure that the hose shut-off valves (Figure 9-19) are closed.
3. Finger-tighten the low-side manifold hose to the suction side of the system.
4. Finger-tighten the high-side manifold hose to the discharge side of the system.

Figure 9-15 Remove protective caps from service valves: (A) Schrader type; (B) stem type. (Courtesy of BET, Inc.)

Figure 9-16 CFC-12 service hoses equipped with Schrader valve depresser pin.

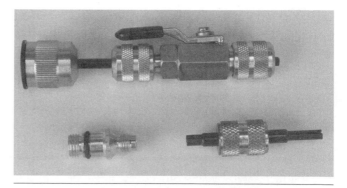

Figure 9-17 CFC-12 adapters for Schrader access valves.

Figure 9-18 Make sure that manifold hand shut-off valves are closed. (Courtesy of Uniweld Products)

Figure 9-19 Make sure that hose shut-off valves are closed. (Courtesy of Uniweld Products)

 SERVICE TIP: The CFC-12 high-side fitting on most late-model car lines requires that a special adapter be connected to the hose (Figure 9-20) before being connected to the fitting.

5. If retrofitting an older vehicle or heavy-duty, off-road equipment air conditioning system having shut-off type service valves (Figure 9-21), use a service valve wrench to rotate the stem two turns clockwise (cw).

6. Connect the service hose to the CFC-12 recovery system.

Special Tools

Recovery system

Refrigerant Recovery

Until the early 1990s, service technicians vented refrigerant into the atmosphere. Refrigerant was inexpensive and the cost of recovery would probably have been greater than the cost of the refrigerant. The Clean Air Act (CAA) Amendments of 1990 changed that practice. The CAA enacted by the Environmental Protection Agency (EPA) required that, after July 1, 1992, no refrigerants may be intentionally vented.

CAUTION: Adequate ventilation must be maintained during this procedure. Do not discharge refrigerant near an open flame as a hazardous toxic gas may be formed.

Unintentional venting in the performance of repairs is permitted under the CAA.

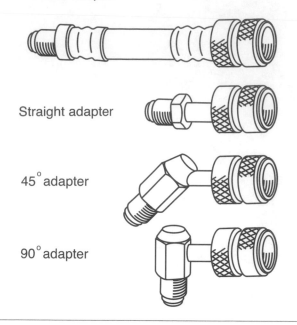

Flexible adapter

Straight adapter

45° adapter

90° adapter

Figure 9-20 Special high-side hose adapters. (Courtesy of Ford Motor Company)

Figure 9-21 Use a wrench to turn the shut-off type service valve. (Courtesy of BET, Inc.)

Prepare the System

 WARNING: Certain system malfunctions, such as a defective compressor, may make this step impossible.

1. Start the engine and adjust its speed to 1,250–1,500 **rpm**.
2. Set all air conditioning controls to the MAX cold position with the blower on HI speed.
3. Operate for 10–15 minutes to stabilize the system.

Recover Refrigerant

Special Tools

Recovery cylinder

Recovery system

1. Return the engine speed to normal idle to prevent dieseling.
2. Turn off all air conditioning controls.
3. Shut off the engine.
4. If not integrated in the recovery system, use a service hose and connect the recovery system to an approved recovery cylinder.
5. Open all hose shut-off valves.
6. Open both low- and high-side manifold hand valves.
7. Open the recovery cylinder shut-off valves, as applicable.
8. Connect the recovery system into an approved electrical outlet and turn on the main power switch.

WARNING: If an extension cord is used, make certain that it has an electrical rating sufficient to carry the rated load of the recovery system.

9. Turn on the recovery system compressor switch.
10. Operate the vacuum pump until a vacuum pressure is indicated (Figure 9-22).
11. If the recovery system is not equipped with an automatic shut-off, turn off the compressor switch after achieving a vacuum (step 10).
12. Be sure that the vacuum holds for a minimum of five minutes.

Figure 9-22 Operate pump until a vacuum is noted.

    a. If the vaccum does not hold, repeat the procedures starting with step 9 and continue until the system holds a stable vacuum for a minimum of two minutes.

    b. If the vacuum holds, proceed with step 13.

13. Close all valves: at the recovery cylinder, recovery system, service hoses, manifold, and compressor.

14. Disconnect all hoses previously connected.

CAUTION: Some recovery systems have automatic shut-off valves. Be certain they are operating properly before disconnecting the hoses to avoid refrigerant loss that could result in personal injury.

Repair or Replace Components

Flushing is generally not recommended unless the component has first been removed from the vehicle.

    1. Determine what repairs, if any, are required.

    2. If an oil change is required, proceed with step 3; if not, proceed with step 4.

    3. Remove the necessary components to drain the oil from the component (Figure 9-23).

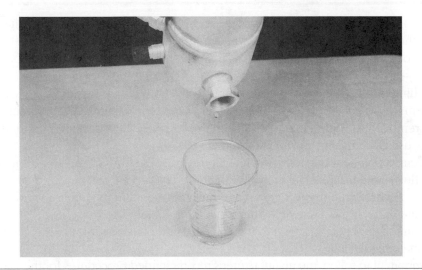

Figure 9-23 Draining oil from an accumulator.

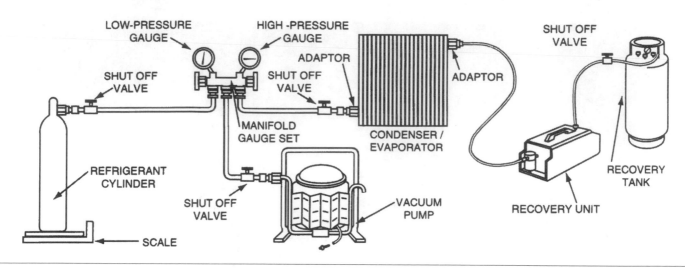

Figure 9-24 Typical setup for flushing a component.

4. Flush the individual components while they are out of the vehicle. A typical setup for this procedure is shown in Figure 9-24.
5. Replace components such as the accumulator, receiver-drier, and/or condenser, if required. It may be necessary to replace the receiver-drier or accumulator-drier if the desiccant is not compatible with HFC-134a refrigerant.

> **CAUTION:** It is recommended that the receiver-drier or accumulator-drier be replaced any time the air conditioning system is opened for major repairs.

6. Add or replace electrical fail-safe components, such as the refrigerant containment and high-pressure switch, if required.
7. Perform any other modifications and/or procedures required by the specific vehicle manufacturer.
8. Replace/reinstall all components serviced in steps 3 and 7.
9. If not accomplished by the requirements of steps 5, 6, or 7, repair any problems determined in step 1.

Flush the System?

Flushing the air conditioning system is not generally recommended. Because of the screens and strainers in the system, little if any debris will be removed. Also, most liquids (moisture and lubricant) in the low areas in the system—such as in the bottom of the evaporator, muffler, receiver, or accumulator—are not removed by flushing.

If flushing is to be performed, the flushing agent should be refrigerant—the same type used in the system. In the case of CFC-12, this is an expensive procedure. Also, system components should be removed for individual flushing after excess lubricant has been drained from them. Refrigerants used for flushing must be recovered.

There are a number of systems and techniques available for flushing an air conditioning system. Some flush systems are attachments for the recovery/recycle machine and other systems are self-contained units. Some use refrigerant as a flushing agent and others use various fluids, even methylhydrate or naphtha (flammable fluids). Nitrogen is often used as a propellant for the cleaning fluid. Some suggest adding a filter to the liquid line after flushing to catch any remaining debris before the metering device.

It is important to follow the manufacturer's recommended procedures for the particular flushing system being used. It is also important not to neglect system lubrication after flushing, regard-

Figure 9-25 Use adapters to change service ports from CFC-12 (A) to HFC-134a (B) service. (Courtesy of BET, Inc.)

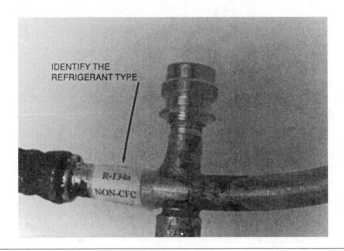

Figure 9-26 Affix decals to identify refrigerant type. (Courtesy of BET, Inc.)

less of the method or system used. Any lubricant flushed out of the system must be replaced with clean, fresh lubricant of the proper type for the refrigerant being used.

Prepare the System for HFC-134a

1. Charge the system with the proper type and quantity of lubricant as recommended by the vehicle manufacturer for HFC-134a refrigerant.

 WARNING: If the system was flushed, charge oil directly into the compressor to provide lubrication at startup.

2. Change service ports from CFC-12 to HFC-134a access type (Figure 9-25).
3. Check for leaks.
4. Affix decals to identify refrigerant type for future service (Figure 9-26).

Only PAG or POE lubricant should be used in an HFC-134a refrigerant system.

Evacuating the System

Whenever it is serviced, an automotive air conditioning system should be evacuated to the extent that the refrigerant has been removed. There are some who claim that moisture cannot be removed from an automotive air conditioning system with a "standard" vacuum pump. SAE standard J1661, however, requires that a vacuum pump be capable of achieving a vacuum level of 29.2 in. Hg (2.7

SYSTEM VACUUM		TEMPERATURE	
in.Hg	kPa (ABS)	°F	°C
0.00	101.33	212	100
27.75	7.35	104	40
28.67	4.23	86	30
29.32	2.03	64	18
29.62	1.01	45	7
29.74	0.61	32	0
29.82	0.34	6	−14
29.91	0.03	−24	−31

Figure 9-27 Boiling point of water (H_2O) in a vacuum at sea level atmospheric pressure.

kPa absolute) adjusted to altitude. The boiling point of water (H_2O) at this level is 69°F (20.6°C) at sea level atmospheric pressure (Figure 9-27). That means that moisture cannot be removed from an air conditioning system when the ambient temperature is below, say, 70°F (21.1°C) .

If a vacuum pump is to be used to remove moisture from an automotive air conditioning system, a quality two-stage, high-vacuum pump is recommended for adequate performance over a long period of time. Even the best vacuum pump, however, requires regular maintenance to ensure optimum performance.

Frequent oil changes are perhaps the single most important factor in a preventive maintenance (PM) program. A vacuum pump cannot handle moisture without some of it condensing in the lubricant. If this moisture is not removed by changing the oil it can attack metal components within the pump. This will result in lockups or loss of pumping efficiency and/or capacity. For the average service shop, oil changes should be a normal part of a the daily equipment maintenance program. It would be well, however, to change the oil after an extended pump down, especially after pumping down a system known to be wet. Specific instructions included with a vacuum pump should be followed for changing the oil.

Speed at Which a System Is Dehydrated

Several factors influence the "pumping speed" of a high-vacuum pump and thus the time required to remove all moisture from a refrigerant system. Some of the most important factors include:

- ❏ Size of the system, in cubic feet
- ❏ Amount of moisture to be removed
- ❏ Ambient temperature
- ❏ Internal restrictions within the system (Schrader valves and metering device)
- ❏ External restrictions (between the system and vacuum pump)
- ❏ Size of the pump
- ❏ Condition of the pump (clean, fresh oil)

The elimination of restrictions in an air conditioning system is not generally possible. The size of valves, manifold, and metering device cannot be altered during evacuation. The service lines, however, can be enlarged as well as shortened, and the Schrader valve cores can be removed during the evacuation process. Photo Sequence 16 illustrates the procedure for removing Schrader valve cores for evacuation and charging procedures.

How Vacuum Is Measured

In the automotive air conditioning industry, vacuum is generally measured with a standard Bourdon tube compound gauge. This type of gauge is suitable for standard vacuum reading, say 29 in.

Photo Sequence 16
Removing and Replacing a Schrader Valve Core in a Service Valve

The following procedure may be considered typical for replacing a Schrader valve core in an HFC-134a air conditioning system service valve. This procedure, with minor variations, may also be used to replace a Schrader valve core in a CFC-12 air conditioning system service valve as well. Always follow specific procedures included by the equipment manufacturer when they differ from those given in this text.

P16-1 Install the high side service valve access fitting.

P16-2 Install the low side service valve access fitting.

P16-3 With the recovery machine connected, open the appropriate gas and/or liquid valve (follow specific instructions for machine being used).

P16-4 Turn on the main power.

P16-5 Open the low- and/or high-side valve.

P16-6 Allow recovery equipment to operate until both gauges indicate 0 psig (0 kPa), or less. Follow procedures outlined in Photo Sequence 11.

P16-7 Remove the service valve access fitting from the leaking Schrader valve core.

P16-8 Using a valve core tool, remove the leaking Schrader valve core.

P16-9 Install a new valve core using the valve core tool.

Photo Sequence 16
Removing and Replacing a
Schrader Valve Core in a Service Valve (continued)

P16-10 Open the low side hand valve.

P16-11 Start the vacuum pump.

P16-12 After 5 minutes, open the high side hand valve and proceed with procedures as outlined in Photo Sequence 10.

Hg. It cannot be used, however, to read millimeters or microns. For this reason, it is not suitable for use with high-vacuum pumps.

Electronic thermistor vacuum gauges (Figure 9-28) are available for use with high-vacuum pumps. They can accurately read a vacuum as low as 1 micron by using a sensing tube mounted at some point in the vacuum service line. The readout can be an analog meter scale, digital display, or a light-emitting diode (LED) sequential display. One advantage of a thermistor vacuum gauge is that it is sensitive to water vapor and other condensables and can give a good indication of the actual vacuum level within a system. A thermistor vacuum gauge, though not essential, is a worthwhile companion instrument for high-vacuum dehydration of an automotive air conditioning system.

The location of the vacuum gauge will affect its reading in relation to the actual vacuum in the system. The closer the gauge is to the vacuum source, the lower the reading. When taking a

A Micron is a unit of linear measurement equal to 1/25,400 or an inch.

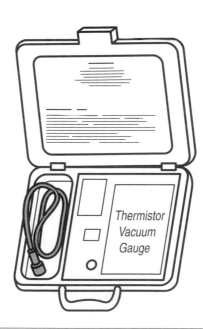

Figure 9-28 A thermistor vacuum gauge. (Courtesy Robinair SPX Corporation)

final reading of the vacuum created in an air conditioning system, one should isolate the vacuum pump with a vacuum valve and allow the pressure in the system to equalize.

If the pressure does not equalize, it is an indication of a leak. If it does equalize, but only at a higher pressure, it is an indication that moisture remains in the system. If this is the case, more pumping time is required.

The following service procedure for evacuating the system may be used for the independent vacuum pump (Figure 9-29) or the dedicated charging station (Figure 9-30). The vacuum pump may be used for either CFC-12 or HFC-134a refrigerant systems. The charging station that is pictured contains a vacuum pump, manifold and gauge set, and calibrated charging cylinder and is for CFC-12 only. It is compatible with all CFC-12 recovery and recycling systems. Robinair (and others) also produce a similar dedicated charging station for HFC-134a refrigerant that is compatible with all HFC-134a recovery and recycle systems.

Prepare the System

NOTE: Before performing any service procedure, ensure that both the low-side (compound) and high-side (pressure) gauges are zero calibrated.

1. Make sure that the high- and low-side manifold hand valves are in the closed position.
2. Make sure that the service hose shut-off valves are closed.
3. Remove the protective caps and covers from all service access fittings.
4. Connect the R-134a manifold and gauge set to the system in the same manner as was previously outlined for the R-12 manifold and gauge set.
5. Place the high- and low-side compressor service valves, if equipped, in the **cracked position**.
6. Remove the protective caps from the inlet and exhaust of the vacuum pump.

▲ **WARNING:** Make sure the port cap is removed from the exhaust port to avoid damage to the vacuum pump.

7. Connect the center manifold hose to the inlet of the vacuum pump.
8. Open all service hose shut-off valves.
9. Start the vacuum pump.

Follow the vacuum pump manufacturer's operating instructions if they differ from those given in this manual.

If, after five minutes, there is not a reasonable vacuum noted, a leak is indicated.

Figure 9-29 A typical high-vacuum pump. (Courtesy of Robinair Division of SPX Corporation)

Figure 9-30 A typical CFC-12 charging station. (Courtesy of Robinair Division of SPX Corporation)

Figure 9-31 The compound gauge should indicate 20 in. Hg (33.8 kPa absolute) or below.

Figure 9-32 The high-side gauge should drop below zero.

10. Open the low-side manifold hand valve.
11. Observe the low-side (compound) gauge needle. The needle should indicate a slight vacuum.
12. After five minutes, the compound gauge should indicate 20 in. Hg (33.8 kPa absolute) or less (Figure 9-31).
13. The high-side (pressure) gauge needle should be slightly below the zero index of the gauge.
14. If the high-side gauge does not drop below zero (Figure 9-32), unless restricted by a stop, a system blockage is indicated.
    a. If the system is blocked, discontinue the evacuation. Repair or remove the obstruction.
    b. If the system is clear, continue the evacuation with step 15.
15. Open the high-side manifold hand valve.
16. Operate the pump for 15 minutes and observe the gauges. The system should be at a vacuum of 24–26 in. Hg (20.3–13.5 kPa absolute) minimum if there is no leak.
17. If the system is not down to 24–26 in. Hg (20.3–13.5 kPa absolute), close the low-side hand valve and observe the compound gauge.
    a. If the compound gauge needle rises, indicating a loss of vacuum, there is a leak that must be repaired before the evacuation is continued.
    b. If no leak is evident, continue with the pump-down.
18. Pump for a minimum of 30 minutes, as required by SAE J1661. A longer pump down is much better, if time permits. For maximum performance, a triple pump down is recommended by many.
19. After pump down, close the high- and low-side manifold hand valves.
20. Shut off the vacuum pump.
21. Close all valves (service hose, vacuum pump, and compressor, if equipped).
22. Disconnect the manifold hoses.
23. Replace any protective caps previously removed.

Charging an R-134a Air Conditioning System

It may be noted that HFC-134a is an **ozone friendly** refrigerant and, as such, poses no known threat to the environment. Nonetheless, the EPA requires that this refrigerant also be recovered. This law became effective in the middle of November 1995.

Procedure

1. Place the vehicle in a draft-free work area. This is an aid in detecting small leaks.
2. Close all valves (service valves, if equipped, manifold gauge, service hose shut-off valves, and refrigerant cylinder or charging station shut-off valve).
3. Connect the manifold and gauge set to the system following procedures previously outlined.
4. Connect the service hose to the refrigerant source. If a charging station is used, it is very important that the instructions provided by the manufacturer of the equipment are followed.

> ⚠️ **WARNING:** Do not open the manifold and gauge set hand valves until instructed to do so. Early opening could contaminate the system with moisture-laden air.

5. Open the service hose shut-off valves.
6. Open the system service valves, if equipped.
7. Observe the gauges.
   a. Confirm that the system is in a vacuum. If it is, proceed with step 8.
   b. If it is not, follow the procedure outlined for evacuating the system before proceeding.
8. Dispense one "pound" can (Figure 9-33) of HFC-134a refrigerant into the system.
   a. Invert can for liquid dispensing (Figure 9-34).
   b. Open the high-side manifold hand valve.
   c. Empty the contents of the can into the system.
   d. Close the manifold high-side valve.
   e. Rotate the clutch armature several revolutions by hand to ensure that no liquid refrigerant is in the compressor.
9. Attach the electronic thermometer probes (Figure 9-35) to the inlet and outlet of the evaporator. Be sure that the end of the probe makes good contact with the metal tubes of the evaporator.
10. Open all windows.
11. Place a **jumper** wire across the terminals of the temperature or pressure control, usually found on the accumulator.

The system should now be under a deep vacuum.

Special Tools
Electronic thermometer

A dual-probe electronic thermometer is ideal for measuring superheat.

The jumper prevents compressor short-cycling during charging procedures.

Figure 9-33 Dispense one can of refrigerant into the system.

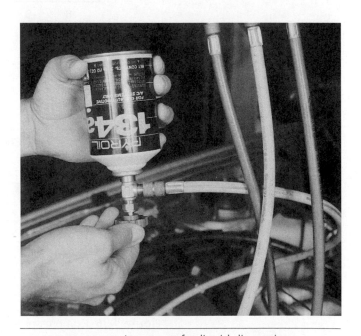

Figure 9-34 Invert can for liquid dispensing.

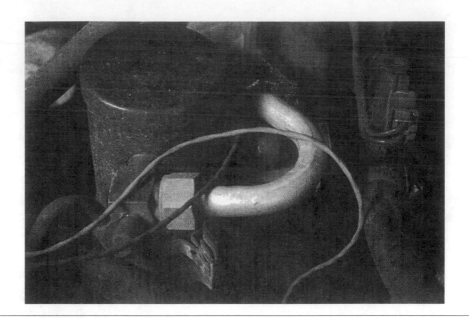

Figure 9-35 Attach the electronic thermometer probes to the inlet and outlet of evaporator.

12. Start the engine.

13. Set all air conditioner controls to HI.

14. Allow the engine to reach normal operating temperature.

15. Note and record the temperature of the two thermometers. Calculate the difference in temperature between the inlet and outlet tubes of the evaporator.

16. Wait a few minutes and record the temperatures again to confirm readings.

17. Note and record the ambient temperature. Compare it with the chart in Figure 9-36 or 9-37 as applicable.

AMBIENT TEMPERATURE (°F)						AMOUNT OF HFC-134a TO ADD (OUNCES)
60	70	80	90	100	110	
Evaporator Inlet To Outlet Temperature Difference						
-8	-8	-8	-8	-8	-8	0
-7	-7	-7	-7	-7	-7	2
-6	-6	-6	-6	-6	-6	4
-5	-5	-5	-5	-5	-5	6
+13	+13	+13	+17	+20	+25	8
+21	+25	+29	+33	+37	+42	12
+40	+45	+50	+55	+60	+65	14

A

AMBIENT TEMPERATURE (°C)						AMOUNT OF HFC-134a TO ADD (mL)
16	21	27	32	38	43	
Evaporator Inlet To Outlet Temperature Difference						
-5	-5	-5	-5	-5	-5	0
-4	-4	-4	-4	-4	-4	59
-3	-3	-3	-3	-3	-3	118
-3	-3	-2	-1	0	0	177
+7	+7	+7	+9	+11	+14	237
+10	+14	+16	+18	+21	+23	335
+22	+25	+28	+31	+33	+36	414

B

Figure 9-36 Compressor discharge pressure vs. ambient temperature chart: (A) English; (B) metric.

NOTICE: RETROFITTED TO R-134a

RETROFIT PROCEEDURE PERFORMED TO SAE J1661
USE ONLY R-134a REFRIGERANT AND SYNTHETIC
OIL TYPE: ____1____ PN: ____2____ OR
EQUIVALENT, OR A/C SYSTEM WILL BE DAMAGED

REFRIGERANT CHARGE/AMOUNT: ____3____
LUBRICANT AMOUNT: ____4____ PAG ☐ ESTER ☐ 5

RETROFITTER NAME: ____6____ DATE: ____7____
ADDRESS: ____8____
CITY: ____9____ STATE: ____10____ ZIP: ____11____

1 Type: manufacturer of oil (Saturn, GM, Union Carbide, Etc).

2 PN: Part number assigned by manufacturer.

3 Refrigerant charge / amount: Quanity of charge installed

4 Lubricant amount: Quanity of oil installed (indicate ounces, cc. ml).

5 Kind of oil installed (check either PAG or ESTER).

6 Retrofitter name: Name of facility that performed the retrofit.

7 Date: Date retrofit is performed.

8 Address: Address of facility that performed the retrofit.

9 City: City in which the facility is located.

10 State: State in which the facility is located.

11 Zip: Zip code of the facility.

Figure 9-37 A retrofit label.

18. Follow the temperature differential chart (step 15) to determine how much refrigerant must be added to the system to ensure a proper charge.

 WARNING: As a general rule, the capacity of a retrofit system for refrigerant HFC-134a capacity is about 90 percent of the original capacity for CFC-12 refrigerant.

19. Continue charging, as required. Tap a "pound" can of HFC-134a. With the can upright, open the manifold low-side valve. Dispense the contents of the can into the system. Close the low-side manifold valve. Repeat this step, as required.

20. Turn off the air conditioner.

21. Stop the engine.

22. Remove the jumper wire from the temperature/pressure switch (see step 11).

23. Close all valves (manifold, hose shut-off, and service, if equipped).

24. Recover refrigerant from service hoses.

25. Disconnect all hoses from the system.

26. Replace all protective covers and caps.

Remember to replace the connector to the switch.

CUSTOMER CARE: When performing underhood service such as refrigerant retrofit, make a visual inspection of the engine cooling system. Advise the customer of any problems noticed that may lead to early failure of the cooling or heating system. These problems may include leaks, rotted or cracked radiator or heater hoses, or frayed or worn belt(s). In bringing these problems to the customer's attention, the customer is made aware of pending problems. Nothing is more frustrating than having a breakdown due to other failures just after having extensive (and expensive) repairs.

When customers are made aware of potential problems they will generally approve repairs. While some may put off repairs, most will be thankful that your inspection may have prevented an expensive and inconvenient breakdown in the future.

In any event, the customer has been made aware of pending problems that are not covered by the current repair warranty. If the customer chooses not to have the repairs made, make a proper notation on the shop order form so it may be a matter of record.

Conclusion

Some final points:

❏ At the end of a successful retrofit, affix the proper label in a conspicuous place under the hood. The label (Figure 9-37) should at least contain the following information:
   1. Date of retrofit.
   2. Company and/or technician name and address
   3. Type and amount of refrigerant (HFC-134a) in pounds (lb.), ounces (oz.), or milliliters (mL)
   4. Type and amount of lubricant (PAG or POE) in ounces (oz.) or milliliters (mL)
❏ Do not remove the HFC-134a fitting adapters from the CFC-12 fittings. Once installed they are to become a permanent part of the air conditioning system.
❏ Do not overcharge the air conditioning system with refrigerant. The typical HFC-134a charge of refrigerant is about 90 percent of the original CFC-12 refrigerant charge. Refer to the chart in Figure 9-38 for the 90-percent rule.

CFC-12		HFC-134A	
OUNCES	MILLILITERS	OUNCES	MILLILITERS
48	1420	43.2	1278
44	1302	39.6	1171
40	1183	36.0	1065
36	1065	32.4	958
32	947	28.8	852
30	887	27.0	799
28	828	25.2	745
26	769	23.4	692
24	710	21.6	639
22	651	19.8	586
20	592	18.0	532
18	532	16.2	479
16	473	14.4	426
14	414	12.6	373

Figure 9-38 The 90 percent rule for HFC-134a versus CFC-12 refrigerant charge.

CASE STUDY

A customer complained about an inoperative air conditioning system. Questioning the customer revealed that the system had not worked since the end of last summer. "It was going to get cool in a few weeks and I would not need the air conditioner. I decided to put it off until spring."

A visual inspection of the system by the technician did not reveal any oil spots or ruptured hoses indicating a leak. When the manifold set was connected, the gauges revealed system pressure was equal on both gauges. Further, the temperature/pressure chart indicated that the system pressure was within acceptable limits for CFC-12 refrigerant for the ambient temperature.

The technician noticed that the lead wire to the clutch coil had been disconnected. Assuming that it had been intentionally disconnected, the technician reconnected it. Further questioning of the customer, however, revealed no knowledge of a disconnected wire.

Shortly after starting the engine and turning the air conditioner on, cool air was noted coming from the driver-side vent. The manifold gauges indicated proper pressures. A thermometer inserted in the passenger-side vent also indicated proper temperature.

Further discussion with the customer revealed that the problem apparently had begun while on vacation. The belts had been replaced and the mechanic must have pulled the wire loose during the repairs. The customer didn't realize the problem for several days after the repairs since the climate was mild and the air conditioner was not turned on. The customer had not considered that the problem may have occurred during repairs. The customer suffered through the close of one summer and the start of another simply because of a mechanic's error, and putting off repairs.

Terms to Know

Access valve	Ozone friendly	rpm
Adapter	Piercing pin	Saddle valve
Cracked position	Pressure switch	Society of Automotive Engineers
Depressing pin	Purity test	Tank
Jumper	Restricted	

ASE-Style Review Questions

1. *Technician A* says that a shut-off type service valve has a front-seated position.
   Technician B says that a Schrader-type service valve has a front seated position.
   Who is right?
   A. A only **C.** Both A and B
   B. B only **D.** Neither A nor B

2. After stabilizing the air conditioning system, the engine speed is returned to normal.
   Technician A says this is to reduce air flow across the condenser.
   Technician B say this is to increase the cooling capacity of the evaporator.
   Who is right?
   A. A only **C.** Both A and B
   B. B only **D.** Neither A nor B

3. *Technician A* says that the system is purged of refrigerant if the manifold gauges read a slight vacuum.

Technician B says that the system is purged of refrigerant even if the manifold gauges read a slight pressure.
Who is right?
A. A only **C.** Both A and B
B. B only **D.** Neither A nor B

4. *Technician A* says that a minimum evacuation of 30 minutes is required.
   Technician B says an evacuation of one or two hours provides a better pump down.
   Who is right?
   A. A only **C.** Both A and B
   B. B only **D.** Neither A nor B

5. The recommended minimum efficiency of a vacuum pump at sea level atmospheric pressure is being discussed:
   Technician A says that atmospheric pressure has no effect on efficiency.
   Technician B says that the greater the vacuum

achieved, the better the efficiency.
Who is right?

A. A only   **C.** Both A and B

B. B only   **D.** Neither A nor B

6. *Technician A* says that POE lubricant is not compatible with a CFC-12 system.
   Technician B says that XH9 desiccant used in HFC-134a systems should not be used in a CFC-12 system.
   Who is right?

   A. A only   **C.** Both A and B

   B. B only   **D.** Neither A nor B

7. All of the following statements are true, EXCEPT:

   A. Flushing the air conditioning system is not recommended when retrofitting.

   B. HFC-134a retrofit capacity is about 90 percent of the original CFC-12 capacity.

   C. A refrigerant identifier will identify flammable refrigerants.

   D. A refrigerant identifier will identify HFC-134a at any purity.

8. The pressure required for leak testing is being discussed:

Technician A says that a minimum of 60 psig (414 kPa) is required.
Technician B says that a maximum of 40 psig (276 kPa) is required.
Who is right?

A. A only   **C.** Both A and B

B. B only   **D.** Neither A nor B

9. *Technician A* says that special leak detectors are available for use with HFCs.
   Technician B says that there are leak detectors available that will detect CFCs as well as HFCs.
   Who is right?

   A. A only   **C.** Both A and B

   B. B only   **D.** Neither A nor B

10. If the compressor cycles while charging:
    Technician A says to place a jumper across the temperature switch.
    Technician B says to place a jumper across the low-pressure switch.
    Who is right?

    A. A only   **C.** Both A and B

    B. B only   **D.** Neither A nor B

ASE Challenge

1. Under the Clean Air Act (CAA), any person who performs service to a motor vehicle air conditioner (MVAC) must:

   A. Be properly certified by an approved agency

   B. Use properly certified recovery equipment

   C. Both A and B

   D. Neither A nor B

2. Which of the following refrigerants may be vented?

   A. CFC-12   **C.** Both A and B

   B. HFC-134a   **D.** Neither A nor B

3. All of the following are removed from an air conditioning system during evacuation, EXCEPT:

   A. Air   **C.** Refrigerant

   B. Moisture   **D.** Lubricant

4. When retrofitting, the proper HFC-134a charge of refrigerant is about _____ of the original CFC-12 refrigerant charge.

   A. 95 percent   **C.** 85 percent

   B. 90 percent   **D.** 80 percent

5. The equipment shown in the above illustration may be used to:

   A. Evacuate and charge an air conditioning system

   B. Recover refrigerant from an air conditioning system

   C. Recycle refrigerant that has been recovered from an air conditioning system

   D. All of the above

Table 9-1 ASE TASK

Maintain and verify correct operation of certified equipment.

Identify and recover A/C system refrigerant.

Recycle or properly dispose of refrigerant.

Label and store refrigerant.

Test recycled refrigerant for noncondensable gases.

Problem Area	Symptoms	Possible Causes	Classroom Manual	Shop Manual
RECOVERY/ RECYCLE EQUIPMENT	Will not operate	1. Defective fuse or circuit breaker (main)		321
		2. Not connected to electrical service		321
		3. Defective fuse or circuit breaker (equipment)		321
		4. Improper use		321
	Erratic operation in recovery or recycle cycle	1. Improper hook-up		321
		2. Valve(s) closed		321
		3. Recovery tank full		321
		4. Defective pressure control		321
REFRIGERANT CONTAMI- NATION	Improper pressure while operating	1. Unknown type of refrigerant	195–199	147
		2. Air in system	89	111
		3. Moisture in system	84–85	249
		4. Excess lubricant in system	89	110–131
	Improper pressure not running	1. Unknown type refrigerant in system	195–199	147
		2. Air in system	89	111

JOB SHEET 30

30

Name _____ Date _____

Determining Refrigerant Purity in a Mobile Air Conditioning System by Verbal Communication

Upon completion of this job sheet you should be able to use good judgment regarding refrigerant purity.

ASE Correlation

This job sheet is related to the ASE Heating and Air Conditioning Systems Test's content area: *Refrigerant Recovery, Recycling, Handling, and Retrofit,* Task: *Identify, and recover A/C system refrigerant.*

Tools and Materials

Vehicle with air conditioning system charged with refrigerant

Describe the vehicle being worked on:

Year _____ Make _____ Model _____

VIN _____ Engine type and size _____

Procedure

Procedure

1. Question the customer:
   a. Has the vehicle air conditioner been serviced recently? _____
   b. When was it serviced? _____
   c. By whom? _____
   d. What type service was performed? _____
   e. What type refrigerant, if any, was used? _____
   f. What problems are you experiencing? _____

   Responses by the customer to the above questions help determine if the system should now be serviced. If there are any safety concerns, such as flammable refrigerant, DO NOT service the air conditioning system. Give a brief summary of your interpretation of the customers responses and tell why you:

2a. Decided to service the air conditioning system:

   _____

   _____

2b. Decided not to service the air conditioning system:

   _____

   _____

☑ **Instructor's Check** _____

JOB SHEET 31

31

Name _____ Date _____

Determining Refrigerant Purity in a Mobile Air Conditioning System by Testing

Upon completion of this job sheet you should be able to use a refrigerant tester to test for refrigerant purity.

ASE Correlation

This job sheet is related to the ASE Heating and Air Conditioning Systems Test's content area: *Refrigerant Recovery, Recycling, Handling, and Retrofit,* Task: *Identify, and recover A/C system refrigerant.*

Tools and Materials

Vehicle with air conditioning system charged with refrigerant
Refrigerant purity tester
Instruction manual

Describe the vehicle being worked on:

Year _____ Make _____ Model _____

VIN _____ Engine type and size _____

Procedure

1. Determine, following Job Sheet 30, if you wish to proceed with refrigerant testing. Briefly explain:

   _____

   _____

2. Following procedures included with the purity tester, connect the tester to the air conditioning system to draw a sample of refrigerant. Describe your procedure:

   _____

   _____

3. Was there an audible signal? _____ What would an audible signal indicate?

   _____

   _____

4. What is indicated on the readout? _____ What does this reading mean?

   _____

   _____

5. Based on the results of this test, what procedure will you use to recover the refrigerant?

_____

_____

_____

_____

☑ **Instructor's Check** _____

JOB SHEET 32

32

Name _____ Date _____

Identifying Retrofit Components

Upon completion of this job sheet you should be able to identify those components that must be replaced during retrofit procedures.

Tools and Materials

Vehicle with CFC-12 air conditioning system to be retrofitted for HFC-134a
Service manual
Factory-approved retrofit kit

Describe the vehicle being worked on:

Year _____ Make _____ Model _____

VIN _____ Engine type and size _____

Procedure

Write a short report about component replacement during retrofit from CFC-12 to HFC-134a procedures. Explain why or why not the following components should be replaced:

1. Receiver or Accumulator:

_____

_____

2. Hose or hoses:

_____

_____

3. Evaporator:

_____

_____

4. Condenser:

_____

_____

5. Pressure control switch:

_____

_____

6. Control thermostat:

_____

_____

7. Compressor:

_____

_____

8. Condenser fan and/or motor:

_____

_____

9. Evaporator blower and/or motor:

_____

_____

10. Other:

_____

_____

☑ **Instructor's Check** _____

JOB SHEET 33

33

Name _____ Date _____

CFC-12 to HFC-134a Retrofit

Upon completion of this job sheet you should be able to retrofit a CFC-12 air conditioning system to an HFC-134a air conditioning system.

Tools and Materials

Vehicle with CFC-12 air conditioning system to be retrofitted for HFC-134a
Factory-approved retrofit kit
CFC-12 refrigerant recovery equipment
HFC-134a refrigerant charging equipment
Hand tools, as required

Describe the vehicle being worked on:

Year _____ Make _____ Model _____
VIN _____ Engine type and size _____

Procedure

After each step, write a brief summary of your procedure:

1. Remove the CFC-12 refrigerant by recovering it for future use.

NOTE: DO NOT vent refrigerant to the atmosphere.

_____

_____

2. Remove and replace any defective air conditioning system components.

_____

_____

3. Remove as much of the mineral oil as possible.

_____

_____

4. Add and/or replace components as required in the retrofit procedures.

_____

_____

5. Add and/or replace lubricant as required in the retrofit procedures.

_____

_____

6. Install HFC-134a service valve fittings and label.

_____

_____

7. Evacuate the air conditioning system.

_____

_____

8. Leak test the air conditioning system.

_____

_____

9. Charge the air conditioning system with HFC-134a refrigerant.

_____

_____

10. Performance test the air conditioning system.

_____

_____

11. What problems, if any, were encountered?

_____

_____

☑ **Instructor's Check** _____

Diagnosis and Service of System Controls

Upon completion and review of this chapter you should be able to:

❑ Discuss the methods used to diagnose fuse and circuit breaker defects.

❑ Recognize and identify the components of the climate control system.

❑ Identify and service the different types of blower motors.

❑ Understand and practice the methods used to diagnose compressor clutch malfunctions.

❑ Identify and troubleshoot the different types of pressure- and temperature-actuated controls.

❑ Understand the function of, and be able to troubleshoot, the components of an automatic temperature control system.

Introduction

The control **system** of an automotive air conditioning system, at first, may seem to be very complex. And it is a complex system of many single wires. Compare the control system schematic to a road map. As one looks at a road map and notes the many highways and byways, it, too, looks complex. It is, however, but one route that is of interest at any one time. All the other routes are unimportant for any particular journey. The same is true, for the most part, when diagnosing any control system or **subsystem**: though the "map" may seem very complex, most of it will prove to be of no interest.

The schematic in Figure 10-1 is a composite of several car line schematics and, while it may be representative of any make or model automobile, it should not be considered typical for any specific make or model. For specific information, manufacturers' shop and service manuals must be consulted.

Refer to the schematics of this text as you are led through a systematic approach to diagnosis, troubleshooting, and repair procedures for today's modern automotive air conditioning system.

Basic Tools

Basic mechanic's tool set

Fender cover

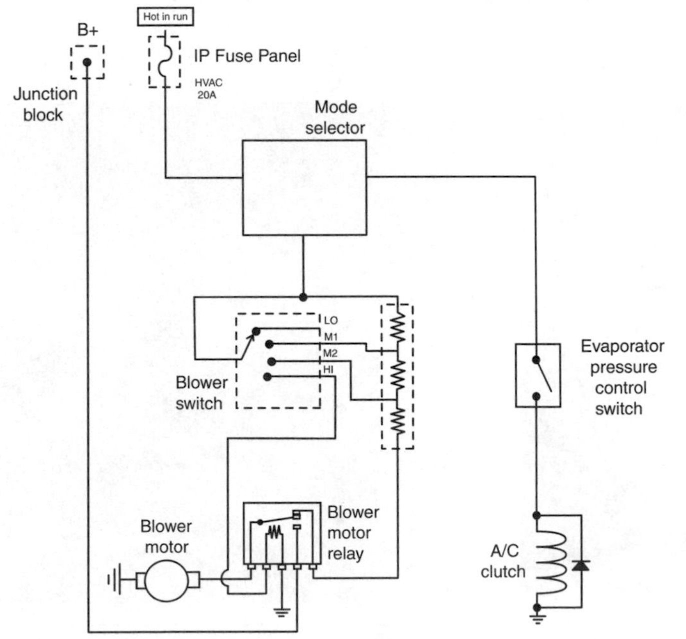

Figure 10-1 A typical automotive air conditioning system schematic.

Fuses and Circuit Breakers

An electrical schematic often requires several pages in a service manual.

Classroom Manual
Chapter 10, page 215

Take care not to short fuses or circuit breakers to ground when testing them.

Special Tools

Nonpowered test lamp

Voltmeter

Make sure that the test lamp is not burned out.

Special Tools

Powered test lamp

Ohmmeter

Note that there are several fuses, a circuit breaker, and a **fusible link** in the schematic. The purpose of these devices is to provide optimum protection of all of the circuits at all times. A fuse or fusible link is generally used in circuits that are **hot** all the time. That is, circuits that are not interrupted when the ignition switch is **open**. This provides a positive nonrestorable interruption of power should an **overload** occur when the vehicle is unattended. A circuit breaker, on the other hand, is generally used only in circuits that are interrupted when the ignition switch is open (off).

There are several methods that may be used to check a fuse or circuit breaker: in-vehicle testing with a **voltmeter** or nonpowered test lamp and out-of-vehicle testing with an **ohmmeter** or powered test lamp.

To test a fuse or circuit breaker in the vehicle, use a voltmeter or test lamp as follows:

1. Connect one lead of the test lamp or voltmeter to body ground (–).
2. Touch the other lead to the hot side of the fuse or circuit breaker in the fuse block or holder (Figure 10-2). If the lamp does not light or if voltage is not indicated, power is not available and the problem is elsewhere. If the lamp lights or if voltage is indicated, power is available. Proceed with step 3.
3. Touch the lead to the other side of the fuse or circuit breaker (Figure 10-3). If the lamp does not light or if voltage is not indicated, the fuse is blown or the circuit breaker is defective. Proceed with step 4. If the lamp lights or voltage is indicated, the problem is elsewhere and further testing is indicated.
4. Test protected components for **shorts** or overloads, then replace the fuse or circuit breaker.

To test a fuse or circuit breaker that has been removed from the vehicle, follow this procedure:

1. Set the ohmmeter in the 1× scale, touch the leads together, and zero the meter or make sure that the test lamp battery is good.
2. Touch the two leads of the ohmmeter or test lamp to either side of the fuse or circuit breaker (Figure 10-4). If the ohmmeter indicates a low resistance or if the test lamp lights, the fuse or circuit breaker is good. If there is no resistance indicated on the ohmmeter or if the test lamp does not light, the fuse is blown or the circuit breaker is defective.

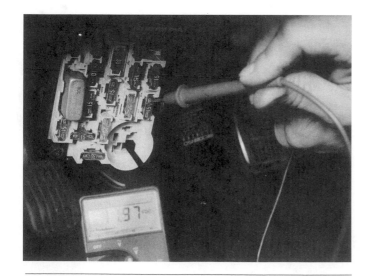

Figure 10-2 Touch the other lead to the hot side of the fuse block or holder.

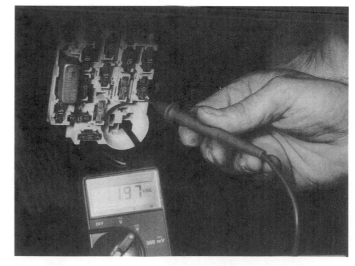

Figure 10-3 Touch the lead to the other side of the fuse.

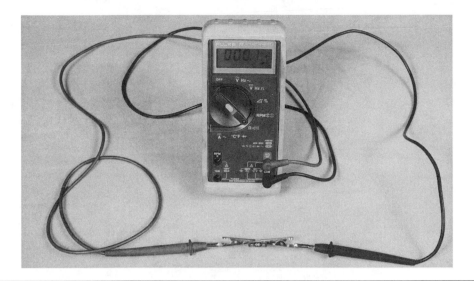

Figure 10-4 Touch the two leads to either side of the fuse.

Master Control Heads

The master control head (Figure 10-5) is found in the instrument panel where it is easily accessed by the driver or front-seat passengers. Some **dual systems** also have a control panel in the rear of the vehicle (Figure 10-6) for the comfort and convenience of the rear-seat passengers.

A minimicroprocessor is found in the control head to input temperature and humidity data selected by the operator to the programmer. Most electronic temperature control heads have provisions for self-testing, known as on board diagnostics (OBD). This system provides a number, letter, or alphanumeric code to provide the technician information relative to the problem. Manufacturers' charts must be consulted to interpret a particular code. For example Ford's "09" or "88," a no-trouble code, corresponds to General Motors' ".7,0." Another example, code "14," indicates "control head

Classroom Manual
Chapter 10,
pages 216

Readout of diagnostic codes is given in manufacturers' service manuals.

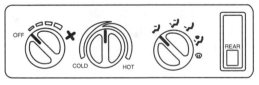

Heater control

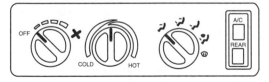

Manual A/C control

ATC control module

Figure 10-5 A typical master control head.

Figure 10-6 A typical rear control panel.

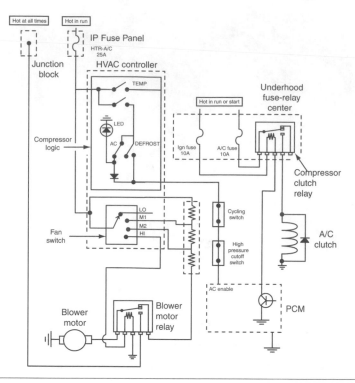

Figure 10-7 A typical electronic climate control (ECC) system schematic.

defective" in a Ford system while "36" indicates "ATC head communications failure" in a Chrysler system.

Note, too, that some electronic climate control (**ECC**) programmers (Figure 10-7) have an "ECC diagnostic connector" provision for the connection of an external read-out.

Thermostat

Classroom Manual
Chapter 10,
pages 217–220

Special Tools

Powered test lamp
Ohmmeter

Special Tools

Powered test lamp
Ohmmeter

The thermostat (Figure 10-8) cycles the air conditioning compressor electromagnetic clutch on and off as determined by a preset temperature. There are two types of thermostat: fixed and variable. Testing either type of thermostat is a relatively simple matter if it has been removed from the vehicle. Proceed as follows:

Variable-Type Thermostat

The variable-type thermostat is generally found on aftermarket air conditioning systems.

1. Connect an ohmmeter (1× scale) or powered test lamp to the two terminals of the thermostat (Figure 10-9).
2. While observing the ohmmeter or test lamp, rotate the thermostat from fully clockwise (cw) to fully counterclockwise (ccw). If a low resistance was noted or if the test lamp was lighted, the thermostat is probably all right. If no resistance was noted or if the test lamp did not light, the thermostat is defective.
3. Repeat step 2 several times to ensure stable and consistent results.

Fixed-Type Thermostat

A fixed-type thermostat that has no provisions for temperature adjustment is generally found on factory-installed air conditioning systems. Two beakers of water are required: one with ice (32°F or 0°C) and the other heated to about 120°F (49°C). The thermostat is tested as follows:

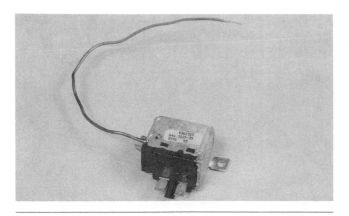

Figure 10-8 A typical thermostat.

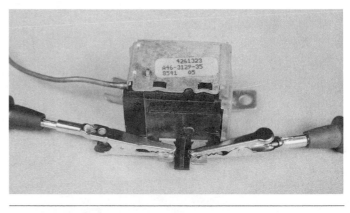

Figure 10-9 Connect an ohmmeter to the two terminals of a thermostat.

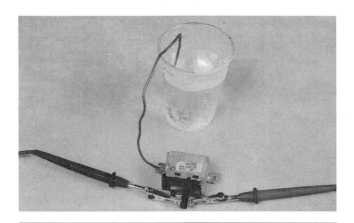

Figure 10-10 Immerse the cap tube into an ice bath.

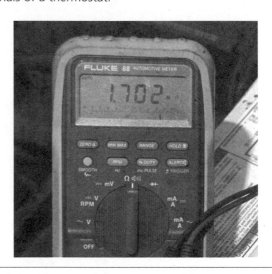

Figure 10-11 Did the resistance decrease?

1. Connect the ohmmeter or test lamp in the same manner as with the adjustable thermostat test.
2. Is low resistance noted or is the lamp lit? Generally, at ambient temperature, the thermostat will be closed. If yes, proceed with step 3. If no, proceed with step 5.
3. Immerse the capillary tube end or remote bulb into the ice bath (Figure 10-10).
4. Did the resistance increase or the lamp go out? A reduction in temperature below the set point should open the thermostat contacts. If yes, proceed with step 5. If no, the contacts are stuck closed and the thermostat is defective.
5. Immerse the capillary tube in the hot bath.
6. Did the resistance decrease (Figure 10-11) or the lamp light? If yes, the thermostat is probably all right. If no, the thermostat is probably defective with contacts stuck open.

Systems with a fixed thermostat usually maintain the desired in-car temperature by tempering cooled and heated air in the plenum section of the duct system.

Blower Motor

The blower **motor** often receives its power through a low blower **relay** or a high blower relay. To determine whether a blower motor is inoperative, it is necessary to bypass the other components in the control circuit. Follow the schematic (Figure 10-12) to troubleshoot a blower motor.

1. First, make certain the ground wire (1) has not been disconnected.
2. Disconnect the hot lead (2) at the blower motor.

Classroom Manual
Chapter 10,
pages 221–222

Special Tools

Jumper wire (fuse
protected)

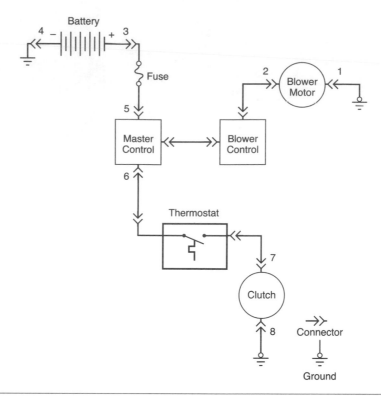

Figure 10-12 A typical blower motor schematic.

For protection against accidental shorts, the jumper wire should be protected with an in-line fuse.

Classroom Manual
Chapter 10, pages 222–224

3. Using a jumper wire, connect from the battery positive (+) terminal (3) to the blower motor. If the motor runs, it is all right and the problem is elsewhere. If the motor does not run, it is probably defective and must be replaced.

Electromagnetic Clutch

The electromagnetic clutch (Figure 10-13) starts compressor action when wanted and stops it when it is not wanted. The clutch either works or it does not work. It may be noisy when it works, a sign that it needs attention before it fails. If it does not work, the problem may be that it is burned, slipping, will not engage, or will not disengage.

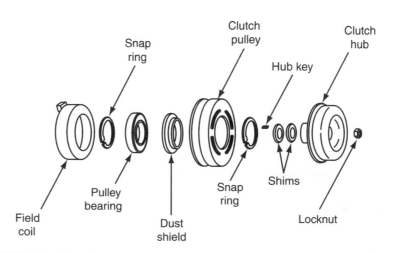

Figure 10-13 Details of a typical electromagnetic clutch.

Noisy

PROBLEM	REMEDY
1. Slipping belt	Tighten belt
2. Misaligned belt	Align belt
3. Clutch slipping	See "Clutch Will Not Engage"
4. Rotor/pulley snap ring missing	Replace snap ring
5. Rotor/pulley snap ring improperly installed	Properly install new snap ring
6. Rotor-to-armature **air gap** too small	Properly adjust air gap
7. Improper field coil snap ring	Install proper snap ring
8. Field coil snap ring installed improperly	Reinstall snap ring
9. Damaged bearing	Replace clutch assembly

The thickness of a snap ring is just as important as its diameter.

Burned Clutch

A burned clutch is often noted by charred paint, blued steel, melted bearing seals, broken springs, or a charred field coil. To prevent a recurrence, all of the problems leading to a burned clutch should be addressed before replacing it. These problems and the recommended remedies are:

PROBLEM	REMEDY
1. Compressor shaft seal leak	Replace shaft seal
2. Compressor thru bolt leak	Repair as required
3. Oil leak: engine, power steering, transmission	Repair or replace as required
4. Missing rotor/pulley snap ring	Make sure snap ring is installed
5. Improperly installed rotor/pulley snap ring	Make sure snap ring is properly installed
6. Improper field coil snap ring	Make sure snap ring is proper
7. Improperly installed field coil snap ring	Make sure field coil snap ring is installed properly
8. Mismatched components	Replace with matched components

Clutch Will Not Engage

PROBLEM	REMEDY
1. Excessive air gap	Adjust air gap
2. Poor electrical connection(s)	Repair as required
3. Undersized wiring	Use minimum 18-gauge wire
4. Damaged wiring	Repair as required
5. Defective clutch relay	Replace relay
6. Electrical component failure	Replace defective component
7. Shorted field coil	Replace field coil
8. Open field coil	Replace field coil

Many component malfunctions are due to a defective ground connection.

Clutch Will Not Disengage

PROBLEM	REMEDY
1. Improper air gap	Adjust air gap
2. Rotor/pulley snap ring not installed	Install snap ring
3. Rotor/pulley snap ring improperly installed	Reinstall snap ring
4. Electrical problem	Correct problem as required

Classroom Manual
Chapter 10,
page 225

A pressure cycling switch may open (close) on low or high pressure depending on application.

When replacing a pressure switch, make certain that the new replacement is the same pressure range as the old defective one.

Classroom Manual
Chapter 10,
page 227

Pressure Switches and Controls

There are many types of pressure switch controls used in the automotive air conditioning system. These are the low-pressure cutoff switch, high-pressure cutoff switch, compressor discharge pressure switch, and pressure cycling switch.

Some pressure switches (Figure 10-14) can be replaced without having to remove the refrigerant from the system. Others, on the other hand, require that the refrigerant be recovered before the old switch can be removed. Those that do not require refrigerant removal have a valve depressor located inside the threaded end of the pressure switch (Figure 10-15). This pin presses on the Schrader-type valve stem as the switch is screwed on and allows system pressure to be expressed on the switch.

Replacement is rather simple. Remove the old pressure switch and install a new one. A word of caution, however. If it is not known if the switch is equipped with a Schrader-type service port, the refrigerant must first be removed from the system.

At atmospheric pressure, a low- or high-pressure switch should be normally closed (nc). If there is a low-pressure switch, a vacuum pump may be used to see at what low pressure (if any) it opens. Similarly, a nitrogen source may be substituted for the vacuum pump to test the operation of a high-pressure switch. A test lamp is used to determine if and when the pressure switch opens and/or closes.

Coolant Temperature Warning Switches

There are two types of coolant temperature warning systems: telltale lamp and gauge system. Testing the telltale lamp system is rather straightforward. If the lamp(s) is/are good and the wiring is sound, an inoperative system is generally due to a defective sending unit. It is a relatively simple matter to substitute a new sending unit if a defective unit is suspected. The gauge system requires the use of a tester, however. A diagnostic chart of a typical temperature gauge/sending unit test is shown in Figure 10-16.

Photo Sequence 17
Typical Procedure for Testing a Temperature or Pressure Switch

Note: Internal resistance of a pressure transducer varies by design and manufacturers specifications must be referenced to determine proper values based on powertrain control module (PCM) requirements, generally 0.15–4.85 volts with a 5.0-volt input.

P17-1 Typical location of a thermal switch.

P17-2 Typical location of a pressure transducer.

P17-3 Typical location of a low-pressure switch.

Photo Sequence 17
Typical Procedure for Testing a
Temperature or Pressure Switch (continued)

P17-4 Typical location of a low-pressure clutch cycling switch used for temperature control.

P17-5 Disconnect the wires from the switch to be tested.

P17-6 Connect a digital volt ohmmeter to the switch terminals.

P17-7 The resistance of a normally closed (nc) switch should be 0 ohms (Ω).

P17-8 If the resistance is higher than 0 Ω the switch is defective.

P17-9 The resistance of a normally open (no) switch should be $\infty\Omega$.

P17-10 Low resistance indicates a defective switch (Note 1).

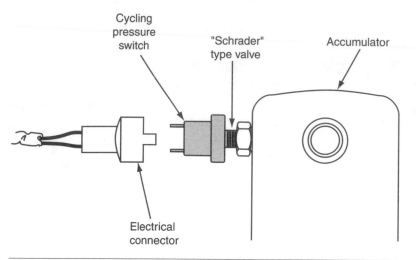

Figure 10-14 Some pressure switches can be replaced without removing the refrigerant.

Figure 10-15 A valve depressor located inside the threaded portion of the pressure switch.

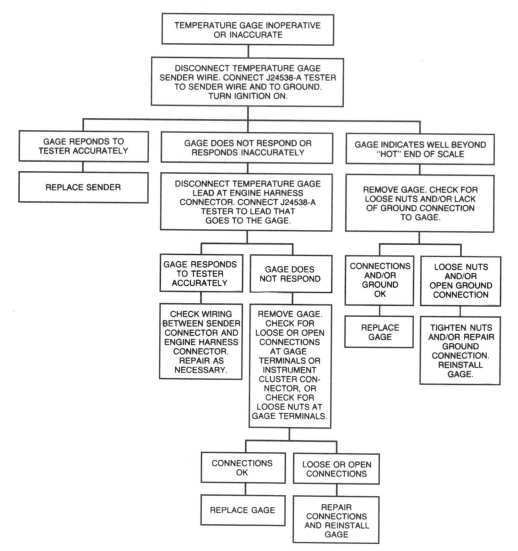

Figure 10-16 Diagnostic chart for a typical temperature gauge/sending unit test. (Courtesy of General Motors Corporation, service operations)

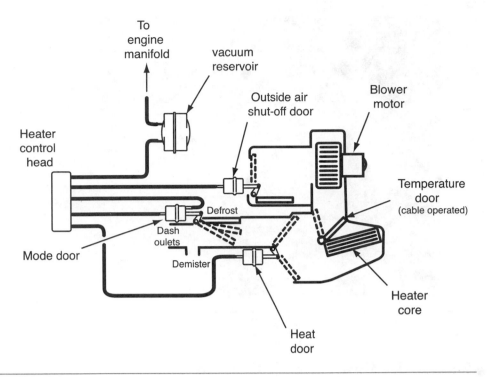

Heater control head

To engine manifold

vacuum reservoir

Outside air shut-off door

Blower motor

Temperature door (cable operated)

Defrost

Dash oulets

Mode door

Demister

Heater core

Heat door

Figure 10-17 A typical vacuum control schematic.

Vacuum Switches and Controls

Vacuum-controlled vacuum actuators, often called motors, are used to position the A/C-defog valve, up-mode valve, down-mode valve, and the inside air valve (Figure 10-17).

Scan Tool

A **scan tool** (Figure 10-18) is a microprocessor that is designed to communicate with the vehicle's computer. When connected to the computer, through diagnostic connectors, a scan tool accesses diagnostic trouble codes (DTCs), runs tests to check system operations, and monitors the activity of the system. Both trouble codes and test results are displayed on a light-emitting diode (LED) screen or are printed out on the scanner printer.

A scan tool receives its information from several sources. Some scan tools use cartridges containing programmable read-only memory (PROM) chips. Each chip contains all of the information needed to diagnose problems in a specific model line, and the appropriate cartridge for whatever vehicle is being worked on is inserted into the tool accordingly.

LED displays are generally only large enough to display four short lines of information, limiting the technician's ability to compare test data. Most scan tools, however, store the test data in a random access memory (RAM) that can be accessed by a printer, personal computer, or an engine analyzer to retrieve the information.

Trouble codes set by the computer help the technician identify the cause of the problem. Most diagnostic work on computer control systems should be based on a description of symptoms to help locate any technical service bulletins that refer to the problem. One can also use the symptom description to locate the appropriate troubleshooting sequence in the manufacturer's service manuals.

Classroom Manual Chapter 10, pages 228–233

The average person is comfortable at 78–80°F (25.6–26.7°C) at a relative humidity (rh) of 45–50 percent.

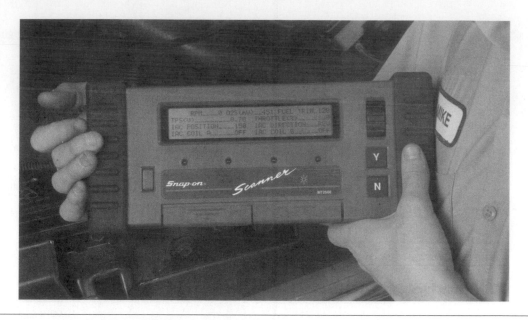

Figure 10-18 A typical handheld scan tool.

Sensor Testing With a Scan Tool

Most scan tools will display the voltage values or switch position of many sensors. Access to this information differs depending on the scan tool used. For example, when using the DRB-II (Figure 10-19), if the tool display reads BODY COMP MENU, select STATE DISPLAY. The display will change to BODY COMP STATE. By selecting SENSORS, the value of selected sensors can be

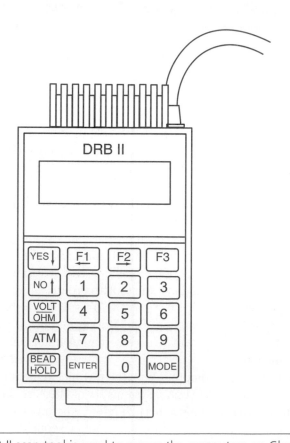

Figure 10-19 The DRB-II scan tool is used to access the computers on Chrysler vehicles.

viewed. If the technician is interested in a switch position, select INPUTS/OUTPUTS and the display will indicate the various positions of various switches used as inputs to the computer.

Breakout State

A **breakout box** (Figure 10-20) is a device that, when connected between the module and the wiring harness, allows the technician to "see" the exact information the computer is receiving and sending.

The breakout box taps directly into the sensor or actuator circuit providing the technician with the exact voltage signal being sent or received. A breakout box connected into the system, then, allows a digital multimeter (DMM) to be used to measure the voltage signals and resistance values of the circuit (Figure 10-21). The diagnostic manual provided with the breakout box should

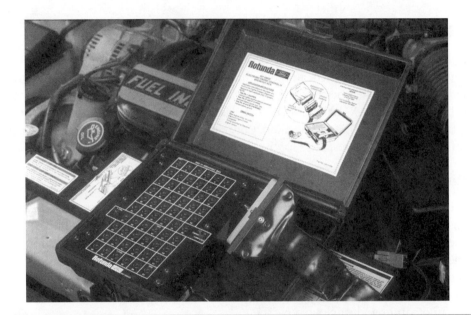

Figure 10-20 A breakout box provides test points for voltmeter and ohmmeter connections.

Figure 10-21 Using the breakout box to test a circuit.

be used as a guide through a series of test procedures. Comparing the test results with specifications will lead to the problem area.

Automatic (Electronic) Temperature Controls

Comfort of those in the passenger compartment is maintained by mixing cooled and ambient or heated air in the plenum section of the heater-air conditioning duct system. In the **AUTO** mode, the operator sets the desired comfort level, often humidity as well as temperature. Both the quality as well as the quantity of air delivered to the passenger compartment is then controlled automatically. The blower speed and air delivery can, however, be manually controlled if desired.

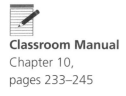

Classroom Manual
Chapter 10,
pages 233–245

Control Panel

The control panel in Figure 10-22 is typical for an automatic temperature control (ATC) system. The control panel may be used to select a predetermined temperature level that will be automatically maintained at all times. The operator, if desired, can override the automatic provisions of the control head by selecting MAX A/C or MAX heat. Manual modes are also available for selecting BILEVEL, DEF, VENT, and DEFOG operation. Check the heater and A/C function test chart (Figure 10-23) for the proper system response for the various control settings.

Programmer

The programmer is generally identified by the electrical and vacuum lines attached to it. It primarily contains a small circuit board that controls a small reversible dc motor that adjusts the air mix valve to blend cold and warm air. It also contains four vacuum **solenoids** that control the various vacuum mode actuators. The programmer also provides data for blower speed selection and operation to provide selected in-vehicle temperature conditions.

Typically, to remove the programmer (Figure 10-24), gain adequate access by removing the right-side sound barrier and/or glove box. Then:

1. Remove the brackets and/or covers to gain access to the programmer.
2. Remove the threaded rod from the programmer.
3. Remove the vacuum connector retaining nut.
4. Remove the vacuum and electrical connectors from the programmer.
5. Remove the programmer from the vehicle.
6. To replace, reverse the preceding procedure.

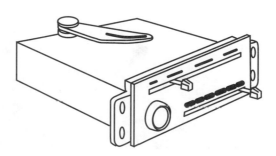

Figure 10-22 A typical automatic temperature control panel.

STEP	CONTROL SETTINGS			SYSTEM RESPONSE				
	MODE CONTROL	TEMP CONTROL	FAN CONTROL	BLOWER SPEED	HEATER OUTLET	A/C OUTLETS	DEFROSTER OUTLETS	SEE REMARK
1	OFF	60	DOES NOT FUNCTION	OFF	NO AIR FLOW	NO AIR FLOW	NO AIR FLOW	A
2	AUTO	60	LO	LOW	MINIMUM AIR FLOW	AIR FLOW	NO AIR FLOW	A
3	AUTO	60	LO TO HI	LO TO HI	SMALL AIR FLOW	AIR FLOW	NO AIR FLOW	D
4	BI–LEVEL	60	LO TO HI	LO TO HI	AIR FLOW	AIR FLOW	SMALL AIR FLOW	A,D
5	AUTO	90	HI	HIGH	AIR FLOW	NO AIR FLOW	SMALL AIR FLOW	A,B,C,D
6	DEF FRT	90	HI	HIGH	SMALL AIR FLOW	NO AIR FLOW	AIR FLOW	A,D
7	DEF REAR	DOES NOT CHANGE SYSTEM RESPONSE						A

REMARKS:

A. THE WORD 'AUTO' MUST APPEAR IN L/H UPPER CORNER OF DISPLAY WHEN IN AUTOMATIC MODE. MODE ARROWS IN DISPLAY MUST INDICATE FLOW FROM APPROPRIATE OUTLET (EXAMPLE: EITHER HEATER, A/C OR BOTH).
LED ANNUNCIATORS MUST LIGHT ABOVE FIXED MODE BUTTONS WHEN MODE IS SELECTED.

B. LISTEN FOR REDUCTION OF AIR NOISE DUE TO RECIRCULATION VALVE (DOOR) CLOSING.

C. DURING THE TRANSITION OF THE AIR FLOW BETWEEN THE A/C OUTLETS AND THE HEATER OUTLET, THERE MAY BE A BRIEF PERIOD WHEN APPROXIMATELY ONE HALF OF THE AIR WILL BE DIRECTED OUT THE DEFROSTER OUTLETS.

D. CHECK FOR AIR FLOW AT SIDE WINDOW DEFOGGER OUTLETS (AIR FLOW IN ANY MODE POSITION EXCEPT OFF).

Figure 10-23 A typical heater and air conditioner function test chart. (Courtesy of General Motors Corporation, service operations)

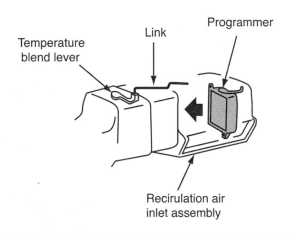

Figure 10-24 A typical programmer.

Blower Control

For any given ECC signal, the blower control module (**BCM**) has a predetermined blower motor voltage value. A signal from the BCM to the programmer causes a variable voltage signal to be sent to the power module.

Variable Resistor Test. (Figure 10-25) The control assembly variable resistor, also referred to as a **sliding resistor** or blower speed control, can be tested in or out of the vehicle by the following procedure:

1. Disconnect the electrical connector to the variable resistor.

2. Connect an ohmmeter's test leads across the terminals.

3. Set the comfort control lever to the 65 selection and note the resistance. The resistance in this position should be less than 390 ohms (390 Ω).

4. Slowly move the comfort control lever toward the right while observing the ohmmeter. When the ohmmeter indicates 930 Ω, the lever should be near the 75 setting.

   NOTE: The ohmmeter should have indicated a smooth increase in resistance.

5. Move the lever to the 85 setting. The resistance value should increase smoothly to at least 1,500 Ω.

If the resistance values are not within these specifications or the increase in resistance is not smooth, the control assembly must be replaced.

Classroom Manual
Chapter 10, page 239

The compressor clutch is used to turn the compressor on and off.

Power Module

The power module (Figure 10-26) controls the speed and operation of the evaporator blower motor. The power module receives a weak voltage signal from the programmer. It amplifies this weak signal to a stronger signal in proportion to its input. Since the power module is transistorized, it provides infinitely variable blower speeds. In all but high speed, the power module generates a great amount of heat. To carry this heat away, a large heat sink extends into the air stream of the evaporator.

Clutch Control

The compressor clutch is controlled by the powertrain control module (**PCM**) from inputs to the PCM, such as engine coolant temperature as well as rpm, and to the body computer module (**BCM**), such as outside temperature, in-car temperature, sun load temperature, and air conditioning system high- and low-side temperatures.

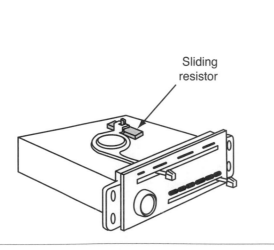

Figure 10-25 Testing the control assembly sliding resistor. Resistor values should be within specifications and change smoothly.

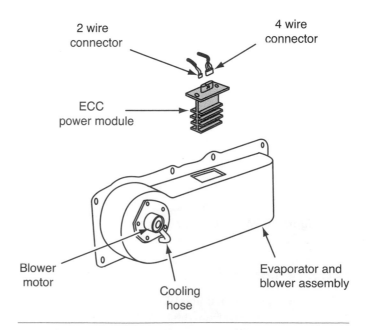

Figure 10-26 A typical power module.

Photo Sequence 18
Procedure for Testing PCM-Controlled Air Conditioners

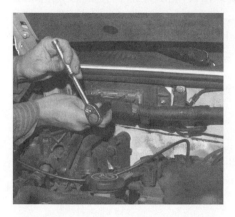

P18-1 Locate the powertrain control module (PCM) and gain access to its wiring harness. Disconnect the PCM.

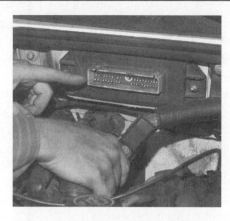

P18-2 Check for a poor connection at the (PCM).

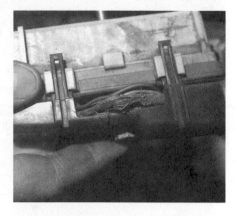

P18-3 Inspect the wiring harness for damage.

P18-4 Connect a digital voltmeter to the relay driver circuit at the PCM harness connector.

P18-5 Turn the ignition switch to ON. Do not start the engine.

P18-6 Observe the voltmeter while moving connectors and wiring harness relating to the relay.

P18-7 Note any change in voltage while moving the relay driver wiring harness indicatng a wiring harness fault.

Clutch Diode

A diode may be thought of as an electrical check valve; it provides current flow in one direction and blocks current flow in the opposite direction.

The clutch diode (Figure 10-27) is found connected across the electromagnetic clutch coil in many systems. Its purpose is to prevent unwanted electrical **spikes** that could damage delicate minicomputer electrical systems and subsystems as the clutch is engaged and disengaged.

Testing the Diode. A diode is thought of as an electrical check valve; it has high resistance to the flow of electricity in one direction and low resistance in the other direction. An ohmmeter may be used to check a diode. Proceed as follows:

1. Remove the diode to be tested from the circuit. Observe the polarity for proper replacement.
2. Connect the ohmmeter leads to the diode leads.
3. Note the reading: resistance is high/low.
4. Reverse the ohmmeter leads as connected in step 2.
5. Note the reading resistance is now high/low.
6. Disconnect the ohmmeter.
7. Compare the readings of steps 3 and 5.

> ✓ **SERVICE TIP:** The diode is good if the resistance reading in step 3 is high and in step 5 is low or if step 3 is low and step 5 is high. The diode is defective if the resistance readings in steps 3 and 5 are the same, or nearly the same.

8. Return the diode to service or replace it, as required. Note the proper polarity as in step 1.

Low– and High–Pressure Controls

![notebook icon]

Classroom Manual
Chapter 10, page 225

Low- and high-side pressure controls are provided for system protection as well as for safety. A high-pressure control, which is normally closed (nc), is designed to open and stop compressor action before the system pressure can rise to a point that could cause damage, such as rupturing a hose or component. This pressure, which is generally 350–400 psig (2,413–2,758 kPa), is considered adequate protection for the average air conditioning system.

A low-pressure switch (Figure 10-28), often referred to as a **low-refrigerant switch**, is located in the low-pressure side of the system. Though generally found on the accumulator, it could be anywhere from the evaporator inlet to the compressor inlet. This normally closed (nc) switch is generally electrically in line with the compressor clutch coil or the clutch relay. It is designed to open if the system pressure should drop to about 10 psig (690 kPa) or below.

An electrical circuit to the BCM monitors the condition of the low-pressure switch as well as high-pressure switches as applicable.

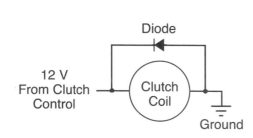

Figure 10-27 A clutch diode prevents electrical spikes.

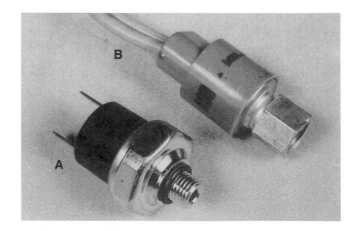

Figure 10-28 Typical low-pressure switches with (A) male threads and (B) female threads.

Evaporator Thermistor

A **thermistor** is actually a resistor that changes value depending on its temperature. It is ideal for use in solid-state temperature control applications, such as electronic automatic temperature control (**EATC**) systems. The thermistor (Figure 10-29) is covered in more detail under the following heading "Sensors."

Sensors

A sensor is a general name given to a transducer, which is short for "transfer inductor." In automotive terms, a sensor is a device that is capable of sensing a change in pressure, temperature, or other controlled variables.

Sun load Sensor. The **sun load sensor** (Figure 10-30) is generally found under the defrost grille at about the center of the windshield. It is a thermistor that is sensitive to the heat load of the sun on the vehicle.

The BCM compares the sun load values with in-car temperature values to determine how much cooling is required in order to maintain selected in-vehicle temperature conditions.

Ambient Temperature Sensor. The **ambient temperature sensor** (ATS), also referred to as the **outside temperature sensor** (OTS), is a thermistor found in a protective housing just behind the radiator grille (Figure 10-31). Data from the ATS is processed by the BCM and displayed on the ECC. Through a rather complicated process, the ATS provides information regarding outside ambient temperature that is essential for the proper operation of an electronic automatic temperature control (EATC) system.

To test the ATS, first remove it from its socket and then measure its resistance using an ohmmeter. At an ambient temperature between 70–80°F (21–27°C), the sensor resistance should be between 225–235 Ω. If the resistance is not within this range, the ambient sensor is defective and must be replaced.

NOTE: Because ohmmeter battery current flow through the sensor and body heat will affect the readings, do not hold the sensor in your hand or leave the ohmmeter connected for longer than 5 seconds (Figure 10-32).

In-Car Temperature Sensor. The **in-car temperature sensor**, a thermistor, is located inside an aspirator. To provide an accurate temperature reading, a small sample of air is drawn through the aspirator across the in-car temperature sensor.

As a thermistor's temperature increases, its resistance decreases.

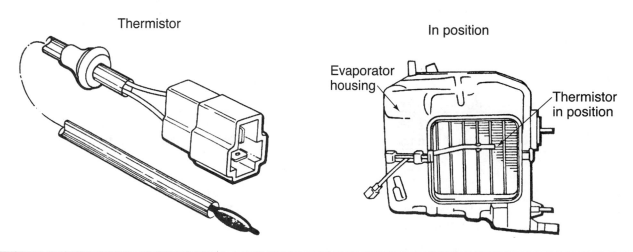

Thermistor

In position

Evaporator housing

Thermistor in position

Figure 10-29 A thermistor (A); in position (B). (From Green/Green/Dwiggins, *Australian Automotive Air Conditioning*, 1E © 1993 by Thomas Nelson Holdings, Ltd.)

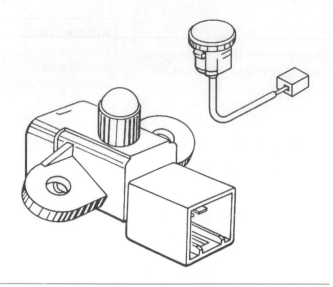

Figure 10-30 A typical sun load sensor. (From Green/Green/Dwiggins, *Australian Automotive Air Conditioning*, 1E © 1993 by Thomas Nelson Holdings, Ltd.)

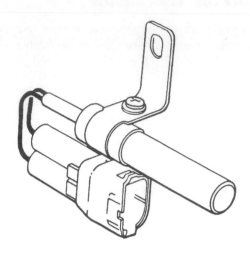

Figure 10-31 The outside temperature sensor is located behind the grille. (From Green/Green/Dwiggins, *Australian Automotive Air Conditioning*, 1E © 1993 by Thomas Nelson Holdings, Ltd.)

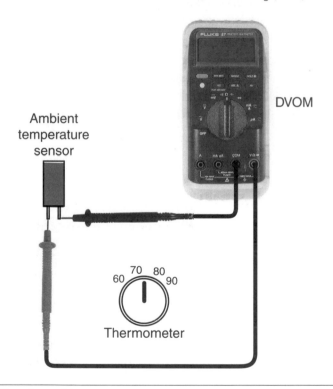

Figure 10-32 Test the ambient temperature sensor at room temperature. Avoid touching the sensor during testing. Also, do NOT connect the ohmmeter for longer than 5 seconds.

The resistive value of the in-car temperature sensor is sent to the BCM and is used by the ECC for calculations to maintain the preselected in-vehicle temperature conditions.

The following procedure may be used to test the in-car temperature sensor:

1. Disconnect its electrical connector. Do not disconnect the aspirator tubes or remove the temperature sensor from the panel.
2. Place a test thermometer into the air inlet grill near the sensor.

3. Set the blower motor speed control to MED.
4. Depress then pull out the NM-A/C button. This will turn off the compressor and close the water valve.
5. Operate the blower while quickly measuring the resistance of the sensor.

NOTE: Do not leave the ohmmeter connected to the sensor terminals for longer than 5 seconds or inaccurate readings will result. Ohmmeter battery current flow through the sensor, a thermistor, and body heat will affect resistance.

Resistance of the in-car temperature sensor should be 1,100–1,800 Ω at an ambient thermometer between 70–80°F (21–27°C). The sensor must be replaced if the resistance is not within these specifications.

Coolant Temperature Sensor.   The coolant temperature sensor (CTS) is also a thermistor. It is located in a passage of the coolant intake manifold. The resistance of the CTS, monitored by the PCM, sends a signal representing the coolant temperature value to the BCM.

Aspirator

The **aspirator** is an assembly device that houses the in-car temperature sensor (Figure 10-33). A quick method for testing the aspirator assembly to verify that it is providing enough air flow to the in-car temperature sensor is to set the controls for HI blower speed while in the heat mode of operation. Place a piece of paper, large enough to cover the aspirator air inlet, over the inlet (Figure 10-34). The suction of the aspirator should be great enough to hold the paper against the inlet grill. If not, refer to the aspirator system diagnostic chart in Figure 10-35.

To aspirate is to draw by suction.

Heater Flow Control Valve.   The heater flow control valve (Figure 10-36) is opened or closed by a signal from the ECM to provide in-vehicle temperature control. If the valve is found to be defective, it must be replaced. Photo Sequence 19 illustrates a typical procedure for replacing a heater flow control valve.

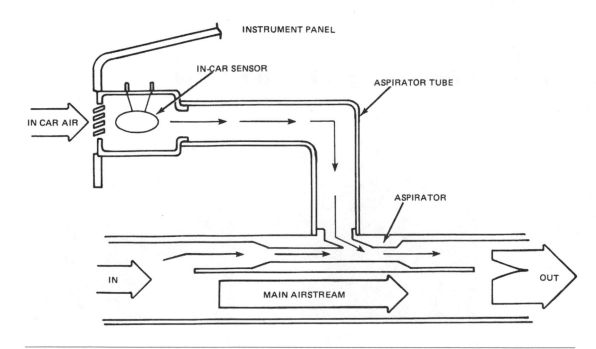

Figure 10-33 A typical aspirator assembly.

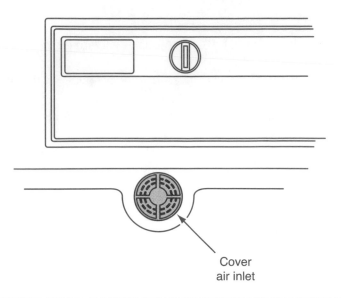

Cover
air inlet

Figure 10-34 When performing the aspirator paper test, the vacuum should hold the paper against the grill.

Classroom Manual
Chapter 11, page 267

Actuators

An actuator is a device that transforms a vacuum or electrical signal to a mechanical motion. It is the component that performs the actual work commanded by the computer. An actuator may be an electric or vacuum motor, relay, switch, or solenoid that typically performs an on/off, open/close, or push/pull operation.

Testing Actuators.   Most systems allow for testing of the actuator through the scan tool or FCC panel while in the correct mode. Actuators that are duty cycled by the computer are more accurately diagnosed through this method. In the earlier example of retrieving trouble codes from the Chrysler system using the DRB-II scan tool, select ACTUATOR TESTS. This will allow activation of selected actuators to test their operation.

Photo Sequence 19
Typical Procedure for Replacing a Heater Flow Control Valve

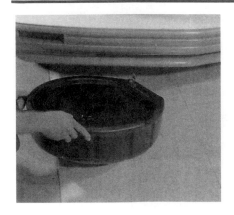

P19-1 Drain the coolant from the system and recycle it.

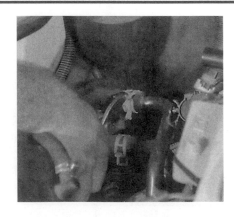

P19-2 Loosen and remove the inlet hose clamp at the heater core. Slip the heater hoses off their fittings.

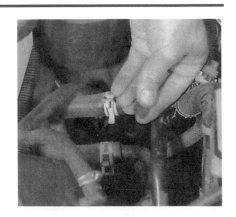

P19-3 Loosen and remove the bolt or nut that holds the inlet pipe bracket in place.

Photo Sequence 19
Typical Procedure for Replacing a Heater Flow Control Valve (continued)

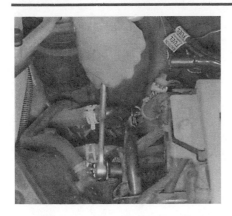

P19-4 Loosen and remove the retaining bolt for the bracket of the heater control valve.

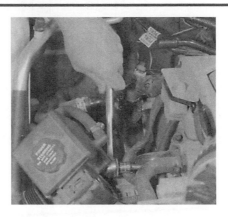

P19-5 Remove all other heating control valve retaining bolts and any parts that may interfere with the removal of the control valve.

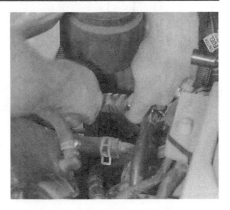

P19-6 Remove the control valve and inspect the hoses connected to it.

P19-7 Clean the coolant pipes and hoses. Make sure all damaged parts are replaced. Then replace the heater control valve.

P19-8 Install and tighten all heater control valve and valve bracket retaining nuts and bolts.

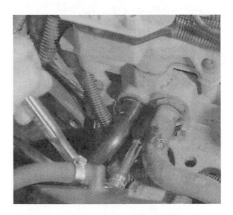

P19-9 Install and tighten the retaining nut or bolt for the inlet pipe.

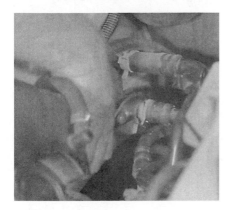

P19-10 Install new clamps and connect the heater hoses to the heater core.

P19-11 Fill the cooling system with fresh coolant and to the correct level. Bleed the system, if necessary.

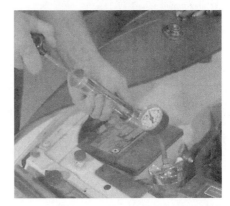

P19-12 Pressure test the system and check for leaks. Then run the engine and allow it to reach normal operating temperature. Then shut if off and retest for leaks.

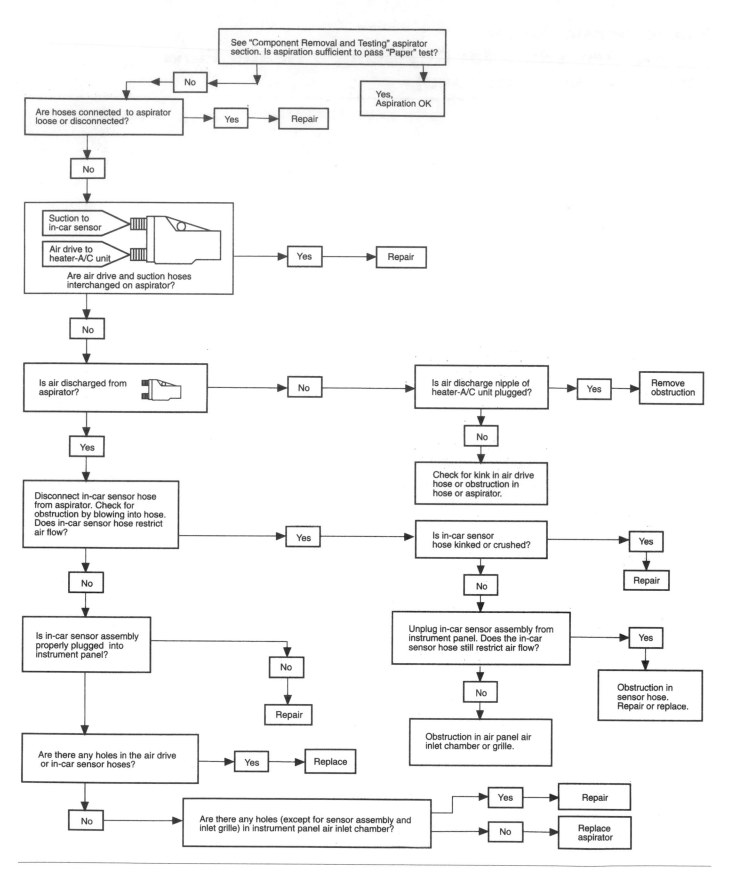

Figure 10-35 Aspirator diagnostic test chart. (Courtesy of Chrysler Corporation)

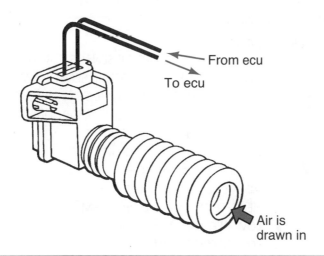

Figure 10-36 A typical heater flow control valve.

Servo Motors

A **servo motor** is a vacuum or electric motor that is used to control the position of the mode and blend air doors in an automotive heating and air conditioning case/duct system.

Servo Motor Test.   The servo motor must be removed from the vehicle for testing. Follow the procedures outlined in the service manual for removing the motor. While operating the motor, check for smooth operation and observe the test light. If the motor briefly jams, the test light illumination level will increase. If the test light flickers while the motor is operating, the motor is not moving in a smooth fashion. The motor must move to the full clockwise (cw) and full counterclockwise (ccw) position. If the cause of a problem cannot be corrected, the servo motor will have to be replaced.

Trouble Codes

Most body control modules (BCMs) are capable of displaying the fault codes that were stored in memory. The procedure used to retrieve the codes varies greatly and reference must be made to the appropriate service manual for the correct procedure.

Only some systems retain the code when the ignition is turned off. Those that do not require test driving the vehicle in an attempt to duplicate the fault. Once the fault is detected by the computer, the code must be retrieved before the ignition switch is turned off. The trouble code, however, does not necessarily indicate the faulty component. It only indicates that circuit of the system that is not operating properly (Figure 10-37). For example, the code displayed may be F11, indicating an air conditioning system high-side temperature sensor problem. This does not mean, however, that the sensor is defective. It means that the fault is in that circuit, which includes the wiring, connections, and BCM as well as the sensor. To locate the problem, follow the diagnostic procedure in the service manual for the code received (Figure 10-38). There are two types of code that can be displayed: intermittent code and hard fault code.

Hard and Intemittent Codes

Some BCMs store trouble codes in their memory until they are erased by the technician or until a predetermined number of engine starts have occurred. Usually, the first set of fault codes to be displayed represent all of the fault codes that are stored in memory, including both hard and intermittent codes.

BCM DIAGNOSTIC TROUBLE CODES

CODE	NOTES	PROBLEM
F10	1	OUTSIDE TERMPERATURE SENSOR CIRCUIT
F11	1 - 2	A/C HIGH SIDE TEMPERATURE SENSOR CIRCUIT
F12	1 - 3	A/C LOW SIDE TEMPERATURE SENSOR CIRCUIT
F13	1	IN-CAR TEMPERATURE SENSOR CIRCUIT
F30	1	CCP TO BCM DATA CIRCUIT
F31	1	FDC TO BCM DATA CIRCUIT
F32	1 - 4	ECM TO BCM DATA CIRCUIT
F40	1	AIR MIX DOOR PROBLEM
F43	1	HEATED WINDSHIELD PROBLEM
F46	2	LOW REFRIGERANT WARNING
F47	2 - 5	LOW REFRIGERANT PROBLEM
F48	2 - 5	LOW REFRIGERANT PRESSURE
F49	1	HIGH TEMPERATURE CLUTCH DISENGAGE
F51	1	BCM PROM ERROR

NOTES: 1 Does not turn on any light
2 Turns on SERVICE AIR COND light
3 Disengages A/C clutch
4 Turns on cooling fans
5 Switches from AUTO to ECON

Figure 10-37 Body Control Module (BCM) diagnostic trouble codes lead the technician to the problem. (Courtesy BET, Inc.)

The second set of fault codes to be displayed are only hard codes. The codes that are displayed in the first set but not displayed in the second set are intermittent codes.

Most diagnostic charts cannot be used to locate intermittent faults. This is because the testing at various points of the chart requires that the fault be present to locate the problem. Intermittent problems are often caused by poor electrical connections. Diagnosis, then, should start with a good visual inspection of the connectors, especially those involved with the trouble code.

Visual Inspection.   One of the most important checks to be made before diagnosing a BCM-controlled system is a complete visual inspection. The inspection can identify faults that could otherwise waste time in unnecessary diagnostics. Inspect the following:

1. Sensors and actuators for physical damage
2. Electrical connections to actuators, control modules, and sensors
3. All ground connections
4. Wiring for signs of broken or pinched wires or burned or chaffed spots indicating contact with sharp edges or hot exhaust manifolds
5. Vacuum hoses for breaks, cuts, disconnects, or pinches

NOTE: Check wires and hoses that are hidden under other components.

Entering BCM Diagnostics

There are perhaps as many methods of entering BCM diagnostics as there are vehicle makes and models. One thing that most have in common, however, is that a scan tool be plugged into the diagnostic connector for the system being tested. Always refer to the correct service manual for the vehicle being serviced and use only the methods identified for retrieving trouble codes. Once

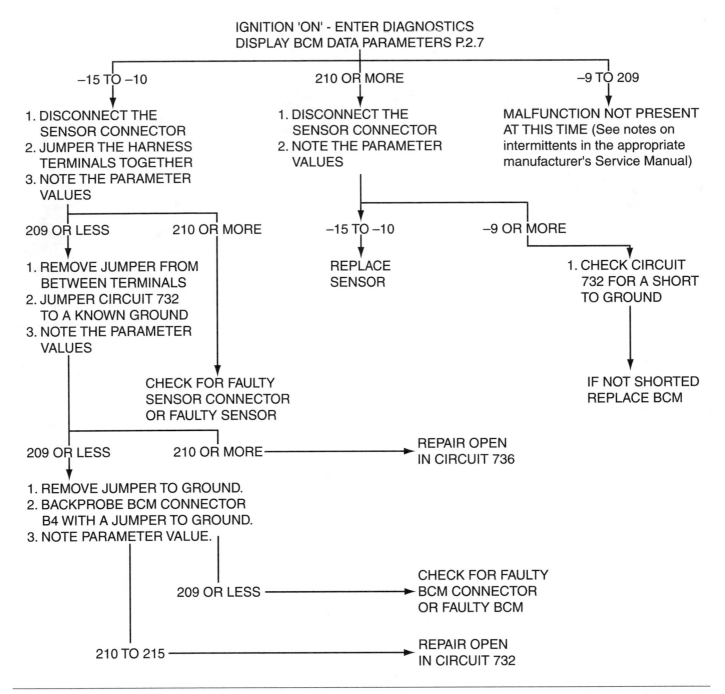

IGNITION 'ON' - ENTER DIAGNOSTICS
DISPLAY BCM DATA PARAMETERS P.2.7

−15 TO −10

1. DISCONNECT THE
   SENSOR CONNECTOR
2. JUMPER THE HARNESS
   TERMINALS TOGETHER
3. NOTE THE PARAMETER
   VALUES

209 OR LESS   **210 OR MORE**

1. REMOVE JUMPER FROM
   BETWEEN TERMINALS
2. JUMPER CIRCUIT 732
   TO A KNOWN GROUND
3. NOTE THE PARAMETER
   VALUES

CHECK FOR FAULTY
SENSOR CONNECTOR
OR FAULTY SENSOR

209 OR LESS   **210 OR MORE** ──────→ REPAIR OPEN
                                            IN CIRCUIT 736

1. REMOVE JUMPER TO GROUND.
2. BACKPROBE BCM CONNECTOR
   B4 WITH A JUMPER TO GROUND.
3. NOTE PARAMETER VALUE.

209 OR LESS ──────→ CHECK FOR FAULTY
                         BCM CONNECTOR
                         OR FAULTY BCM

210 TO 215 ──────→ REPAIR OPEN
                        IN CIRCUIT 732

210 OR MORE

1. DISCONNECT THE
   SENSOR CONNECTOR
2. NOTE THE PARAMETER
   VALUES

−15 TO −10   **−9 OR MORE**

REPLACE
SENSOR

1. CHECK CIRCUIT
   732 FOR A SHORT
   TO GROUND

IF NOT SHORTED
REPLACE BCM

−9 TO 209

MALFUNCTION NOT PRESENT
AT THIS TIME (See notes on
intermittents in the appropriate
manufacturer's Service Manual)

Figure 10-38 A typical diagnostic chart used to locate the cause of General Motors' trouble code. (Courtesy of BET, Inc.)

the trouble codes are retrieved, consult the appropriate diagnostic chart for instructions on isolating the fault. It is also important to check the codes in the order required by the manufacturer.

Chrysler's DRB-II

The following procedure for using the DRB-II scanner is meant as a general guide only. It is intended to complement, not to replace, the service manual. Improper methods of trouble code retrieval may result in damage to the computer.

Chrysler uses several modules that share information with the body controller through a multiplex system (Figure 10-39). Connecting the DRB-II into the diagnostic connector will access information concerning the operation of most vehicle systems. A typical procedure for entering body controller diagnostics using the DRB-II scanner is as follows:

1. Locate the diagnostic connection using the component locator (Figure 10-40).
2. Insert the correct program cartridge into the DRB-II scanner.
3. Connect the DRB-II to the vehicle by plugging its connector into the vehicle's diagnostic connector.
4. Turn the ignition switch to the RUN position. After the power-up sequence is completed, the copyright date and diagnostic program version should be displayed.
5. The display will change to a selection menu. The entire menu is not displayed; press the down arrow until the desired selection is found. In this example, press the down arrow twice.
6. Select 4 (SELECT SYSTEM) to enter the diagnostic test program. The display will change to a menu for selecting the system to be tested. Use of the down arrow reveals additional choices. Push the down arrow until the BODY option is shown.
7. Enter body system diagnostics by selecting 3 (BODY). The display will change to indicate that the BUS test is being performed (Figure 10-41). If the message is different than that shown in the figure, there is a problem in the CCD bus that must be corrected. No further testing is possible until this problem is corrected.
8. After a few seconds, the display will change and ask for input concerning the body style of the vehicle. Use the down arrow and scroll through the choices available.
9. Enter the number indicating the body style being diagnosed.

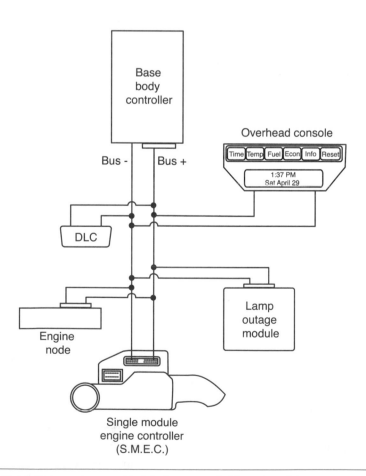

Figure 10-39 Multiplex system used to interface several different modules.

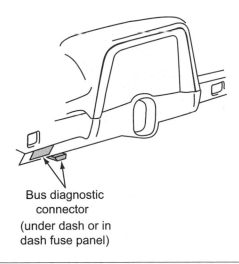

Bus diagnostic
connector
(under dash or in
dash fuse panel)

Figure 10-40 Diagnostic connector location.

Bus test in progress

Bus operational

Figure 10-41 This message must appear before proceeding with diagnostics.

10. The display will then ask that a module be selected. Select BODY COMPUTER.
11. The display will indicate the name of the module selected, along with the version number of the module. Then, after a few seconds, the display will indicate BODY COMP MENU.
12. Use the down arrow key to scroll the menu selection if needed. Press 2 (READ FAULTS). The DRB-II will either display that no faults were detected or provide the fault codes.

The first screen will indicate the number of fault codes found, the code for the first fault, and a description of the code. Scroll down the entire list of codes retrieved.

Retrieving Cadillac BCM Trouble Codes

These procedures may vary among models, years, and the type of instrument cluster installed. Refer to the appropriate manufacturer's service manual for the vehicle being tested.

Cadillac allows access to trouble codes and other system operation information through the electronic climate control (ECC) panel. The body control module (BCM) and the electronic control module (ECM) share information with each other so both system codes are retrieved through the ECC. The following procedure may typically be followed to enter diagnostics:

1. Place the ignition switch in the RUN position.
2. Depress the OFF and WARMER buttons on the ECC panel simultaneously (Figure 10-42). Hold the buttons until all display segments are illuminated.

Cadillac uses the onboard ECC panel to display trouble codes, whereas other General Motors (GM) vehicles use a Tech I scan tool. Beginning in 1996, the Tech II scan tool was used to retrieve codes on certain models. That same year, Cadillac switched to the use of the Tech I scan tool to retrieve class 2 data. When diagnosing GM systems, make sure to follow the procedures specifically designated by GM for the vehicle being tested.

Diagnosis should not be attempted if all segments of the display do not illuminate. A problem may be misdiagnosed as the result of receiving an incorrect code. For example, if two segments of a display fail to illuminate, a code 24 could look like code 21 (Figure 10-43).

When the segment check is completed, the computer will display any trouble codes in its memory. An "8.8.8" will be displayed for about 1 second, then an "..E" will appear. This signals the beginning of engine controller trouble codes. The display will show all engine controller trouble

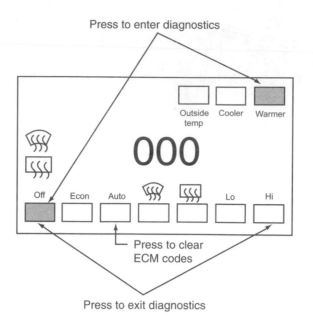

Press to enter diagnostics

Press to clear ECM codes

Press to exit diagnostics

Snapshot: Econ and Cooler   Increment: Hi

Snapshot review: Econ and Warmer   Decrement: Lo

Figure 10-42 The buttons on the ECC panel allow the technician to access information from the computer when it is in the diagnostic mode.

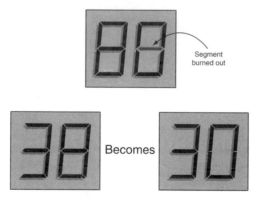

Segment burned out

Becomes

Figure 10-43 Burned out segments give a false code.

codes beginning with the lowest number and progressing through the higher numbers. All codes associated with the engine controller will be prefixed with an "E"; if there are no codes, however, "..E" will not be displayed.

Once all "E" codes are displayed, the computer will display BCM codes. The BCM codes are prefixed by an "F." An "..F" will precede the first set of codes displayed. The first set will be all codes stored in memory since the last 100 engine starts. An ".F.F" will appear to signal the separation of the first pass and the second. The second set of trouble codes will be all hard codes.

When all codes are displayed, ".7.0" will be displayed, indicating that the system is ready for the next diagnostic feature to be selected. To erase the BCM trouble codes, press the OFF and LOW buttons simultaneously until "F.O.O" appears. Release the buttons and ".7.0" will reappear. Turn off the ignition switch and wait at least 10 seconds before reentering the diagnostic mode.

When in the diagnostic mode, to exit the system without erasing the trouble codes, press AUTO on the ECC panel and the temperature will reappear in the display.

Diagnosing SATC and EATC Systems

Semi-automatic or automatic control of the interior (cabin) temperature is made possible through the use of electronic components such as microprocessors, thermistors, and potentiometers that control vacuum and electric actuators. A failure in any of these components will result in inaccurate or no temperature control. Today's technician must possess a basic knowledge and understanding of the operating principles of both the semi-automatic temperature control (SATC) and electronic automatic temperature control (EATC) systems and must be proficient at diagnosing and servicing these systems.

These troubleshooting procedures include that portion of the system that controls its operation. Procedures to troubleshoot the other components of the system are found elsewhere in this manual.

SATC System Diagnosis

A semi-automatic temperature control (SATC) system controls both the operating mode and blower fan speed of the air conditioning system. Faults within the evaporator and heater systems will have an adverse effect on the operation and control of the SATC system.

To properly troubleshoot and service a SATC system, a schematic and specifications for the vehicle being diagnosed is essential. Motors and compressor clutch circuits of the SATC system are tested in the same manner as manual temperature control (MTC) systems.

The air delivery control of SATC systems differs among manufacturers and requires specific diagnostic procedures. Most such systems have specific tests to troubleshoot each particular system, and reference should always be made to each model's service manual. A typical example of one system and its procedures follows.

Chrysler SATC Troubleshooting.   To perform some of the service manual tests on Chrysler's SATC system, it may be necessary to place the blend air door in one of three different positions. When locking the blend door in the full and minimum positions, set the controls for LOW blower speed and BILEVEL mode. Follow these procedures for setting the door position:

1. Set the blend door in the full reheat position by disconnecting the in-car sensor and turn the system ON.
2. Obtain the minimum reheat position by connecting a jumper wire between the red terminal wire from the variable resistor and ground (Figure 10-44), and then turn the system ON. DO NOT connect the jumper wire to the sensor side of the red terminal.

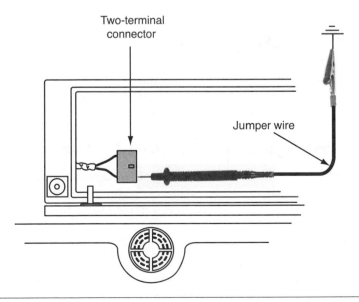

Figure 10-44 Jumper wire connections to set the blend door to the minimum position.

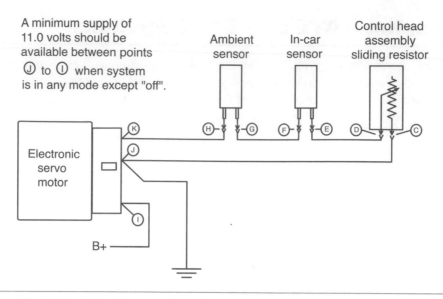

A minimum supply of 11.0 volts should be available between points J to I when system is in any mode except "off".

Ambient sensor

In-car sensor

Control head assembly sliding resistor

Electronic servo motor

B+

Figure 10-45 Chrysler SATC system electrical test points.

3. a. Set the blend door in the middle position by first disconnecting the negative battery cable and removing the ground screw on the passenger side cowl.

b. Next connect a jumper wire from the blower ground wire (black wire with tracer) to a good ground.

c. Then reconnect the battery negative cable.

d. Finally, move the temperature control lever until the blend door moves to the middle position.

e. Then disconnect the jumper wire.

Refer to the test point diagram in Figure 10-45 for the particular vehicle being serviced. Use a voltmeter to measure the voltage between points A and B and points J and I while the system is placed in any mode other than OFF. The voltage values at these test points should be 11 volts or more.

Before performing the continuity tests (Figure 10-46 and Figure 10-47) disconnect the in-car sensor and the power feed connector. Follow the continuity test procedures to determine any system defects.

EATC System Diagnosis

Proper diagnostics of electronic automatic temperature control (EATC) systems depend on system design. There are two basic system designs: those that use their own microprocessor and those that incorporate the controls of the system into the body control module (BCM).

Separate Microprocessor-Controlled Systems

Most EATC systems that use a separate microprocessor for diagnosis have the microprocessor contained in the control assembly (Figure 10-48). Also, most of these systems provide a means of self-diagnostics and have a method of retrieving trouble codes. Typical examples of such systems follow.

Chrysler EATC Troubleshooting. Before entering self-diagnostics, start the vehicle and allow it to reach normal operating temperature. Ensure that all exterior lights are off and press the PANEL button. If the display illuminates, the self-diagnostic mode can be entered. If, however, the display does not illuminate, check the fuses and circuits to the control assembly. If the fuses and circuits are good, replace the ATC computer.

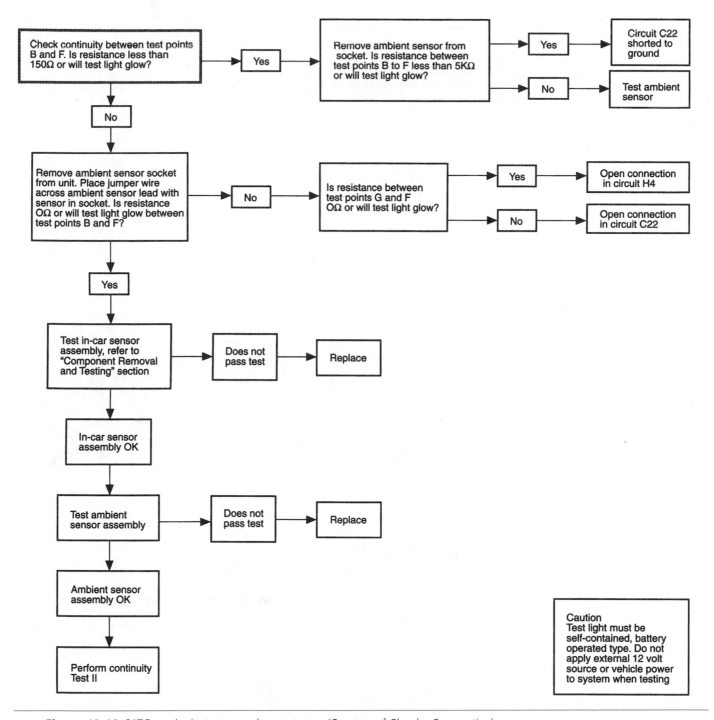

Figure 10-46 SATC continuity test procedure part one. (Courtesy of Chrysler Corporation)

If the display illuminates, the self-diagnostic mode may be entered by pressing the BILEVEL, FLOOR, and DEFROST buttons simultaneously (Figure 10-49). If no trouble codes are present, the self-test program will be completed within 90 seconds and display a "75."

During the process of running the self-diagnostic tests, make four observations that the computer is not able to make by itself:

1. When the test is first initiated, all of the display symbols and indicators should illuminate.
2. The blower motor should operate at its highest speed.
3. Air should flow through the panel outlets.
4. The air temperature should become hot then cycle to cold.

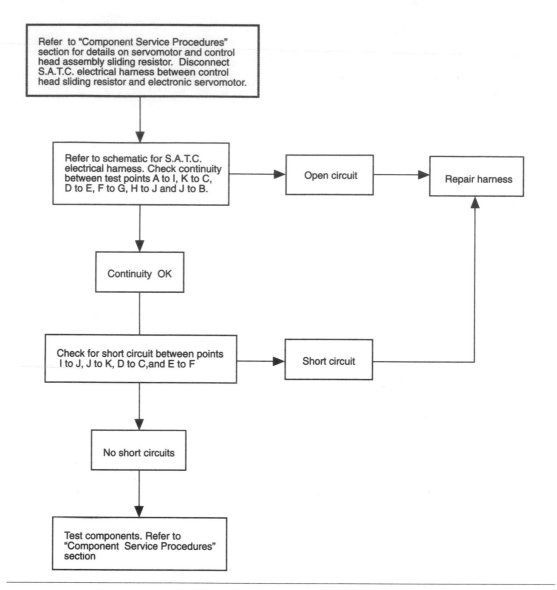

Figure 10-47 SATC continuity test procedure part two. (Courtesy of Chrysler Corporation)

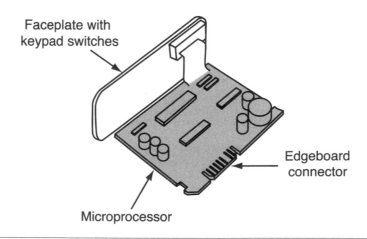

Figure 10-48 Most EATC systems use a separate microprocessor located in the control assembly.

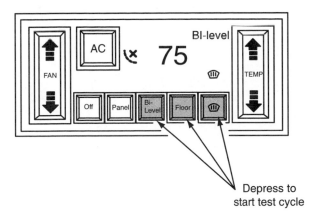

Depress to
start test cycle

Figure 10-49 Use the panel buttons to enter diagnostics.

The diagnostic flow chart may be used to determine the correct test to perform if any of these functions fail (Figure 10-50). The proper procedures for an observed failure are found in the table in Figure 10-51.

If a fault is detected in the system, a trouble code will be flashed on the display panel. To resume the test, record the trouble code then press the PANEL button. Refer to the service manual to diagnose the trouble codes received.

Ford EATC Troubleshooting. To correctly diagnose Ford's EATC system, the exact system description as well as the exact procedures for trouble code retrieval are required. This is because Ford uses different versions of EATC systems that have different diagnostic capabilities. The following is thus only a typical example of performing the self-test:

1. Turn the ignition switch to the RUN position.
2. Place the temperature selector to the "90" setting and select the OFF mode.
3. Wait 40 seconds while observing the display panel. If the VFD display begins to flash, there is a malfunction in the blend actuator circuit, the actuator, or the control assembly. If the LED light begins to flash, this indicates there is a malfunction in one of the other actuator circuits, the actuators, or the control assembly.
4. If no flashing of displays occurs, place the temperature selection to "60" and select the DEF mode.
5. Wait 40 seconds while observing the VFD and LED displays. If there are no malfunctions in the actuator drive or feedback circuits, the displays will not flash.
6. Regardless of whether or not flashing displays were indicated, continue with self-diagnostics. Press the OFF and DEFROST buttons at the same time.
7. Within 2 seconds, press the AUTO button.

Once the self-diagnostics is entered, if an "88" is displayed, there are no trouble codes present. If there are any trouble codes retrieved, they will be displayed in sequence until the COOLER button is pressed. Always exit self-test mode by pressing the COOLER button before turning the ignition switch to the OFF position. Refer to the trouble code chart in Figure 10-52. When service repairs have been performed on the system, rerun the self-test to confirm that all faults have been corrected.

GM ETCC Troubleshooting. General Motors (GM) uses several different versions of the microprocessor-controlled electronic touch climate control (ETCC) system. Depending on the GM division and system design, the door controls can be either by vacuum or by electric servo motor. Methods of entering diagnostics also vary between divisions and models. For this reason, the correct service manual for the system being serviced is needed to perform correct diagnostic procedures. Knowledge of one ETCC system type is no guarantee that you will be able to service other ETCC systems without the use of the proper service manual.

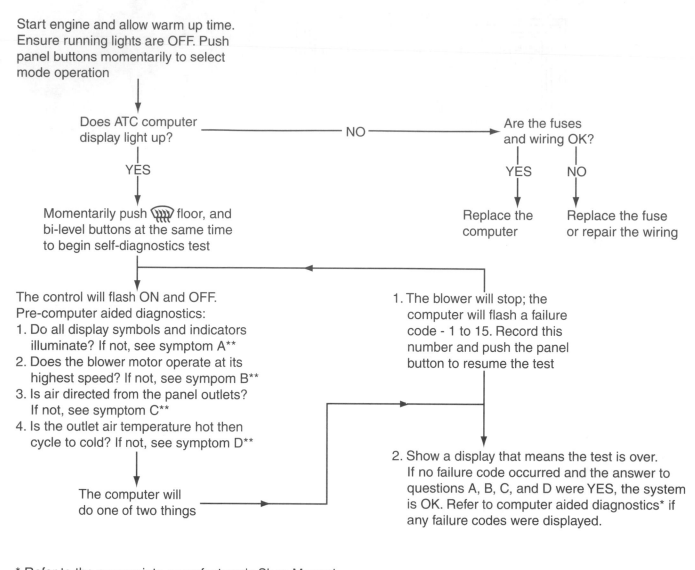

Start engine and allow warm up time. Ensure running lights are OFF. Push panel buttons momentarily to select mode operation

Does ATC computer display light up? —————— NO —————→ Are the fuses and wiring OK?

YES

YES   NO

Momentarily push 🌀 floor, and bi-level buttons at the same time to begin self-diagnostics test

Replace the computer   Replace the fuse or repair the wiring

The control will flash ON and OFF. Pre-computer aided diagnostics:
1. Do all display symbols and indicators illuminate? If not, see symptom A**
2. Does the blower motor operate at its highest speed? If not, see sympom B**
3. Is air directed from the panel outlets? If not, see symptom C**
4. Is the outlet air temperature hot then cycle to cold? If not, see symptom D**

1. The blower will stop; the computer will flash a failure code - 1 to 15. Record this number and push the panel button to resume the test

The computer will do one of two things

2. Show a display that means the test is over. If no failure code occurred and the answer to questions A, B, C, and D were YES, the system is OK. Refer to computer aided diagnostics* if any failure codes were displayed.

* Refer to the appropriate manufacturer's Shop Manual
** Refer to Figure 10-51
NOTE: Technician must be prepared to answer questions during test.

Figure 10-50 A typical Chrysler Electronic Automatic Temperature Control (EATC) diagnostic flow chart. (Courtesy BET, Inc.)

Many GM EATC systems can be checked for proper operation by using a functional chart (Figure 10-53). In addition, troubleshooting charts that correspond with fault symptoms (Figure 10-54) are a great help.

BCM-Controlled EATC Systems

Because the BCM-controlled EATC system incorporates many different microprocessors within its system, diagnostics can be very complex (Figure 10-55). Faults that seem to be unrelated to the EATC

DIAGNOSTIC CHART

"NO"	PROBABLE CAUSE	PROCEDURE
A	1. CONTROL	a. REPLACE CONTROL MODULE
B	1. WIRING PROBLEM	CAUTION: Stay clear of the blower motor wheel. The power/vacuum heat sink is hot (12 volts) DO NOT operate the module longer than 10 minutes with the unit removed from the housing.
	2. POWER VACUUM MODULE	b. Ensure that connections are secure at blower motor and Power/Vacuum Module
		c. If diagnostic test gave code 8 or 12 refer to fault code page in the Service Manual. If no codes, check blower motor fuse.
		d. Disconnect blower motor and check voltage. A reading of 3 to 12 volts for 1 to 8 bar segment on the display is correct. If correct, replace the motor.
		e. If voltage is not correct measure the voltage to vehicle (not motor) ground. Should be 12 volts with ignition ON. If OK replace Power/Vacuum module.
C	1. VACUUM LEAKAGE	a. Service if any codes are found.
	2. POWER VACUUM LEAKAGE	b. Check all connections.
		c. Disconnect vacuum connector and connect it to a manual control to test each port. To test check valve select Panel Mode, disconnect the engine vacuum and determine if the mode changes quickly.
		d. Try a new Power/Vacuum Module.
D	1. REFRIGERATION SYSTEM	a. Complete diagnostic test. Refer to Fault Code Page in Shop Manual if error occurs.
	2. HEATER SYSTEM	b. If a temperature difference of 40°F (22.2°C) or more is noted during the test the Blend-Air Door is engaged in the Servo Motor Actuator. A lower temperature difference indicates a Blend-Air Door operation problem.
	3. BLEND-AIR DOOR	c. Check heater system. 85°F setting is full heat; 65°F is full cool.
		d. Check air conditioning system.

NOTE: See appropriate Shop Manual for this procedure

Figure 10-51 If the Technician's answer was NO to any of the Self Diagnosis Test questions of Figure 10-50, this Diagnostic Chart may be used to isolate the fault. (Courtesy BET, Inc.)

CODE	SYMPTOM	POSSIBLE SOURCE
1	Blend actuator is out of position. VFD flashes.	• Open circuit in one or more of the actuator leads • Actuator output arm jammed. • Actuator inoperative. • Control assembly inoperative.
2	Mode actuator is out of position. LED flashes.	• Same as 1.
3	Pan/Def actuator is out of position. LED flashes.	• Same as 1.
4	Fresh air/recirculator actuator is out of position. LED flashes.	• Same as 1.
1, 5	Blend actuator output shorted. VFD flashes.	• Outputs A or B shorted to ground, or to the supply voltage or together. • Actuator inoperative. • Control assembly inoperative.
2, 6	Mode actuator output shorted. LED flashes.	• Same as 1, 5.
3, 7	Pan/Def actuator output shorted. LED flashes.	• The actuator output is shorted to the supply voltage. • Actuator inoperative. • Control assembly inoperative.
4, 8	Fresh air/recirculator actuator output shorted. LED flashes.	• Same as 3, 7.
9	No failures found. See Supplemental Diagnosis.	
10, 11	A/C clutch never on.	• Circuit 321 open. • BSC inoperative. • Control assembly inoperative.
10, 11	A/C clutch always on.	• Circuit 321 shorted to ground. • BSC inoperative. • Control assembly inoperative.
12	System stays in full heat. In-car temperature must be stabilized above 60°F for this test to be valid.	• Circuit 788, 470, 767, or 790 is open. • The ambient or in-car temperature sensor is inoperative.
13	System stays in full A/C.	• Remove control assembly connectors. Measure the resistance between pin 10 of connector #1 and pin 2 of connector #2. • If the resistance is less than 3K ohms, check the wiring and in-car and ambient temperature sensors. • If the resistance is greater than 3K ohms, replace the control assembly.
14	Blower always at maximum speed.	• Turn off ignition. Remove connector #2 from control assembly. Using a small screwdriver remove terminal #5 from the connector. Replace connector, tape terminal end and turn on ignition. • If blower still at maximum speed, then check circuit 184 and the BSC. • If the blower stops then the control assembly is inoperative.
15	Blower never runs.	• Circuit 184 shorted to the power supply. • BSC inoperative. • Control assembly inoperative.

Figure 10-52 Trouble code chart for Ford EATC system. (Courtesy of Ford Motor Company)

ELECTRONIC CLIMATE CONTROL (ECC) FUNCTIONAL TEST

The air conditioning system diagnostics should begin with a Functional Test. The Functional Test should be performed in the order given in the following chart. If the answer to any test question is NO, turn to the specific Trouble Tree for an analysis of the malfunction. Do not omit steps in the test; to do so may result in an inability to isolate a specific malfunction.

If the Functional Test has been used to select the proper Trouble Tree it is not necessary to perform the test at the beginning of each tree. Also, after a malfunction has been isolated and repaired exit the Trouble Tree at the point of repair.

Check the fuses and stop hazard operation to verify the stop hazard fuse. Warm the engine before beginning the Functional Test, and check the LED above each button as the test is performed.

Proceed as follows:

TEST	SYSTEM CHECKS	CONTROL SETTING	TROUBLE TREES
1	DO THE MPG AND CONTROL HEAD DISPLAY? (MPG is optional)	ALL	1
2	DO COOLER AND WARMER PUSH BUTTONS OPERATE?	ALL	1
3	SET TEMPERATURE TO 60°F (42.2°C):		
	A. DOES THE BLOWER MOTOR OPERATE?	LO-AUTO-HI	2A
	B. IS THERE LOW BLOWER OPERATION?	LO	2B
	C. IS THERE HI BLOWER OPERATION?	HI	2C
	D. IS THERE AIR FROM A/C OUTLETS?	ECON-LO-AUTO-HI	3
	E. DOES COMPRESSOR CLUTCH		
	1. ENGAGE?	LO-AUTO-HI	4
	2. DISENGAGE?	OFF-ECONO	5
	F. IS THE A/C OUTLET AIR COLD	LO-AUTO-HI	6
	1. IS THERE ONLY HEAT?		
	2. IS COOLING SUFFICIENT?		7
	G. DOES RECIRC DOOR FULLY OPEN? (Allow 1-2 minutes)	AUTO-HI	8
4	SET TEMPERATURE TO 90°F (72.2°C):		
	A. IS THERE ADEQUATE HEAT AT OUTLETS?	AUTO	9
5	SET TEMPERATURE TO 85°F (67.2°C):		
	A. IS OUTLET AIR WARM OR HOT?	LO-AUTO	10
6	DOES FRONT DEFROSTER OPERATE?	FRT DEF	11
7	DOES REAR DEFROSTER OPERATE?	RR DEF	12
8	DOES REAR DEFROSTER TURN OFF?	RR DEF OFF	13

Figure 10-53 General Motors' Electronic Automatic Temperature Control (EATC) functional test. (Courtesy BET, Inc.)

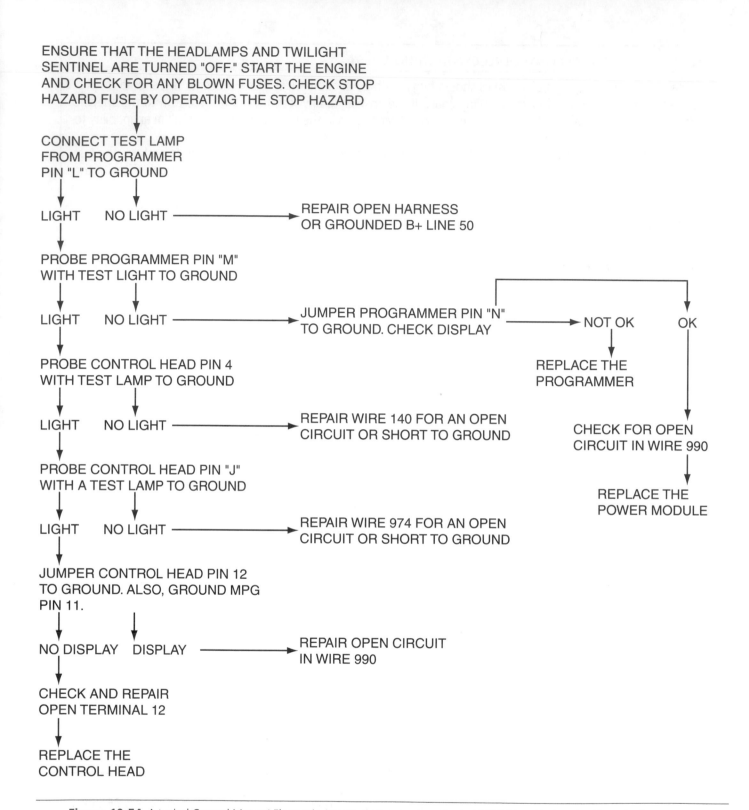

ENSURE THAT THE HEADLAMPS AND TWILIGHT
SENTINEL ARE TURNED "OFF." START THE ENGINE
AND CHECK FOR ANY BLOWN FUSES. CHECK STOP
HAZARD FUSE BY OPERATING THE STOP HAZARD

CONNECT TEST LAMP
FROM PROGRAMMER
PIN "L" TO GROUND

LIGHT   NO LIGHT ————————————→ REPAIR OPEN HARNESS
                                         OR GROUNDED B+ LINE 50

PROBE PROGRAMMER PIN "M"
WITH TEST LIGHT TO GROUND

LIGHT   NO LIGHT ————————————→ JUMPER PROGRAMMER PIN "N"   NOT OK   OK
                                         TO GROUND. CHECK DISPLAY

                                                                   REPLACE THE
                                                                   PROGRAMMER

PROBE CONTROL HEAD PIN 4
WITH TEST LAMP TO GROUND

LIGHT   NO LIGHT ————————————→ REPAIR WIRE 140 FOR AN OPEN   CHECK FOR OPEN
                                         CIRCUIT OR SHORT TO GROUND   CIRCUIT IN WIRE 990

PROBE CONTROL HEAD PIN "J"
WITH A TEST LAMP TO GROUND   REPLACE THE
                                                                    POWER MODULE

LIGHT   NO LIGHT ————————————→ REPAIR WIRE 974 FOR AN OPEN
                                         CIRCUIT OR SHORT TO GROUND

JUMPER CONTROL HEAD PIN 12
TO GROUND. ALSO, GROUND MPG
PIN 11.

NO DISPLAY   DISPLAY ————————————→ REPAIR OPEN CIRCUIT
                                       IN WIRE 990

CHECK AND REPAIR
OPEN TERMINAL 12

REPLACE THE
CONTROL HEAD

Figure 10-54 A typical General Motors' Electronic Automatic Temperature Control (EATC) troubleshooting flow chart.
(Courtesy BET, Inc.)

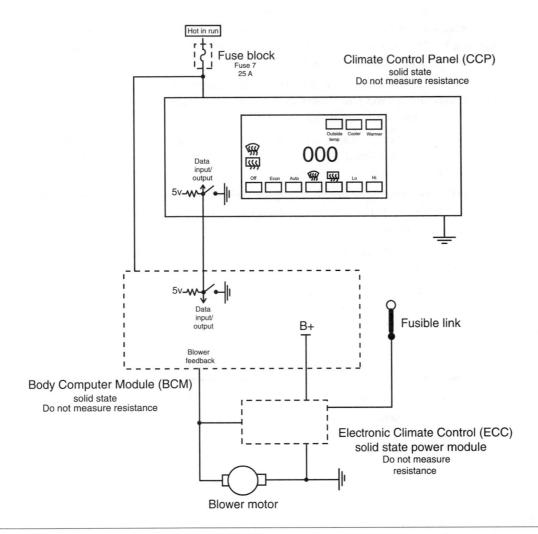

Figure 10-55 The BCM-controlled EATC system has several modules that use multiplexing to share information.

system may cause the system to malfunction. Since it was first introduced in 1986, BCM-controlled EATC systems have become increasingly popular on many GM vehicles. Each model year brings forth revisions and improvements in the system that also require different diagnostic procedures. In addition, system logic, as used by the different GM divisions, has changed through the years.

It is not possible to generally describe the diagnostic procedures required to service the many systems now in use. For this reason, one must have the correct service manual for the system being diagnosed. There are several different methods used to retrieve trouble codes and it is important to follow the correct procedure. In all systems, powertrain control module (PCM) codes are displayed first followed by BCM codes. Codes associated with the EATC system can be in either set of codes.

Once the codes have been retrieved, refer to the correct diagnostic chart. This test will pinpoint the fault in a logical manner. After all repairs to the system are complete, follow the service manual procedure for erasing codes and for resetting the system. Rerun the diagnostic test to confirm that the system is operating property.

CASE STUDY

A customer complains that an abnormal noise is coming from under the hood of her car. The service writer asks the customer the usual questions: "When did the noise start? When does it make the noise? How often is the noise noticeable?"

The customer answers the questions and notes that the noise seems to be growing louder. She first noticed the noise a few days ago.

After noting the mileage, the service writer checks the computer for the service record. According to the records, no major work has been performed on the car. Also, it seems to have been serviced regularly and is well maintained.

On starting the car for a test drive, the noise is immediately noted. The service writer raises the hood and, using a mechanic's stethoscope, is able to pinpoint the noise at the air conditioner compressor.

"It couldn't be the compressor," the customer says. "I just had that repaired a week ago." The customer then explains that the repairs were made by an independent dealer in another city while the customer was out of town.

An inspection of the compressor clutch by the technician reveals that the wrong field coil snap ring was installed. The snap ring, which was too thin, allowed the field coil to barely touch the rotor, creating a noise.

Fortunately for the customer, no major damage was done to the clutch and a proper snap ring corrected the problem.

Terms to Know

Air gap	In-car temperature sensor	Short
Ambient temperature sensor	Low-refrigerant switch	Sliding resistor
Aspirator	Motor	Solenoid
AUTO	Ohmmeter	Spike
BCM	Open	Subsystem
Breakout box	Outside temperature sensor	Sun load
Dual system	Overload	System
EATC	PCM	Thermistor
ECC	Relay	Voltmeter
Fusible link	Scan tool	
Hot	Servo motor	

ASE-Style Review Questions

1. Servicing the automatic temperature control system is being discussed:
   Technician A says that one must have the manufacturer's wiring diagrams to properly diagnose system malfunctions.
   Technician B says that one must have the manufacturer's shop manuals to properly remove and replace defective components.
   Who is right?
   A. A only
   B. B only
   C. Both A and B
   D. Neither A nor B

2. *Technician A* says that a blown fuse *always* indicates a defective component.
   Technician B says that an open circuit breaker *must* be replaced.
   Who is right?
   A. A only
   B. B only
   C. Both A and B
   D. Neither A nor B

3. *Technician A* says that a fuse can be tested in the vehicle with the use of an analog or a digital ohmmeter.
   Technician B says that a fuse can be tested out of the vehicle with a powered or a nonpowered test lamp.
   Who is right?
   A. A only   **C.** Both A and B
   B. B only   **D.** Neither A nor B

4. *Technician A* says that a trouble code may be displayed on some master control heads.
   Technician B says that there are provisions for connecting an external scan tool on some systems.
   Who is right?
   A. A only   **C.** Both A and B
   B. B only   **D.** Neither A nor B

5. *Technician A* says that a thermostat is used for temperature control in some systems.
   Technician B says that a low-pressure control is used for temperature control in some systems.
   Who is right?
   A. A only   **C.** Both A and B
   B. B only   **D.** Neither A nor B

6. *Technician A* says that some thermostats may be adjusted.
   Technician B says that some pressure controls may be adjusted.
   Who is right?
   A. A only   **C.** Both A and B
   B. B only   **D.** Neither A nor B

7. Blower motor speed control is being discussed:
   Technician A says that blower motor speed control is automatic in a temperature-controlled system.
   Technician B says that blower motor speed may be manually selected in a temperature-controlled system.
   Who is right?
   A. A only   **C.** Both A and B
   B. B only   **D.** Neither A nor B

8. The electromagnetic clutch is being discussed:
   Technician A says the clutch will slip if the air gap is too close.
   Technician B says that the clutch will slip if the belt is loose.
   Who is right?
   A. A only   **C.** Both A and B
   B. B only   **D.** Neither A nor B

9. Pressure switches are being discussed:
   Technician A says that a compressor discharge pressure switch is a low-pressure switch.
   Technician B says that the pressure switch used for temperature control is a low-pressure switch.
   Who is right?
   A. A only   **C.** Both A and B
   B. B only   **D.** Neither A nor B

10. *Technician A* says that a diode is used in some sensor circuits to prevent computer damage during operation.
    Technician B says that a thermistor is used in the clutch circuit to prevent spikes during operation.
    Who is right?
    A. A only   **C.** Both A and B
    B. B only   **D.** Neither A nor B

ASE Challenge

1. Low-pressure and high-pressure cutoff switches may be used for air conditioning system:
   A. Temperature control   **C.** Either A or B
   B. Protection   **D.** Neither A nor B

2. The following statements regarding the relationships between the cooling system and air conditioning system are all true, EXCEPT:
   A. An overheating engine will affect air conditioning system performance
   B. Ambient air first passes through the radiator, then the condenser
   C. The same blower motor is used for the heater and air conditioner
   D. An air conditioning system places an additional load on the cooling system

3. Electrical testing is being discussed:
   Technician A says an analog ohmmeter should not be used to test an electronic circuit.
   Technician B says a digital ohmmeter may be used to test any electrical or electronic circuit.
   Who is right?
   A. A only   **C.** Both A and B
   B. B only   **D.** Neither A nor B

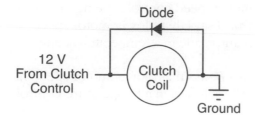

12 V
From Clutch
Control

Clutch
Coil

Diode

Ground

4. The diode shown in the illustration above is used to:
   A. Increase voltage in the clutch coil circuit
   B. Reduce voltage in the clutch coil circuit
   C. Block the flow of current in the electrical system
   D. Prevent voltage spikes in the electrical system

5. Which of the following is NOT a tool used by the air conditioning system technician:
   A. Scan tool
   B. Breakout box
   C. DRB-II scanner
   D. OBD-5 scanner

Table 10-1 ASE TASK

Diagnose the cause of temperature control problems in the heater/ventilation system; determine needed reapairs.

Diagnose temperature control system problems; determine needed repairs.

Diagnose blower system problems; determine needed repairs.

Inspect, test, adjust, or replace climate control temperature and sun load sensors.

Problem Area	Symptoms	Possible Causes	Classroom Manual	Shop Manual
IMPROPER AIR FLOW	No air flow	1. Defective master control	216	348–349
		2. No:		
		a. vacuum, if pneumatic		
		b. power, if electric	227, 233	357
		3. Defective		
		a. motor, if electric		
		b. "pot," if pnuematic		
		c. cable, if manual	227, 233	357
	No fresh air from vents	1. Defective master control; electric, pneumatic, or manual	216	348–349
		2. Defective:		
		a. wiring, if electric		
		b. hose, if pneumatic		
		c. cable, if manual	227, 233	357
		3. Defective actuator:		
		a. motor, if electric		
		b. "pot," if pnuematic		
		c. retainer, if cable	227, 233	357

Table 10-1 ASE TASK (continued)

Problem Area	Symptoms	Possible Causes	Classroom Manual	Shop Manual
IMPROPER AIR FLOW	No cool air from vents	1. Defective master control; electric, pneumatic, or manual	216	348–349
		2. Defective: a. wiring, if electric b. hose, if pneumatic c. cable, if manual	227, 233	77
		3. Defective actuator: a. motor, if electric b. "pot," if pnuematic c. retainer, if cable	227, 233	77
		4. Defective or inoperative air conditioner	84–89	150
		5. Defective (open) heater coolant flow control valve	267	295
		6. Duct disconnected or missing	175–184	281–291
	No warm air from vents	1. Defective master control; electric, pneumatic, or manual	216	348–349
		2. Defective: a. wiring, if electric b. hose, if pneumatic c. cable, if manual	227, 233	77
		3. Defective actuator: a. motor, if electric b. "pot," if pneumatic c. retainer, if cable	175–184	77
		4. Defective (closed) heater coolant flow control valve	267	295
		5. Heater (hoses) disconnected	267–268	424
		6. Duct disconnected or missing	175–184	281–291

Table 10-2 ASE TASK

Diagnose the cause of unusual operating noises of the A/C system; determine needed repairs.

Problem Area	Symptoms	Possible Causes	Classroom Manual	Shop Manual
NOISE	Chattering sound	1. Defective blower motor	221–222	351–352
		2. Blower loose on motor shaft	221–222	351–352
		3. Blower rubbing insulation on mode door gasket	221–222	351–352
		4. Debris in duct	171–184	281
		5. Low compressor: a. lubricant b. refrigerant	129–133, 137–148	187–234
		6. Loose bracket or other part(s)	129–133, 137–148	187–234
	Squealing sounds	1. Loose or glazed belts(s)	133–134	410
		2. Worn belt(s) and or pulley(s)	133–134	410
		3. Defective A/C clutch or idler pulley bearing(s)	134–137	351–352
		4. Defective blower motor	221–222	351–352
		5. Defective A/C compressor	129–133	187
	Grinding sounds	1. Defective A/C clutch	134–137	351–352
		2. Defective A/C compressor	129–133	187
		3. Defective bearing(s) in clutch or idler	134–137	351–352
		4. Defective coolant (water) pump	253	407

Table 10-3 ASE TASK

Diagnose the cause of failure in the electrical control system of heating, ventilating, and A/C systems; determine needed repairs.

Inspect, test, repair, replace, and adjust load sensitive A/C compressor cut-off systems.

Inspect, test, repair, and replace engine cooling/condenser fan motors, relays/modules, switches, sensors, wiring, and protection devices.

Problem Area	Symptoms	Possible Causes	Classroom Manual	Shop Manual
INADEQUATE OR NO COOLING	Little to no air flow from ducts	1. Blown fuse or defective circuit breaker	215	348
		2. Defective blower speed control or resistor	237	351, 360
		3. Defective master control	216	357
		4. Defective relay	263–264	351–352, 360
		5. Defective wiring	222–225	351–352, 360
	Air not cool	1. Defective clutch coil or ground connection	222–225	351–354
		2. Defective low- or high-pressure control	225–226	354
		3. Defective temperature control	217–221	350
		4. Defective master control	216	357
		5. Defective relay	263–264	351–352, 360
		6. Defective sensor	233–234	360
	Air not warm	1. Defective master control	216	357
		2. Defective temperature control	217–221	350
		3. Defective electric coolant flow control valve	267	348–349
		4. Defective relay	263–264	351–352, 360
		5. Defective sensor	233–234	360
INTERMITTENT OR NO COOLING	A/C system cycles rapidly	1. Defective low-pressure control	241	354
		2. Defective high-pressure control	240	354
		3. Thermostat adjustment	217–221	350
	Cycles on high pressure protector	1. Defective or inoperative cooling fan or motor	262	283
		2. Defective high-pressure control	226	354
		3. Engine overheating	227–228	430

Table 10-4 ASE TASK

Inspect, test, repair, and replace A/C compressor clutch components or assembly.

Inspect, test, repair, replace, and adjust A/C-related engine control systems.

Inspect, test, adjust, repair, and replace electric actuator motors, relays/modules, switches, sensors, wiring, and protection devices.

Diagnose compressor clutch control system; determine needed repairs.

Inspect, test, repair, and replace electric and vacuum motors, solenoids, and switches.

Problem Area	Symptoms	Possible Causes	Classroom Manual	Shop Manual
INADEQUATE OR NO COOLING	Little to no air flow from ducts	1. Blown fuse or defective circuit breaker	215	348
		2. Defective blower speed control or resistor	237	351, 360
		3. Defective master control	216	357
		4. Defective relay	263–264	351–354
		5. Defective wiring	222–225	351–354
	Air not cool with A/C selected	1. Defective clutch coil	222–225	351–354
		2. Poor electrical connection	227	352, 354
		3. Defective low-pressure control	241	354
		4. Defective high-pressure control	240	354
		5. Defective temperature control	217–221	350
		6. Defective master control	216	357
		7. Defective relay	263, 264	251–354
		8. Defective sensor	233–234	360
		9. Defective module	239	351, 360
		10. Shorted clutch diode	241	352
	Air not warm	1. Defective master control	216	357
		2. Defective temperature control	217–221	350
		3. Defective electric coolant flow control valve	297	360
		4. Defective relay	263–264	251–354
		5. Defective sensor	233–234	360
		6. Defective module	239	351, 360
INTERMITTENT OR NO COOLING	A/C system cycles rapidly	1. Defective low-pressure control	241	354
		2. Defective high-pressure control	240	354
		3. Thermostat adjustment	217–221	350
		4. Defective module	239	253
		5. Defective relay	237–238	351–354
		6. Loose or defective wiring	222–225	351–354
	Cycles on high pressure protector	1. Defective or inoperative cooling fan or motor	262	416
		2. Defective high-pressure control	226	354
		3. Engine overheating	227–228	430

Table 10-5 ASE TASK

Inspect, test, and replace ATC control panel.

Inspect, test, adjust or replace ATC microprocessor (climate control computer/programmer).

Check and adjust calibration of ATC system.

Problem Area	Symptoms	Possible Causes	Classroom Manual	Shop Manual
NO COOLING	Compressor will not engage	1. A/C clutch relay	263–264	351–354
		2. Low-side temperature sensor	233–234	354
		3. Defective low-pressure switch	241	354
		4. Open clutch coil	222	351–354
		5. Power steering pressure switch (if equipped)	244	354
		6. Defective wiring	222–225	351–354
		7. Defective BCM	237	351, 360
	Compressor always engaged	1. Defective clutch	222	351–354
		2. Defective clutch relay	263–264	351–354
		3. Shorted control signal circuit	234–236	357
		4. Mechanical binding	171–177	295
	High blower speed only	1. Open feedback circuit	234–236	351–354
		2. Defective programmer	237	351–354
		3. Open signal circuit	234–236	357
		4. Defective BCM	237	351, 360
	No blower	1. Blown fuse or circuit breaker	215	348
		2. Defective blower motor	221–222	351–352
		3. Loose or disconnected motor ground	221	351–352
		4. Improper signal to programmer due to open or short	237	351, 360
		5. Power module feed open	239	357
		6. Defective programmer	237	360
		7. Defective BCM	237	360
	Improper air delivery	1. Loss of vacuum source	228–232	291–295
		2. Leak in vacuum circuit	228–232	291–295
		3. Defective programmer	237	351, 360
		4. Defective BCM	237–239	351, 360
	Insufficient heating	1. Air mix valve	171–177	295
		2. Defective (closed) coolant flow control valve	267	350
		3. Air mix valve linkage	171–177	295
		4. Programmer arm adjustment	237	360
		5. Programmer	237	360
	Insufficient cooling	1. Insufficient air flow	171–184	281–295
		2. Refrigeration problems	84–89	146
		3. Air mix valve linkage	171–177	295
		4. Programmer arm adjustment	237	360
		5. Programmer	237	360

JOB SHEET 34

34

Name _____ Date _____

Testing and Replacing
Fuses and Circuit Breakers

Upon completion of this job sheet you should be able to test and replace fuses and circuit breakers.

ASE Correlation

This job sheet is related to the ASE Heating and Air Conditioning Systems Test's content area: *Operating Systems and Related Controls Diagnosis and Repair, 1. Electrical,* Task: *Diagnose the cause of failures in the electrical control system of heating, ventilating, and A/C systems; determine needed repairs.*

Tools and Materials

Vehicle with accessible fuse/circuit breaker panel
Test light (fused, nonpowered)
Test light (fused, powered)
Ohmmeter
Tools, as needed

Describe the vehicle being worked on:

Year _____ Make _____ Model _____

VIN _____ Engine type and size _____

Procedure

Using a nonpowered test light, perform the following tasks. Write a brief description of your procedure and of your findings.

1. Connect one probe of the test light to ground.

   a. Touch the other probe to the HOT side of the fuse panel buss bar.

   _____

   _____

2. b. Touch the other probe to the other side of selected fuses and circuit breakers.

   Fuses: _____

   _____

   Circuit Breakers: _____

   _____

3. Remove a selected fuse and a circuit breaker. Make note of their location and amperage.

   _____

   _____

4. Touch the probes of the self-powered test light to the two ends of the fuse.

_____

_____

5. Touch the probes of the self-powered test lamp to the two terminals of the circuit breaker.

_____

_____

NOTE: For steps 6 and 7, DO NOT connect the ohmmeter to the self-powered test lamp.

6. Hold the same test as step 4 using an ohmmeter.

_____

_____

7. Hold the same test as step 5 using an ohmmeter.

_____

_____

8. Perform any other tests as outlined by your instructor.

Test _____

Results _____

_____

✓ **Instructor's Check** _____

JOB SHEET 35

35

Name _____ Date _____

Test a Blower Motor

Upon completion of this job sheet you should be able to test a blower motor.

ASE Correlation

This job sheet is related to the ASE Heating and Air Conditioning Systems Test's content area: *Operating Systems and Related Controls Diagnosis and Repair, 1. Electrical ,* Task: *Inspect, test, repair, and replace A/C-heater blower motors, resistors, switches, relay/modules, wiring, and protection devices.*

Tools and Materials

Late-model vehicle
Shop manual
Voltmeter
Fused jumper wire
Safety glasses or goggles
Hand tools, as required

Describe the vehicle being worked on:

Year _____ Make _____ Model _____

VIN _____ Engine type and size _____

Procedure

After each of the following steps, write a brief description of your procedure followed by your findings.

Test the blower motor:

1. With the ignition switch in the ON position, turn the blower control to:

   a. HIGH _____

   b. MED-HI _____

   c. MED-LO _____

   d. LOW _____

   Did the blower run in any speed? Explain: _____

   _____

2. Turn the blower control and ignition switch OFF. Disconnect the blower motor and connect the voltmeter: one lead to ground and the other lead to the disconnected wire. Is there a voltage? _____ Why? _____

   _____

3. Turn the ignition switch ON. While observing the voltmeter turn the blower control to:

    a. HIGH _____

    b. MED-HI _____

    c. MED-LO _____

    d. LOW _____

Was there voltage noted in any speed position? Explain: _____

_____

    e. Turn the blower control and ignition switch OFF. _____

4. Connect one end of a fused jumper wire to the battery positive terminal. While wearing safety glasses or goggles, carefully connect the other end of the jumper wire to the blower motor terminal. Was there a "spark"? _____ Did the fuse "blow"? _____ Did the motor "run"? _____ Describe what happened. _____

_____

5. Conclusion. Write a brief summary of your findings.

_____

_____

_____

_____

☑ **Instructor's Check** _____

JOB SHEET 36

36

Name _____ Date _____

Remove and Replace a Blower Motor

Upon completion of this job sheet you should be able to remove and replace a blower motor.

ASE Correlation

This job sheet is related to the ASE Heating and Air Conditioning Systems Test's content area: *Operating Systems and Related Controls Diagnosis and Repair, 1. Electrical,* Task: *Inspect, test, repair, and replace A/C-heater blower motors, resistors, switches, relay/modules, wiring, and protection devices.*

Tools and Materials

Late-model vehicle
Shop manual
Safety glasses or goggles
Hand tools, as required

Describe the vehicle being worked on:

Year _____ Make _____ Model _____
VIN _____ Engine type and size _____

Procedure

Follow the procedures outlined in the manufacturer's shop manual. These procedures are given as a typical guide line for the task. After each step write a brief summary of your procedure.

1. While wearing safety glasses or goggles, carefully remove the battery ground cable.

_____

_____

2. Disconnect the VCM or PCM (if applicable) following procedures given in the service manual. _____

_____

_____

3. Remove any components, such as coolant reservoir, that may prevent blower motor removal. _____

_____

_____

4. Remove electrical connector(s) and ground wire(s), if applicable. _____

_____

_____

5. Remove all retaining screws and fasteners. _____

_____

_____

6. If necessary, use a sharp utility knife to cut through any gasket material that may restrict removal of blower motor assembly. _____

_____

_____

7. Lift blower and motor from case/duct. _____

_____

_____

8. Remove retaining nut or clip from motor shaft and remove blower, if applicable.

_____

9. Slide blower onto new motor shaft and secure with nut or clip. _____

_____

_____

10. Reverse the removal procedure and replace the blower and motor. Replace any gasket material cut in step 6 with black weatherstrip adhesive. Do not use RTV.

_____

_____

11. Make electrical connections, reversing the order of steps 1, 2, and 4.

_____

_____

12. Replace any components removed in step 3.

_____

_____

☑ **Instructor's Check** _____

Diagnosis and Service of Engine Cooling and Comfort Heating

Upon completion and review of this chapter you should be able to:

❏ Identify the major components of the automotive engine cooling and comfort heating system.

❏ Compare the different types of radiators.

❏ Discuss the function of the coolant pump.

❏ Explain the need for a pressurized cooling system.

❏ Describe the advantage of a thermostat in the cooling system.

❏ Understand the procedures used for testing the various cooling system components.

❏ Recognize the hazards associated with cooling system service.

❏ Understand troubleshooting procedures for determining the malfunction of cooling system components.

Introduction

A typical gasoline engine is only about 15 percent efficient; only about 15 percent is used to move the vehicle. That means that 85 percent of all energy developed by the engine is wasted in friction and heat—heat that must be removed.

While the heat of combustion may reach as high as 4,000°F (2,200°C), most of it is expelled when the exhaust valve opens. This results in an actual net engine temperature range from about 750°F (410°C) to about 1,500°F (815°C). This is still a lot of heat and it must be removed. The coolant, a mixture of water (H_2O) and ethylene glycol, is the refrigerant used to transfer this heat from the engine to the radiator.

In the cooling system, heat is absorbed into the refrigerant in the engine and dissipated in the radiator just as heat is removed from the interior of the vehicle by being absorbed into the refrigerant in the evaporator and dissipated in the condenser.

The Cooling System

The cooling system (Figure 11-1) is made up of several components, all of which are essential to its proper operation. They are the radiator, pump, pressure cap, thermostat, cooling fan, heater core, hoses and clamps, and coolant.

The most common cooling system problems are a result of a leaking system. A sound system seldom presents a problem. Leaks are generally easy to find using a pressure tester. Several types are available. Following is a typical procedure:

1. Allow the engine and coolant to cool to **ambient temperature**.
2. Remove the pressure cap. Note the pressure range indicated on the cap (Figure 11-2).
3. Adjust the coolant level to a point just below the bottom of the **fill neck** of the radiator.
4. Attach the pressure tester (Figure 11-3).
5. While observing the gauge, pump the tester until a pressure equal to the cap rating is achieved. If the pressure can be achieved, proceed with step 6. If the pressure cannot be achieved, make a visual inspection for leaks.
6. Let the system stand for five minutes. Recheck the gauge. If the pressure is the same as in step 5, the system is all right. If the pressure has dropped very slightly, repressurize

Basic Tools

Basic mechanic's tool set

Classroom Manual
Chapter 11, page 250

A common leak point is due to loose hose clamps.

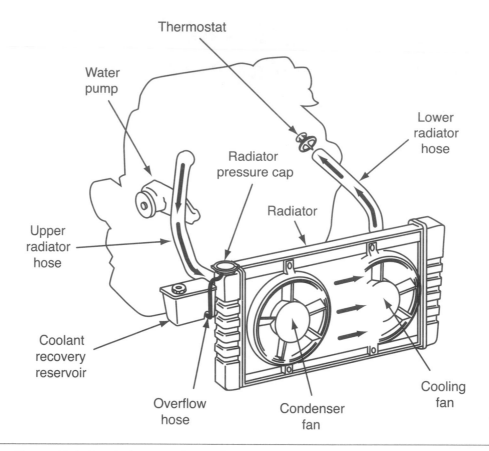

Figure 11-1 A typical automotive cooling system.

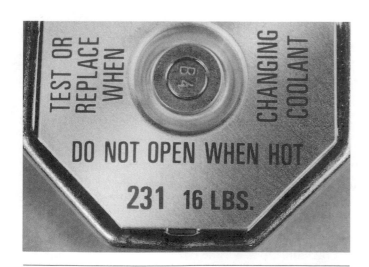

Figure 11-2 A pressure cap showing the pressure rating.

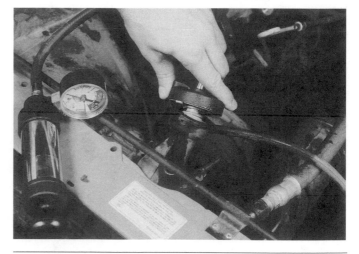

Figure 11-3 Attach a pressure tester to the radiator neck.

the system to the top end of the pressure rating and proceed with step 7. Proceed with checking the radiator cap. The cap should be able to hold the pressure noted on the cap (Figure 11-4). If all right, proceed to step 7. If the pressure dropped considerably, visually check for leaks.

7. Was a leak found? If yes, repair and recheck the system beginning with step 5. If no, repeat steps 5, 6 and 7.

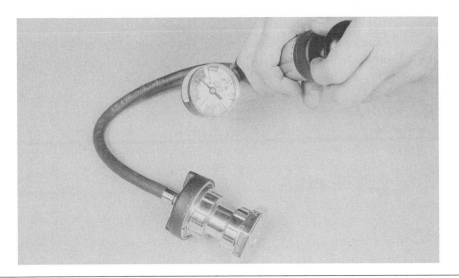

Figure 11-4 Pump the tester until a pressure equal to the cap rating is noted on the gauge.

Radiators

The purpose of the radiator is to **dissipate** heat that is picked up by the coolant in the engine into the air passing through its fins and tubes. This is accomplished by natural or forced means. Natural means are created by **ram air** as the vehicle is in motion. Forced means are created by an engine- or electric motor-driven fan.

Although radiators (Figure 11-5) are often neglected by the vehicle owner, they generally have a long service life. The service life is greatly increased if periodic and scheduled routine maintenance, outlined in the Owner's Manual, is performed. Eventually, however, most radiators will develop a leak or become clogged with rust and/or corrosion due to lack of attention.

For most radiator repairs, the radiator must be removed from the vehicle and "shopped out" to a specialty shop that is equipped with the proper specialized tools, equipment, and knowledgeable service technicians to perform this type of repair. If a radiator is badly damaged, such as would result from a collision, it is often less expensive to replace it with a new unit.

Classroom Manual
Chapter 11, page 251

A heat exchanger is a device that causes heat to move from one medium to another: e.g., fluid to air, air to air, air to fluid, or fluid to fluid.

Figure 11-5 A typical automotive radiator.

Following is a typical procedure for removing a radiator from the vehicle. The actual procedure for any particular make or model vehicle is provided in the manufacturer's specifications. If there is an engine-mounted cooling fan (Figure 11-6), start with step 1. If there is an electric cooling fan (Figure 11-7), start with step 2.

The shroud is a design consideration of the engine cooling system and, as such, should not be removed.

1. If the fan is equipped with a shroud, remove the attachments and slide the shroud toward the engine.
2. Carefully remove the upper and lower coolant hoses from the radiator.
3. Remove the transmission cooler lines (if the vehicle has an automatic transmission) from the radiator. Plug the lines to prevent transmission fluid loss.
4. Remove the radiator attaching bolts and brackets.
5. Carefully lift out the radiator.

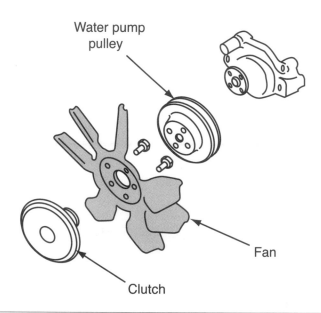

Figure 11-6 A typical engine-mounted belt-driven cooling fan.

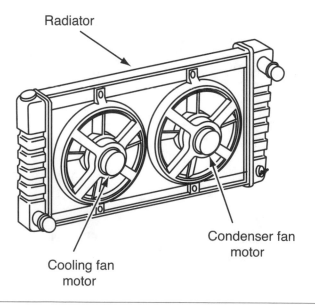

Figure 11-7 A typical electric motor-driven cooling fan.

WARNING: Take care not to damage the delicate fins of the radiator when removing it.

COOLANT PUMP

The coolant pump (Figure 11-8) may be thought of as the heart of the cooling system. Its purpose is to move the coolant through the system as long as the engine is running and the coolant is up to minimum operating temperature. By centrifugal design, the pump cannot be "overloaded" if the coolant flow is restricted. This would be the case when the thermostat is closed, restricting flow, and the impeller is pushing coolant into the engine only. At this time, the coolant passage is back to the pump through a small **bypass** below the thermostat. When the thermostat is open, coolant flow is throughout the cooling system.

To replace a coolant pump, it is often necessary to remove accessories, such as the power steering pump, air conditioning compressor, alternator, or the air pump, to gain access. It is advisable to refer to the specific manufacturer's service manuals when replacing the coolant pump. Following, however, is a typical procedural outline.

1. Remove the radiator as previously outlined if necessary to gain access to the coolant pump.
2. Loosen and remove all belts. If there is an engine-mounted fan, proceed with step 3. If there is an electric fan, proceed with step 4.
3. Remove the fan and fan/clutch assembly.
4. Remove the coolant pump pulley.
5. Remove accessories as necessary to gain access to the water pump bolts.
6. Remove the lower radiator hose from the water pump.
7. Remove the **bypass hose**, if equipped.
8. Remove the bolts securing the water pump to the engine.

WARNING: For reassembly, note the size and length of the bolts removed. The bolts may be both English and metric and all may not be the same length.

9. Tap the water pump lightly, if necessary, to remove it from the engine.
10. Clean the old gasket material from all surfaces. Take care not to scratch the mating surfaces.

A centrifugal pump is a variable-displacement pump. Restricting the flow of coolant does not harm the pump.

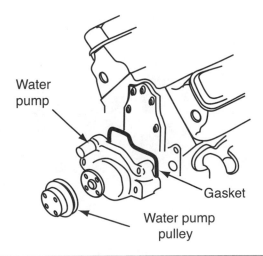

Figure 11-8 A typical coolant pump.

Pressure Cap

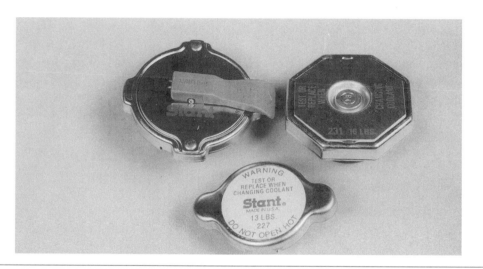

Classroom Manual
Chapter 11, page 255

If a pressure cap fails to hold pressure or if it fails to release high pressure, it must be replaced.

A radiator pressure cap (Figure 11-9) is necessary to maintain the desired engine temperature without coolant loss. A pressure cap is usually designed to operate in the 14–17 psi (97–117 kPa) pressure range.

Pressure caps may be tested using a cooling system pressure tester and an adapter. The requirement is that a pressure cap not leak at a pressure below what it is rated and that it must open at a pressure above what it is rated.

To pressure test a radiator cap, proceed as follows:

1. Attach the adapter to the pressure tester (Figure 11-10).
2. Install the pressure cap to be tested (Figure 11-11).
3. Pump the pressure tester to the value marked on the pressure cap (Figure 11-12).
4. Did the cap hold pressure? If yes, proceed with step 5. If no, replace the cap.
5. Pump to exceed the pressure rating of the cap.
6. Did the cap release pressure? If yes, the cap is good. If no, replace the cap.

Figure 11-9 Typical radiator pressure caps.

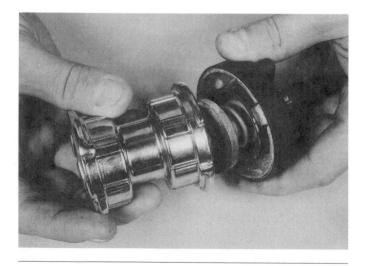

Figure 11-10 Attach the adapter to the pressure tester.

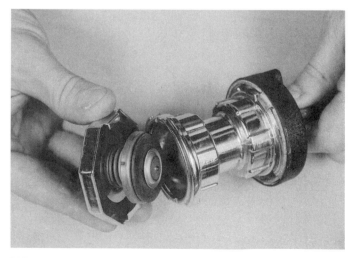

Figure 11-11 Install the pressure cap.

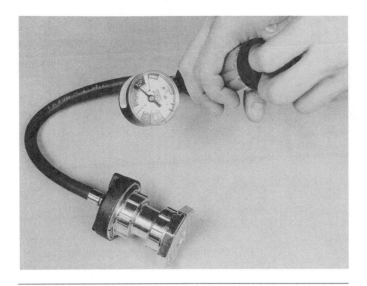

Figure 11-12 Pump the tester to a pressure equal to the cap rating.

Figure 11-13 A typical cooling system thermostat.

Thermostats

The purpose of the thermostat (Figure 11-13) is to trap the coolant in the engine until it reaches its operating temperature. It then restricts the flow of coolant leaving the engine until overall coolant temperature is at or near operating temperature. The thermostat is then fully open to allow unrestricted coolant flow through the system.

A typical procedure for removing a thermostat follows:

Thermostat Removal

1. Reduce the engine coolant to a level below the thermostat.
2. Remove the bolts holding the thermostat housing onto the engine (Figure 11-14). It is not necessary to remove the radiator hose from the housing.
3. Lift off the thermostat housing (Figure 11-15). Observe the pellet-side down position of the thermostat to ensure proper replacement. Do not install the thermostat backward.

Classroom Manual
Chapter 11, page 256

The thermostat operating temperature is a part of the design consideration of the engine and should not be altered.

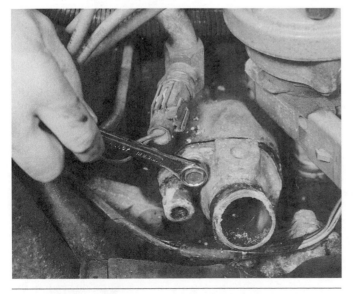

Figure 11-14 Remove the bolts holding the thermostat housing.

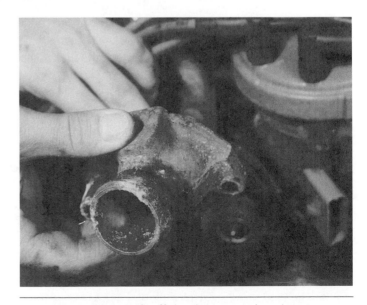

Figure 11-15 Lift off the thermostat housing.

Figure 11-16 Lift out the thermostat.

4. Lift out the thermostat (Figure 11-16).
5. Clean all the old gasket material from the thermostat housing and engine mating surfaces.

There are several ways to test a thermostat. Many, however, believe that if a thermostat is suspect, it should be replaced. The labor cost for the time required to test a thermostat often outweighs the cost of a thermostat.

Following is a procedure for testing a thermostat that has been removed from the engine.

Thermostat Testing

 CAUTION: The water (H_2O) temperature may reach 212°F (100°C). Wear suitable protection.

1. Note the condition of the thermostat.
2. Is the thermostat corroded or open? If no, proceed with step 3. If yes, replace the thermostat.
3. Note the temperature range of the thermostat.
4. Suspend the thermostat in a heatproof glass container filled with water (H_2O).
5. Suspend a thermometer in the container. Neither the thermostat nor the thermometer should touch the container or touch each other.
6. Place the container and contents on a stove burner and turn on the burner.
7. Observe the thermometer. Does the thermostat start to open about 20°F (11°C) below its rating? If yes, proceed with step 8. If no, replace the thermostat.
8. Observe the thermometer. Is the thermostat fully open at its rated temperature? If yes, the thermostat is all right. If no, replace the thermostat.

Pulleys

Classroom Manual
Chapter 11, page 258

Replace any belt that appears to be worn, frayed, or damaged.

Pulleys seldom require attention. The only pulley problem that may occur is damage due to collision or a defective bearing in an **idler** or **clutch pulley**. In all cases, repairs are straightforward; replace the pulley or replace the bearing.

Belts

There are two types of belts used in the automotive engine cooling and air conditioning system: the V-belt (Figure 11-17) and the **serpentine belt** (Figure 11-18).

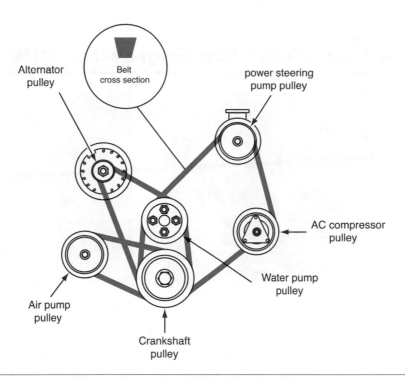

Figure 11-17 A typical V-belt.

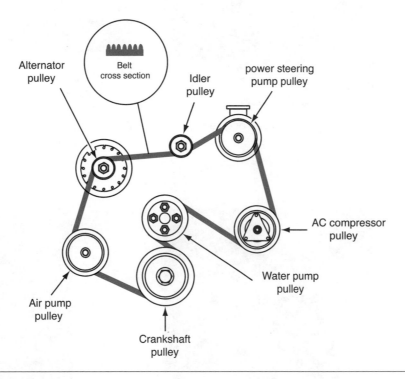

Figure 11-18 A typical serpentine belt.

A belt tension gauge may be used to ensure proper belt tensioning for those systems with manual adjustment. Many late-model vehicles have an automatic belt tensioner, a spring-loaded **idler pulley**. If the belt is manually adjusted, it is suggested that a new belt again be tensioned after about 15 minutes of operation to allow time for initial seating and stretching. The belt should then be checked every 5,000 miles (8,000 kilometers) or so.

Photo Sequence 20
Typical Procedure for Servicing the Serpentine Drive Belt

P20-1 Release the belt tension following specific instructions provided in the manufacturer's shop manual.

P20-2 Remove the belt.

P20-3 Inspect the pulleys for nicks, cracks, bent sidewalls, corrosion, or other damage.

P20-4 Place a straightedge across pulleys to check for alignment.

P20-5 Turn each pulley one-half revolution and repeat step 4.

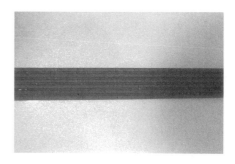

P20-6 Inspect the drive belt. Slight cracks in the belt ribs will not impair performance and is not a cause for replacement.

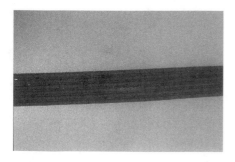

P20-7 If rib sections are missing, the belt should be replaced.

P20-8 Replace the belt by reversing steps 1 and 2.

P20-9 Do not use belt dressing on a serpentine belt. Belt dressing may soften the belt and cause deterioration.

Automatic Belt Tensioner

The drive belts on most latc-model engines are equipped with a spring-loaded automatic **tensioner**. An automatic belt tensioner may be used with all belt configurations, such as with or without power steering and/or air conditioning.

Belt-driven engine accessories are often replaced due to noise or other problem only to learn that the automatic tensioner was at fault. Table 11-1 (on the next page) is an aid in diagnosing drive belt problems.

Replacing A Belt Tensioner

The following procedure is typical for replacing an automatic belt tensioner. Always follow the particular manufacturer's recommended procedures for each vehicle.

1. Attach a socket wrench to mounting bolt of automatic tensioner pulley bolt (Figure 11-19).
2. Rotate tensioner assembly clockwise (cw) until belt tension has been relieved.
3. Remove belt from idler pulley first then remove belt from other pulleys.
4. Disconnect and remove and/or set aside any components hindering tensioner removal.
5. Remove tensioner assembly from mounting bracket.

⚠️ **WARNING:** Because of high spring pressure, do not disassemble automatic tensioner.

6. Remove pulley bolt and remove pulley from tensioner.
7. Install pulley and pulley bolt to tensioner. Tighten bolt to 45 ft.-lb. (61 N·M).
8. Install tensioner assembly to mounting bracket. An **indexing tab** (Figure 11-20) is generally located on the back of the tensioner to align with the slot in the mounting bracket. Tighten the nut to 50 ft.-lb. (67 N·M).
9. Replace any components moved in step 4.
10. Position the drive belt over all pulleys, except idler pulley.

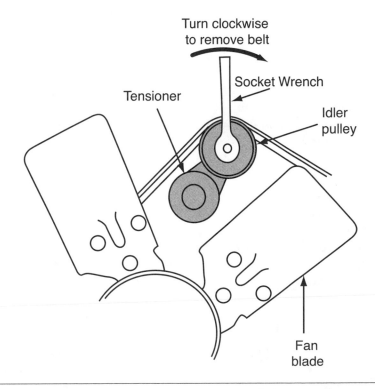

Figure 11-19 Rotate tensioner clockwise (CW) to loosen belt.

Problem	Possible Cause	Possible Remedy
Belt slipping	Belt too loose	Replace and/or tighten belt
	Coolant or oil on belt or pulley	Clean pulleys and replace belt
	Accessory bearing failure (seized)	Replace faulty bearing
	Belt hardened and glazed	Replace belt
Belt squeal when accelerating	Belt glazed or worn	Replace and tension belt
Belt squeak at idle	Belt too loose	Replace and/or tighten belt
	Dirt or paint embedded in belt	Replace and tension belt
	Misaligned accessory pulleys	Align pulleys
	Improper pulley	Replace pulley
Noise (rumble heard or felt)	Belt slipping	Tighten belt
	Defective bearing	Replace bearing
	Belt misalignment	Align belts
	Improper belt	Install proper belt
	Accessory-induced vibration	Locate cause and correct
	Resonant frequency vibration	Vary belt tension and/or replace belt
Belt rolls over (V-belt)	Broken cord in belt	Replace and tension belt
	Belt too loose	Replace and/or tighten belt
	Belt too tight	Replace and/or loosen belt
Belt jumps off	Broken cord in belt	Replace and tension belt
	Belt too loose	Replace and/or tighten belt
	Belt too tight	Replace and/or loosen belt
	Pulleys misaligned	Align pulleys
	Improper pulley	Replace pulley
Belt jumps grooves	Belt too loose	Replace and/or tighten belt
	Belt too tight	Replace and/or loosed belt
	Improper pulley	Replace pulley
	Foreign objects in pulley grooves	Clean grooves or replace pulley
	Pulley misaligned	Align pulley
	Broken belt cordline	Replace belt
Broken belt	Excessive tension	Replace belt and adjust tension
	Belt damaged during installation	Replace belt
	Severe misalignment	Replace and align belt
	Bent or damaged bracket or brace	Repair as required and replace belt
	Pulley and/or bearing failure	Repair as required and replace belt
Rib chunking (rib separation)	Foreign objects imbedded in pulley	Remove objects and replace belt
	Installation damage	Replace belt
Rib or belt wear	Pulley misalignment	Align pulleys
	Abrasive environment	Clean and replace belt
	Rusted pulleys	Clean rust or replace pulleys
	Sharp or jagged pulley groove tips	Replace pulley
	Rubber deteriorated	Replace belt
Belt cracking between ribs	Belt mistracked from pulley groove	Replace belt
	Pulley groove tip has worn away rubber	Replace belt
Backside of belt separated	Contacting stationary object	Correct problem and replace belt
	Excessive heat	Replace belt
	Fractured splice	Replace belt
Cord edge failure	Excessive tension	Replace belt and adjust tension
	Contacting stationary object	Correct problem and replace belt
	Incorrect pulley	Replace pulley and belt
	Incorrect belt	Replace belt

Table 11-1   Troubleshooting Drive and Accessory Belt Problems

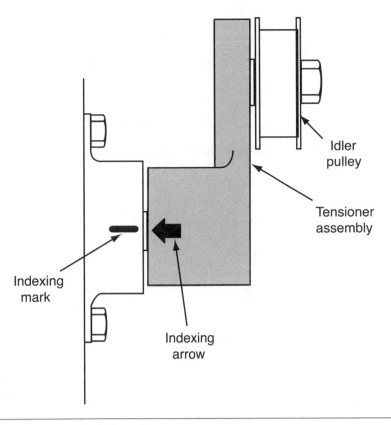

Figure 11-20 Tensioner indexing tab.

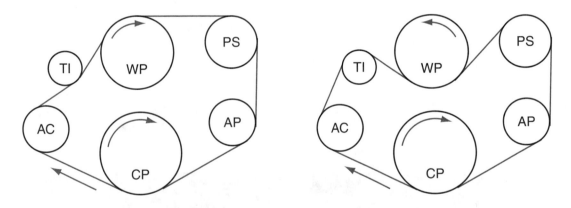

LEGEND:
AC - A/C Compressor Pulley
AP - Alternator Pulley
CP - Crankshaft Pulley
PS - Power Steering Pulley
TI - Tensioner/Idler Pulley
WP - Water Pump Pulley

Figure 11-21 If the belt is not installed correctly, the water pump will be turned in the opposite direction (A) clockwise and (B) counterclockwise.

CAUTION: When installing the serpentine accessory drive belt, the belt must be routed correctly. If not the water pump may rotate in the wrong direction (Figure 11-21), causing the engine to overheat.

11. Using a socket wrench on the pulley mounting bolt of the automatic tensioner, rotate the tensioner clockwise (cw).

12. Place the belt over the idler pulley and allow the tensioner to rotate back into position.

Fans

Classroom Manual
Chapter 11, page 259

If there is any doubt as to the physical condition of the fan, replace it.

Coolant pump-mounted fans (Figure 11-22) occasionally require service. They are damaged due to metal fatigue, collision, road hazards, and abuse. Any condition that causes an out-of-balance pump-mounted fan will result in early pump failure. A check for fan problems is a rather simple task.

1. Remove the belt(s).
2. Visually inspect the fan for cracks, breaks, loose blades, or other damage. Is the fan sound? If yes, proceed with step 3. If no, replace the fan.
3. Hold a straightedge across the front of the fan. Are all blades in equal alignment? If yes, proceed with step 4. If no, replace the fan.
4. Slowly turn the fan while looking for any out-of-true condition or any other damage.
5. Turn the fan fast and look for out-of-true conditions.
6. If the fan fails either test (step 4 or 5), it must be replaced. Due to high operating speeds, it is not recommended that repairs be attempted to an engine cooling fan blade.

Fan Clutch

Classroom Manual
Chapter 11, page 260

If in doubt, replace the fan clutch.

Most air conditioned vehicles with a coolant pump-mounted fan have a fan clutch (Figure 11-23). This device adds about 6 lb. (2.7 kg) to the coolant pump shaft, further increasing the need for a balanced fan.

There are three basic simple tests for troubleshooting a fan clutch. First, make certain that the engine is cold and that it cannot be accidentally started while inspecting the fan clutch.

1. Visually inspect the clutch for signs of fluid loss.
2. Check for a slight resistance when turning the blades.
3. Check for looseness in the shaft bearing.

If the fan clutch fails either of these tests, it should be replaced. There are no repairs for a fan clutch.

Figure 11-22 A damaged coolant pump-mounted fan.

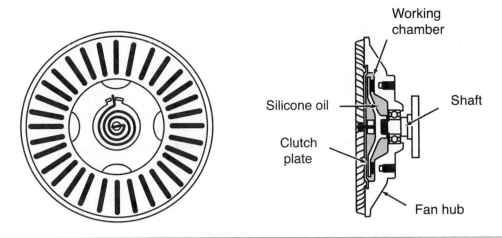

Figure 11-23 A typical fan clutch.

Flexible Fans

Flexible fans are covered in detail in the Classroom Manual. They are subject to the same problems and are tested in basically the same manner as rigid fans.

Classroom Manual
Chapter 11, page 261

Electric Fans

Electric cooling fans (Figure 11-24) are used because there is more precise control over their operation. They may be turned on and off by temperature- and pressure-actuated switches, thereby regulating engine coolant and air conditioning refrigerant temperatures at a more precise level.

Classroom Manual
Chapter 11, page 262

 CAUTION: Electric engine cooling fans may start and operate at any time and without warning. This may occur with the ignition switch off or on.

Follow the schematic in Figure 11-25 for testing and troubleshooting a typical engine cooling fan system.

1. Start the engine and bring the coolant up to operating temperature.
2. Turn on the air conditioner.
3. Disconnect the cooling fan motor electrical lead connector (Figure 11-26).
4. Make sure that the ground wire is not disturbed. If the ground wire is a part of the electrical connector, establish a ground connection with a jumper wire (Figure 11-27).
5. Connect a test lamp from ground to the hot wire of the connector (Figure 11-28). Make sure the lamp is good.

The ground connection must be established to a metal part of the body that is not isolated from electrical ground.

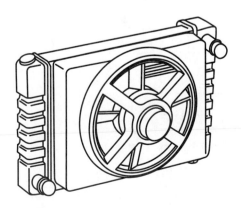

Figure 11-24 A typical electric cooling fan.

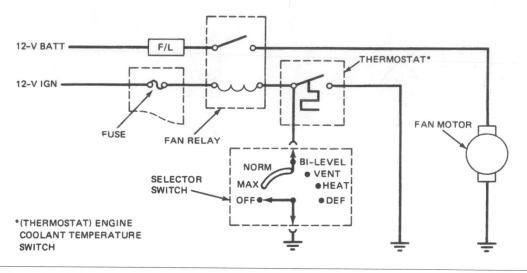

Figure 11-25 An electrical schematic of an engine cooling fan system.

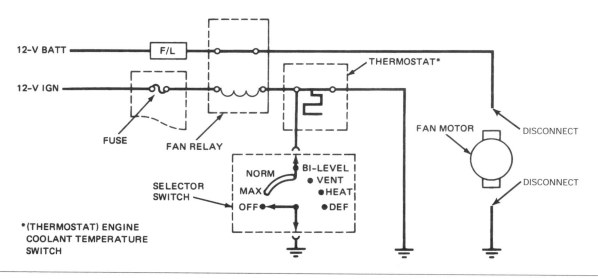

Figure 11-26 Disconnect the cooling fan electric motor.

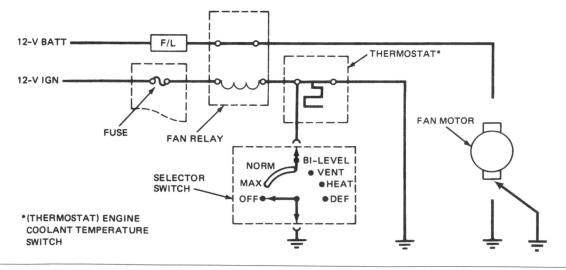

Figure 11-27 Establish ground with a jumper wire.

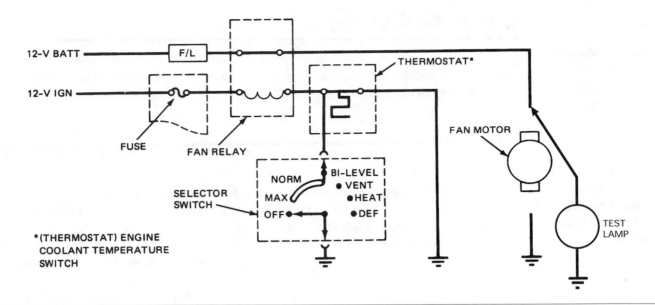

Figure 11-28 Connect a test lamp from ground to the hot wire.

6. Did the lamp light? If yes, proceed with step 7. If no, check for a defective **fan relay** or temperature switch. It is also possible that the engine is not up to sufficient temperature to initiate cooling fan action.

7. Connect a fused jumper wire from the battery positive (+) terminal to the cooling fan connector (Figure 11-29). Make sure the fuse is good.

8. Did the fan start and/or run? If yes, the fan is all right. The system is apparently not warm enough to initiate fan action. If no, proceed with step 9.

9. Again check the fuse in the jumper wire. Is it blown? If yes, the motor is shorted and must be replaced. If no, the motor is open and must be replaced.

Make certain that the engine is up to normal operating temperature.

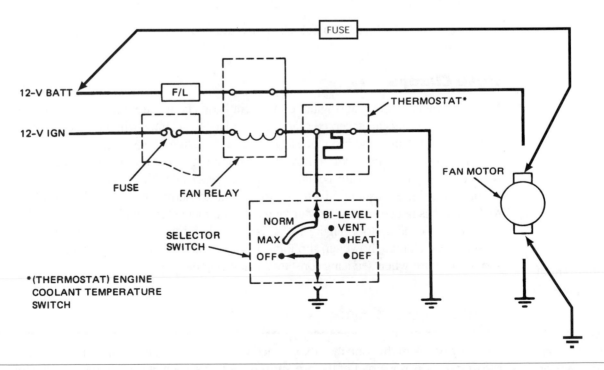

Figure 11-29 Connect a fused jumper wire from the battery positive (+) terminal to the cooling fan connector.

Hoses and Clamps

Classroom Manual
Chapter 11,
pages 264, 267

Engine cooling system hoses and clamps should be replaced every few years. This should become part of a good preventive maintenance program. If done on a periodic schedule, more expensive repairs, such as those caused by an overheating engine, are not as likely to occur.

Hoses

Carefully check all cooling system hoses when a vehicle is being serviced. Following is a simple checklist for this service:

Replace all hoses if
any are found to be
defective.

1. Check for leaks, usually noted by a rust color at the point of the leak.
2. Check for swelling, usually obvious when the engine is at operating pressure.
3. Check for chafing, usually caused by a belt or other nearby component.
4. Check for a soft or spongy hose that would indicate chemical deterioration.
5. Check for a brittle hose indicating repeated heating, usually near the coolant pump.
6. Squeeze the hose. If its outer layer splits or flakes away, replace the hose.
7. Squeeze the lower radiator hose. If the reinforcing wire is missing (due to rust and/or corrosion), replace the hose.

 SERVICE TIP: Some engines have additional hoses such as a small bypass hose between the coolant pump and the engine block, hoses used to carry coolant to heat the throttle body on fuel injected engines, and short hoses used to interconnect coolant carrying components on certain engines (Figure 11-30). Do not overlook these hoses when checking the cooling system.

Replacing a Hose. It is a good practice to replace all of the radiator and heater hoses if any of them are found to be defective. It is not always possible to convince the customer that this should be done.

1. Loosen the hose clamp at both ends of the hose (Figure 11-31).
2. Firmly but carefully twist and turn the hose to break it loose from the coolant pump and/or radiator. Using a box cutter to slice through the hose will help facilitate its removal.
3. Remove the hose (Figure 11-32).

WARNING: Do not use unnecessary force when removing the hose end from the radiator.

Hose Clamps

Many technicians feel that original equipment hose clamps are only good for one-time use. In almost all cases, it is an accepted practice to replace the clamps when replacing a hose. They are inexpensive and are good insurance against an early failure.

At least one manufacturer, however, recommends that a **constant tension hose clamp** (Figure 11-33), used on many cooling systems, be replaced with an original equipment clamp. A number or letter is stamped into the tongue of constant tension clamps for identification. If replacement is necessary, use only an original equipment clamp with matching number or letter.

A special clamp tool (Figure 11-34) is available for use in removing and replacing constant tension hose clamps; however, slip joint pliers may also be used. In either case, eye protection should be worn when servicing constant tension clamps.

Classroom Manual
Chapter 11, page 266

Recovery Tank

The recovery tank is
often referred to as
an overflow tank or
a surge tank.

The only problem that one may experience with a recovery tank is an occasional leak. Since the recovery tank is not a part of the pressurized cooling system, it can often be successfully repaired with hot glue. If it is found to be leaking, proceed as follows:

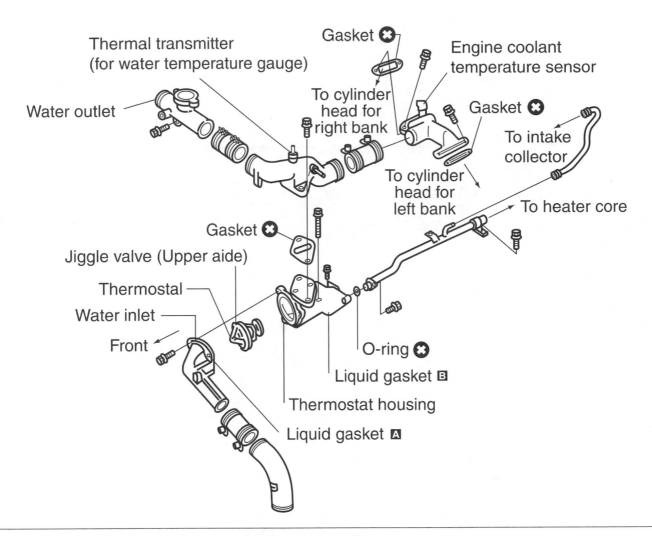

Figure 11-30 Do not neglect interconnecting hoses in the cooling system.

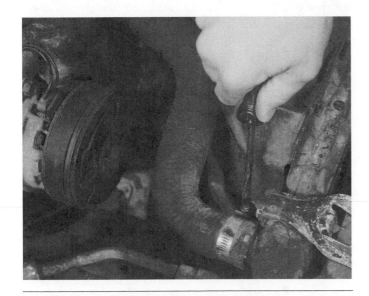

Figure 11-31 Loosen the radiator hose clamp.

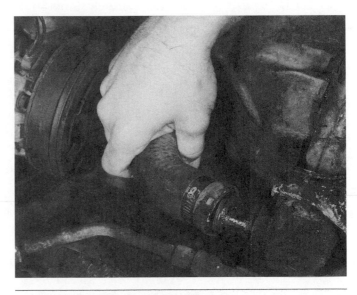

Figure 11-32 Remove the hose.

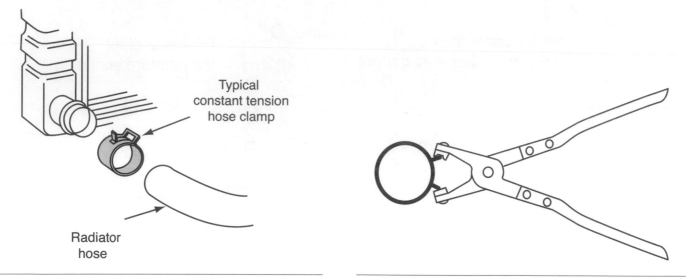

Figure 11-33 A typical constant tension hose clamp.

Typical
constant tension
hose clamp

Radiator
hose

Figure 11-34 A typical hose clamp tool.

1. Remove the tank from the vehicle.
2. Thoroughly clean the tank inside and outside.
3. Use a piece of sandpaper to roughen up the surface at the area of the leak.
4. Use a hot glue gun and make several small beads of hot glue at the point of the leak. Cover the area thoroughly, overlapping each successive bead.
5. If it is accessible, repeat steps 3 and 4 inside the tank.

Heater System

The comfort heater system is actually a part of the engine cooling system. The heater core, a small radiator-type **heat exchanger**, is located in the case/duct system of the heater/air conditioner unit.

Classroom Manual
Chapter 11, page 267

A wet floor carpet can also be caused by a leaking seal around the windshield.

Heater Core

Most failures of the heater core (Figure 11-35) are due to a leak. This is easily detected by noting a wet floor carpet just below the case on the passenger side of the vehicle. Replacement of the heater core, unfortunately, is not so simple. Because of the many different variations of installation, it is necessary to follow the manufacturer's shop manual instructions for replacing the heater core.

Following is a typical procedure only and is not intended for any particular make or model vehicle:

1. Remove the coolant.
2. Remove the access panel(s) or the split heater/air conditioning case to gain access to the heater core.
3. Loosen the hose clamps and remove the heater coolant hoses.
4. Remove the cable and/or vacuum control lines (if equipped).
5. Remove the heater core securing brackets and/or clamps.
6. Lift the core from the case.

▲ **WARNING:** Do not use force. Take care not to damage the fins of the heater core when removing and replacing it.

Control Valve

The control valve is a cable-, vacuum-, or electrically operated shut-off valve used at the inlet of the heater core to regulate coolant flow through the core. Other than a leak, which is usually obvious, the valve fails due to rust or corrosion. To replace the valve:

Figure 11-35 A typical heater core.

Figure 11-36 A typical heater flow control (water valve). (Courtesy of BET, Inc.)

1. Remove the coolant to a level below the control valve.
2. Remove the cable linkage, vacuum hose(s), or electrical connector from the control valve.
3. Loosen the hose clamps and remove the inlet hose from the control valve.
4. Remove the heater control, as applicable (Figure 11-36). Remove the outlet hose from the heater core. Remove the attaching brackets or fasteners from the control.
5. Inspect the hose ends removed. If they are hard or split, cut 0.5–1.0 in. (12.7–25.4 mm) from the damaged ends. Better yet, replace the hoses.

Hoses and Clamps

Heater hoses and clamps are basically about the same as radiator hoses and clamps except they are generally smaller in diameter. It is a practice of some technicians to use a hose that is too large for the application and overtighten the hose clamp to stop the leak. A hose clamp that is too large for the hose is often distorted when tightened sufficiently to secure the hose.

Hoses. Heater hoses are replaced in the same manner as radiator hoses. It is much easier to use the wrong size hose, however. For example, a 3/4-in. hose fits very easily onto a 5/8-in. fitting. The hose clamp then must be overtightened (Figure 11-37) to squeeze the hose onto the fitting sufficiently to prevent a leak. It is not so easy to slide a 5/8-in. hose onto a 5/8-in. fitting. The intent, however, is to use the proper size hose for the application.

It is good practice to replace all heater hoses if any are found to be defective. Following is a typical procedure:

1. Remove the coolant to a level below that of the hoses to be replaced.
2. Loosen the hose clamp at both ends of the hose.
3. Turn and twist the hoses to break them loose.
4. Remove the hose. Do not use unnecessary force when removing the hose end from the heater core.

Clamps. As with cooling system hose clamps, heater hose clamps should be replaced when a hose is replaced. It is most important that the proper size clamp be used for the hose. If the clamp is too large, it will be distorted before being tightened enough to secure the hose onto the fitting (Figure 11-38). When this occurs, it is extremely difficult to stop a leak.

Most control valves are not omnidirectional. Observe proper direction of coolant flow.

If a hose slides onto a fitting very easily, it is probably too large. The fit should require slight resistance.

Classroom Manual Chapter 11, pages 264, 267

Figure 11-37 A hose clamp overtightened to compensate for a hose that is too large for the application.

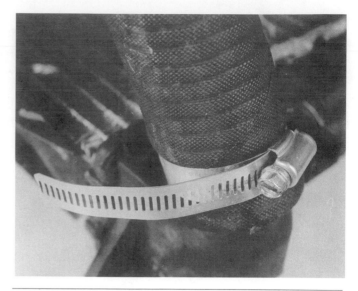

Figure 11-38 A hose clamp that is too large for the hose application.

Antifreeze

> ⚠️ **WARNING:** Antifreeze solution is considered a hazardous material. Dispose of antifreeze in an environmentally safe manner. Refer to all applicable federal, state, and local ordinances and regulations.

Classroom Manual
Chapter 11, page 268

1. Make sure that the engine is cool and the cooling system is not under pressure.
2. Place a clean, dedicated container of adequate size under the drain provision of the cooling system.
3. Open the radiator drain provision and drain the cooling system.

> ✅ **SERVICE TIP:** If the coolant is to be reused, drain it into a clean container. If, however, it is not to be reused, it must be disposed of or recycled in a manner considered to be environmentally safe.

> ⚠️ **WARNING:** Refer to local ordinances and regulations regarding proper disposal procedures for ethyl glycol-type antifreeze solutions.

Preventive Maintenance

Changing antifreeze/coolant annually helps to prevent cooling system failure, the primary cause of engine-related breakdowns. The antifreeze/coolant, depending on mix ratio, can provide protection for the cooling systems from –84–276°F (–64–135.6°C). The generally recommended ratio of 1:1 (50 percent antifreeze and 50 percent water) provides protection against freezing with an ambient temperature as low as –34°F (–36.7°C) and provides protection from rust and corrosion for the cooling system metals. This includes protection for the thin, lightweight aluminum radiator found in many late-model vehicles.

Extended Life

Extended-life antifreeze generally provides freeze-up and boil over protection for up to five years or 150,000 miles (241,350 kilometers). High-mileage drivers as well as those who do not have, or take, the time for regular vehicle preventive maintenance (PM) should rely on the long-lasting protection

Photo Sequence 21
Draining and Refilling the Cooling System

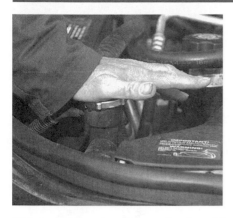

P21-1 Ensure that the engine is cold and slowly remove the radiator cap. **CAUTION:** If the radiator cap is removed from a hot cooling system, serious personal injury may result.

P21-2 Place a drain pan of adequate size under the radiator drain cock.

P21-3 Install one end of a tube or hose on the drain cock and position the other end in the drain pan.

P21-4 Open the radiator drain cock and allow the radiator to drain until the flow stops.

P21-5 Place a drain pan of adequate size under the engine.

P21-6 Remove the drain plug from the engine block and allow the engine block to drain until the flow stops. NOTE: There may be more drainage from the radiator at this time.

P21-7 Close the radiator drain cock and replace the engine block drain plug.

P21-8 Remove the pans and dispose of the coolant a manner consistent with local regulations.

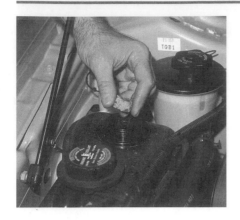

P21-9 Add sealant pellets, if required. Some car lines require that sealant pellets be added to the radiator whenever the cooling system is drained and refilled with fresh coolant. Failure to use the correct sealant pellets may result in premature water pump leakage.

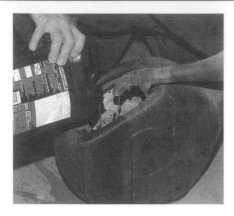

P21-10 Premix the antifreeze with clear water to a 50/50 ratio. NOTE: Distilled water is preferred by many technicians.

P21-11 With a large funnel in the radiator fill hole slowly pour in the coolant mixture. **NOTE:** Refer to manufacturer's specifications for the cooling system capacity.

P21-12 Fill the cooling system to within about 1 in. (25.4 mm) below the fill hole.

P21-13 Start the engine and let the cooling system warm up. **NOTE:** When the thermostat opens the coolant level may drop. After the thermostat opens, add coolant until the level is up to the fill hole.

P21-14 Replace the radiator cap.

P21-15 Check the coolant level in the recovery reservoir and add coolant if needed.

of an extended-life antifreeze/coolant. Such mixtures are silicate and phosphate free, providing extended protection against rust and corrosion to all metals of the cooling system. Extended-life antifreeze, such as Havolin's Extended Life DEX-COOL™, manufactured from ethylene glycol (EG), meets all of the compatibility requirements for the extended-life antifreeze used in most General Motors (GM) vehicles since 1996.

Low Tox

Low-tox antifreeze contains propylene glycol (PG) and is less toxic than ethylene glycol (EG) types. A low-tox antifreeze/coolant mixture offers similar freeze-up and boil over protection while providing rust and corrosion protection for all cooling system parts, including the aluminum used in radiators. Also, low-tox antifreeze/coolant provides an added margin of safety if accidentally ingested by pets or wildlife.

Cooling System Treatment

While it is important to flush and fill the cooling system as recommended by the vehicle manufacturer, not even the most thorough maintenance and the best antifreeze can provide a perfect level of protection. Today's modern engines are more complex than ever and require a good preventive maintenance (PM) program. Some of the simplest, and perhaps most important, vehicle PM is by pouring in the right fluid at the right time. One such fluid, Prestone's Cooling System Treatment, is used to protect aluminum radiators and other metals found in the cooling system. For use in all types of coolants, it also contains patented grafted polymer water pump additives to provide vital water pump lubrication and refreshes the corrosion inhibitor capabilities of older coolant. Prestone recommends that their Cooling System Treatment be added to the coolant recovery tank annually for maximum performance and cooling system protection.

Drain and Flush. If replacing existing antifreeze/coolant with any other type of antifreeze/coolant, first completely drain and flush the cooling system. Be aware that if antifreeze solutions are mixed, the intended protection of either solution may be lost, particularly the added margin of safety afforded by the propylene glycol (PG) formulas.

Both types of antifreeze, EG and PG, are biodegradable. It is the rate at which they degrade, however, that is important. EG, though considered more toxic than PG, degrades the fastest. For more specific information, request a Material Safety Data Sheet (MSDS) from the manufacturer of the particular product that is being used. All used EG and PG coolants must be recycled for reuse or disposed of in a manner consistent with applicable local, state, and national regulations. Both used EG and PG coolants, though considered toxic, are not considered as a hazardous waste by the Environmental Protection Agency (EPA) unless they contain more than five parts per million (5 ppm) lead (Pb). Lead is a hazardous byproduct that leaches out of the solder used to secure the tubes and header tanks.

Coolant Recovery/Recycle. Several companies manufacture antifreeze recovery/recycle/recharge machines. The particular system illustrated is connected to the vehicle's cooling system and injects new or recycled coolant into the cooling system while forcing the old coolant out of the system. This type machine is ideal for field use where electricity is not available. The pump is powered by the vehicle's battery while the engine is running.

Flush the Cooling System

When replacing an existing antifreeze/coolant with an extended-life antifreeze/coolant, the cooling system must first be completely drained and flushed. As mentioned above, this is necessary to gain the full benefits of the longer-lasting formula. For example, if an extended-life antifreeze/coolant is currently being used in a vehicle and a regular-type antifreeze is added to the cooling system the extended life protection will be lost. When adding antifreeze, always add the same type as that previously used in the cooling system. If the type is unknown, it is generally recommended to drain, flush, and refill the cooling system. The procedure that follows is typical:

NOTE: Often instructions are given in the owner's manual and one may be told to open an air bleed valve on the engine or to remove a heater hose to purge air that may have entered the engine during draining.

1. Drain the cooling system following steps 1 through 5 of Job Sheet 37.
2. Close the drain valve.
3. Recycle or dispose of the used coolant according to local laws and regulations.

NOTE: Label the container clearly as used antifreeze. Do not use beverage containers to store antifreeze, new or used. If stored, keep the containers away from children and animals. If the antifreeze is to be disposed, do so promptly and properly.

4. Flush the cooling system to clean the engine block of any scale, rust, or other debris before refilling with new antifreeze/coolant.

NOTE: A cooling system flush may be used to remove stubborn rust, grease and sediment may not be removed by plain water alone.

5. Remove the radiator cap and fill the radiator with cleaner (if used) and water.
6. Run the engine with the heater on HI and the temperature gauge reading normal operating temperature for the time recommended on the flush product label.

NOTE: An infrared (IR) thermometer is a handy tool to use for determining when "normal" engine operating temperature has been reached.

7. Stop the engine and allow it to cool.
8. Again, open the drain valve and drain the cooling system.
9. Close the drain valve and refill radiator with plain water.
10. Run the engine for about 15 minutes at normal engine operating temperature.
11. Stop the engine and allow it to cool, open the drain valve, and redrain the cooling system.
12. Close the drain valve and refill the cooling system following steps 6 through 10 of Job Sheet 37.

NOTE: Check the owner's manual for the cooling system capacity as well as any special service instructions.

13. Once the radiator is filled, run the engine at normal operating temperature with the heater on HI for 15 minutes to mix and disperse the coolant fully throughout the cooling system.
14. Shut off the engine and allow the cooling system to cool.
15. Check the coolant level and concentration. Adjust, if necessary.
16. After a few days driving recheck the cooling system. A hydrometer may be used for EG testing and test strips may be used for PG testing.

NOTE: If additional coolant is required use a premixed 50/50% concentrate. If additional coolant is needed it is suggested that the cooling system be leak tested.

Troubleshooting the Heater and Cooling System

The following procedure is given as a quick reference to enable the service technician to isolate many of the conditions that can cause improper engine cooling system or heater operation. This procedure is given in three parts: engine overcooling, engine overheating, and loss of coolant.

The customer's complaint would usually be for an overheating condition. If the problem is due to a loss of coolant, the customer may complain that coolant or water (H_2O) must be added frequently.

It should be noted that cooling system problems are often caused by, or may cause, air conditioning system problems. Conversely, air conditioning problems may be caused by, or may cause, cooling system problems.

Engine Overcooling

Possible Cause	Possible Remedy
1. Thermostat missing	Replace the thermostat
2. Thermostat defective	Replace the thermostat
3. *Defective temperature sending unit	Replace the sending unit
4. *Defective dash gauge	Replace the dash gauge
5. *Broken or disconnected wire	Repair or replace the wire
6. *Grounded or shorted cold indicator wire (if equipped with a cold lamp)	Repair or replace the wire

*These symptoms indicate overcooling though the engine temperature may be within safe limits.

Engine Overheating

Possible Cause	Possible Remedy
1. Collapsed radiator hose	Replace the radiator hose
2. Coolant leak	Locate and repair the leak
3. Defective water pump	Replace the water pump
4. Loose fan belt(s)	Torque fan belts to specs
5. Defective fan belt(s)	Replace the fan belts
6. Broken belt(s)	Replace the belt(s)
7. Fan bent or damaged	Replace the fan
8. Fan broken	Replace the fan
9. Defective fan clutch	Replace the fan clutch
10. Exterior of radiator dirty	Clean the radiator
11. Dirty bug screen	Clean or remove the screen
12. Damaged radiator	Repair or replace the radiator
13. Engine improperly timed	Service the engine
14. Engine out of tune	Service the engine
15. *Temperature sending unit defective	Replace the sending unit
16. *Dash gauge defective	Replace the dash gauge
17. *Grounded or shorted indicator wire	Repair or replace the wire

*These symptoms indicate overheating though the engine temperature may be within safe limits.

An overheated engine may result in poor to no cooling from the air conditioner.

An air conditioning system malfunction may result in an overheating engine condition.

Loss of Coolant

Possible Cause	Possible Remedy
1. Leaking radiator hose	Replace the radiator hose
2. Leaking heater hose	Replace the heater hose
3. Loose hose clamp	Tighten the clamp
4. Leaking radiator (**external**)	Repair or replace the radiator
5. Leaking transmission cooler (**internal**)	Repair or replace the radiator
6. Leaking coolant pump shaft seal	Repair or replace the coolant pump
7. Leaking gasket(s)	Replace the gaskets
8. Leaking core plug(s)	Replace the core plug(s)
9. Loose engine head(s)	Retorque the head(s)
10. Warped head(s)	Replace the head(s)
11. Excessive coolant	Adjust the coolant level
12. Defective radiator pressure cap	Replace the cap
13. Incorrect pressure cap	Replace the cap
14. Defective thermostat	Replace the thermostat
15. Incorrect thermostat	Replace the thermostat
16. Rust in system	Flush the system and add rust protection
17. Radiator internally clogged	Clean or replace the radiator
18. Heater core leaking	Repair or replace the core
19. Heater control valve leaking	Replace the valve

CASE STUDY

A customer brings his car into the shop because the temperature gauge does not operate. It remains on cold all of the time regardless of engine heat conditions.

The lead wire to the sending unit is disconnected and a test light is used to probe for voltage. The test light comes on when the ignition switch is placed in the ON position. When the lead is connected to ground (–) through a 10-Ω resistor, the dash unit needle moves to the full hot position. This is a normal operation according to the service manual.

The diagnosis is that the sending unit is defective. It is replaced after approval by the customer and the temperature gauge system is returned to normal operation.

Terms to Know

Ambient temperature	External	Indexing tab
Bypass	Fan relay	Internal
Bypass hose	Fill neck	Ram air
Clutch pulley	Heat exchanger	Serpentine belt
Constant tension hose clamp	Idler	Tensioner
Discipate	Idler pulley	V-belt

ASE Style Review Questions

1. Engine overcooling is being discussed:
   Technician A says that a defective thermostat cannot cause this condition.
   Technician B says that a thermostatic control switch stuck in the closed position in an electric cooling fan circuit cannot be the cause.
   Who is right?
   A. A only   C. Both A and B
   B. B only   D. Neither A nor B

2. Assume that the temperature gauge (dash) indicates HOT but the engine coolant is in the proper range.
   Technician A says that the problem may be a grounded or shorted wire sending unit-to-dash gauge.
   Technician B says the problem may be a defective thermostat.
   Who is right?
   A. A only   C. Both A and B
   B. B only   D. Neither A nor B

3. All of the following may cause the back side of a serpentine belt to separate EXCEPT:
   A. Contacting stationary object
   B. Excessive heat
   C. Dractured splice
   D. Pulley misalignment

4. Coolant loss is being discussed:
   Technician A says that a missing thermostat could be the problem.
   Technician B says a heater control valve stuck open may be the problem.
   Who is right?
   A. A only   C. Both A and B
   B. B only   D. Neither A nor B

5. *Technician A* says that air conditioning system problems can cause cooling system problems.
   Technician B says that cooling system problems can cause air conditioning system problems.
   Who is right?
   A. A only   C. Both A and B
   B. B only   D. Neither A nor B

6. Temporary repairs are being discussed. It is decided that some repairs may be made without affecting cooling system performance.
   Technician A says that a loss of coolant due to an overheating engine may be repaired by increasing the pressure range of the coolant pressure cap.
   Technician B says that a loss of coolant due to a leaking heater core may be repaired by disconnecting the heater core.
   Who is right?
   A. A only   C. Both A and B
   B. B only   D. Neither A nor B

7. An overheating condition is being discussed:
   Technician A says that replacing the thermostat with one of a lower temperature rating will reduce the coolant temperature.
   Technician B says that replacing the pressure cap with one of a lower rating will reduce the coolant temperature.
   Who is right?
   A. A only   C. Both A and B
   B. B only   D. Neither A nor B

8. Damaged fan blades are being discussed:
   Technician A says loose blades should be secured by welding.
   Technician B says bent blades should be straightened using proper tools.
   Who is right?
   A. A only   C. Both A and B
   B. B only   D. Neither A nor B

9. Electric cooling fans are being discussed:
   Technician A says that they are independent of the ignition switch and may start and run without notice at any time.
   Technician B says they are generally protected by a shroud and pose no safety problem.
   Who is right?
   A. A only   C. Both A and B
   B. B only   D. Neither A nor B

10. Disposal of hazardous waste is being discussed:
    Technician A says ethyl glycol antifreeze must be disposed of in a manner prescribed by the Environmental Protection Agency (EPA) and/or local ordinances.
    Technician B says similar regulations govern the disposal of other refrigerants, such as CFC-12.
    Who is right?
    A. A only   C. Both A and B
    B. B only   D. Neither A nor B

ASE Challenge

1. Radiators are constructed of all the following materials, EXCEPT:
   A. Magnesium   C. Copper
   B. Aluminum   D. Plastic

2. A properly maintained cooling system removes about ____ percent of the total engine heat.
   A. 25   C. 45
   B. 35   D. 55

3. The least likely problem associated with a cooling system that had the thermostat removed is:
   A. Poor heater performance
   B. Erratic computer engine control
   C. Lower than normal operating temperature
   D. Loss of coolant

4. The most likely use for an engine coolant temperature switch is to electrically energize the:
   A. Compressor clutch
   B. Cooling system fan motor
   C. Blower motor
   D. Coolant "hot" warning light

5. If the crankshaft pulley turns clockwise (cw), all of the following statements about the illustration at right are true, EXCEPT:
   A. The compressor turns clockwise (cw)
   B. The idler pulley turns counter clockwise (ccw)
   C. The water pump turns clockwise (cw)
   D. The alternator pulley turns clockwise (cw)

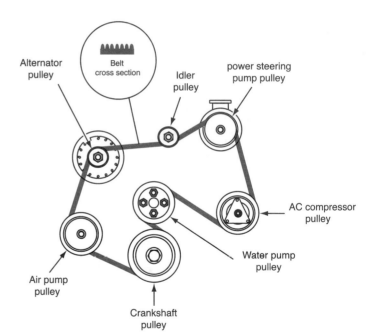

Table 11-1 ASE TASK

Inspect, test, and replace heating, ventilating, and A/C
vacuum actuators (diaphragms, motors) and hoses.

Problem Area	Symptoms	Possible Causes	Classroom Manual	Shop Manual
IMPROPER AIR FLOW	No air flow	Defective vacuum motor	229	357
	No fresh air from vents	Defective actuator: a. motor, if electric b. "pot" if pneumatic c. retainer, if cable	229	357
	No cool air from vents	1. Defective master control: electric, pneumatic, or manual	216	357
		2. Defective: a. hose, if pneumatic b. vacuum motor	229	
		3. Defective or inoperative air conditioner	71–73	93
		4. Defective (open) heater coolant flow control valve	267	422–423
		5. Duct disconnected or missing	175–184	281–286
	No warm air from vents	1. Defective master control: electric pneumatic, or manual	216	357
		2. Defective vacuum a. hose b. motor	229	357
		3. Defective (closed) heater coolant flow control valve	267	422–423
		4. Heater (hoses) disconnected	264	424
		5. Duct disconnected or missing	175–184	281–286

Table 11-2 ASE TASK

Inspect, test, and replace low engine coolant temperature blower control system.

Problem Area	Symptoms	Possible Causes	Classroom Manual	Shop Manual
ENGINE OVERHEATS	Auxiliary blower inoperative	1. Blown fuse or circuit breaker	215	348
		2. Loose or disconnected wiring	222–225	351–352, 360
		3. Defective motor	262	351–352
		4. Defective relay	263	351–352, 360
		5. Defective temperature control	217	350
	Auxiliary fan intermittent	1. Loose wiring	222–225	351–352, 360
		2. Defective motor	262	351–352
		3. Defective relay	263	351–352, 360
		4. Defective temperature control	217	351–352, 360

Table 11-3 ASE TASK

Inspect, test, and replace heater water valve and controls.

Problem Area	Symptoms	Possible Causes	Classroom Manual	Shop Manual
NO HEATING CONDITION	Air from vents not warm/hot	1. Defective master control	216	357
		2. Defective temperature control	217	355
		3. Defective coolant flow control valve	267	422–423
		4. Defective relay	263	351–352, 360
		5. Defective sensor	233	360
		6. Defective module	239	351, 360
		7. Defective thermostat	217–221	409
INTERMITTENT HEATING	Insufficient heat from vents	1. Defective thermostat	217–221	409
		2. Defective module	239	351, 360
		3. Defective relay	263	351–352, 360
		4. Loose or defective wiring	222–225	351–352, 360

JOB SHEET 37

37

Name _____ Date _____

Remove and Replace Coolant

On completion of this job sheet you should be able to remove and replace cooling system coolant.

ASE Correlation

This job sheet is related to the ASE Heating and Air Conditioning Test's content area: *Heating and Engine Cooling Systems Diagnosis and Repair,* Task: *Identify, inspect, recover coolant; flush and refill system with proper coolant.*

Tools and Materials

Late-model vehicle
Shop manual
Two pans
Safety glasses or goggles
Hazardous waste container
Funnel
Rubber hose
Hand tools, as required

Describe the vehicle being worked on:

Year _____ Make _____ Model _____

VIN _____ Engine type and size _____

Procedure

Follow the procedure outlined in the service manual. Photo Sequence 21 may also be used as a guide wherever applicable. Write a brief description of each step.

1. Ensure that the engine is cold and slowly remove the radiator cap.
   CAUTION: If the radiator cap is removed from a hot cooling system, serious personal injury may result.

   _____

   _____

2. Place a drain pan of adequate size under the radiator drain cock and install one end of a tube or hose on the drain cock. Position the other end in the drain pan. Open the radiator drain cock and allow the radiator to drain until the flow stops.

   _____

   _____

3. Place a drain pan of adequate size under the engine. Remove the drain plug from the engine block and allow the engine block to drain until the flow stops.

   _____

   _____

4. Close the radiator drain cock and replace the engine block drain plug.

_____

_____

5. Remove the pans and dispose of the coolant in a manner consistent with local regulations.

_____

_____

6. Add sealant pellets, if required.

_____

_____

7. Premix antifreeze with clear water to a 50/50 ratio.

_____

_____

8. With a large funnel in the radiator fill hole, slowly pour in the coolant mixture. (Refer to the manufacturer's specifications for the cooling system capacity.) Fill to about 1.0 in. (25.4 mm) below the fill hole.

_____

_____

9. Start the engine and let the cooling system warm up. When the thermostat opens, the coolant level may drop. If it drops, add coolant until the level is up to the fill hole.

_____

_____

10. Replace the radiator cap and check the coolant level in the recovery reservoir. Add coolant if needed.

_____

_____

✔ **Instructor's Check** _____

JOB SHEET 38

38

Name _____ Date _____

Leak Test a Cooling System

On completion of this job sheet you should be able to leak test a cooling system.

ASE Correlation

This job sheet is related to the ASE Heating and Air Conditioning Test's content area: *Heating and Engine Cooling Systems Diagnosis and Repair,* Task: *Inspect, test, and replace radiator, pressure cap, coolant recovery system, and water pump.*

Tools and Materials

Late-model vehicle
Shop manual
Safety glasses or goggles
Cooling system leak tester
Hand tools, as required

Describe the vehicle being worked on:

Year _____ Make _____ Model _____

VIN _____ Engine type and size _____

Procedure

Follow the procedures typically outlined in the Shop Manual. Write a brief summary of your procedure and findings following each step. Ensure that the engine is cold for this procedure. Wear eye protection while pressure testing the cooling system.

1. Note cooling system operating pressure on the radiator cap. Verify the pressure by referring to the specifications in the manufacturer's service manual.

_____

_____

2. Remove radiator cap and adjust the coolant level to 0.5 in. (12.7 mm) below the bottom of the fill neck.

_____

_____

3. Attach pressure tester to the filler neck of the radiator. Some radiators require an adapter. _____

_____

_____

4. While observing the gauge, pump the tester until it indicates the same pressure noted in the specifications. What is the specifications pressure? _____
What gauge pressure could you pump in the system ? _____
Explain: _____

5. If the gauge pressure is not correct in step 4, proceed with step 6. If correct, proceed with step 7. Gauge pressure was: Correct _____ Not correct _____

6. Make a visual inspection for coolant leaks. List those areas where a leak is observed.

_____

_____

7. Repair leaks, as required, and repeat leak testing, beginning with step 4.

_____

_____

8. Let pressure stand for 5 minutes. If the pressure is the same as in step 4, proceed with step 9. If not, repeat leak testing beginning with step 4.

_____

_____

9. Release the pressure and remove the tester from the cooling system. Attach the tester to the pressure cap.

_____

_____

10. Pump the tester while observing the gauge. What is the maximum pressure? _____
   a. Does the pressure reach that noted on the radiator cap? _____
   b. Does the excess pressure release when rated pressure is reached? _____
   c. What are your conclusions regarding the pressure cap? _____

☑ **Instructor's Check** _____

JOB SHEET 39

39

Name _____ Date _____

Replace Thermostat

On completion of this job sheet you should be able to replace a cooling system thermostat.

ASE Correlation

This job sheet is related to the ASE Heating and Air Conditioning Test's content area: *Heating and Engine Cooling Systems Diagnosis and Repair,* Task: *Inspect, test, and replace thermostat, bypass, and housing.*

Tools and Materials

Late-model vehicle
Service manual
Safety glasses or goggles
Gasket scraper
Hand tools, as required
Thermostat and gasket, as required.

Describe the vehicle being worked on:

Year _____ Make _____ Model _____
VIN _____ Engine type and size _____

Procedure

Follow procedures outlined in the service manual. Give a brief description of your procedure following each step. Ensure that engine is cold and wear eye protection.

1. Drain coolant to a level below the thermostat. Follow appropriate procedures outlined in Job Sheet 37.

   _____

   _____

2. If required, remove components to gain access to thermostat housing.

   _____

   _____

3. Remove retaining bolts and/or nuts and lift off thermostat housing.

   _____

   _____

4. Remove thermostat. Using a gasket scraper, carefully remove old gasket material.

   _____

   _____

5. Install a new thermostat and gasket. **CAUTION:** Make sure that the thermostat is not installed upside down. Have your instructor initial here at this time. _____

What is the thermostat temperature rating? _____ °F. _____ °C.

6. Replace thermostat housing and retaining bolts and/or nuts. _____
Torque to _____ in.-lb. _____ N·m. Procedure _____

_____

_____

7. Replace components removed in step 2, if any.

_____

_____

8. Replace coolant removed in step 1.

_____

_____

☑ **Instructor's Check** _____

JOB SHEET 40

40

Name _____ Date _____

Troubleshoot an Electric Engine Cooling Fan

On completion of this job sheet you should be able to troubleshoot and repair electric engine cooling fans.

ASE Correlation

This job sheet is related to the ASE Heating and Air Conditioning Test's content area: *Operating Systems and Related Controls Diagnosis and Repair, 1. Electrical,* Task: *Inspect, test, repair, and replace engine cooling/condenser fan motors, relays/modules, switches, sensors, wiring, and protection devices.*

Tools and Materials

Late-model vehicle
Service manual
Fused jumper wire
Hand tools

Describe the vehicle being worked on:

Year _____ Make _____ Model _____
VIN _____ Engine type and size _____

Procedure

Refer to the appropriate service manual for specific procedures for troubleshooting an electric engine cooling/condenser fan motor. Failure to do so can result in serious damage to the control system. The following may be used as a guide only. After each step write a brief report of your procedure.

1. Start the engine and allow the coolant to reach operating temperature. Did the fan turn on? _____ Explain _____

2. Turn on the air conditioner. Did the fan turn on? _____ Explain _____

   _____

   _____

3. Turn the air conditioner OFF and stop the engine. If the fan was running, does it continue to run? _____ Explain

   _____

   _____

4. Carefully disconnect the fan electrical connection. Is this a single- or double-lead connector?

   _____

   _____

5. Refer to the wiring diagram. What is the color of the wire that supplies power to the motor? _____ What is the color of the ground wire? _____

6. Using a fused jumper wire, connect between the fan motor lead and positive battery terminal. DO NOT connect to the motor ground lead. Does the motor operate? _____ Did the fuse "blow"? _____ Explain:

_____

_____

7. If the motor ran in step 6, further testing of the electrical circuit is required. Follow the specific procedure outlined in the service manual. Give your step-by-step procedure and conclusions in the space below.

a. _____

_____

b. _____

_____

c. _____

_____

d. _____

_____

✓ **Instructor's Check** _____

ASE Practice Examination

Final Exam Automotive Heating and Air Conditioning A7

1. What component part of the air conditioning system causes the refrigerant to change from a liquid to a vapor?
   A. Evaporator
   B. Compressor
   C. Condenser
   D. Metering device

2. Which of the following statements about the engine cooling fan are most correct?
   A. The electric engine cooling fan may start and run when the air conditioning system is turned on.
   B. The electric engine cooling fan may start and run when the ignition switch is turned on.
   C. The electric engine cooling fan may start and run when the ambient temperature is high.
   D. The electric engine cooling fan may start and run at any time.

3. A 17-psi radiator pressure cap is replaced with a 7-psi radiator pressure cap. Which of the following is LEAST likely to occur as a result?
   A. The engine will overheat.
   B. The engine will not reach operating temperature.
   C. The coolant will boil over.
   D. Engine performance will be degraded.

4. Latent heat can be measured with:
   A. A spirit thermometer
   B. An electronic thermometer
   C. Either A or B
   D. Neither A nor B

5. Heat required for a change of state, say, from a liquid to a vapor is called:
   A. Sensible heat
   B. Latent heat
   C. Superheat
   D. Subsurface heat

6. All of the following may cause a compressor clutch to slip, EXCEPT:
   A. Overcharge of refrigerant
   B. Loose drive belt
   C. Improper air gap
   D. Low voltage

7. The ohmmeter reading in the illustration below is 0. The MOST probable cause of this problem is that the:
   A. Windings are shorted
   B. Windings are open
   C. Brushes are defective
   D. Motor is seized

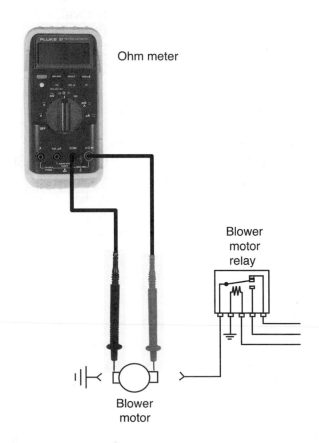

Ohm meter

Blower motor relay

Blower motor

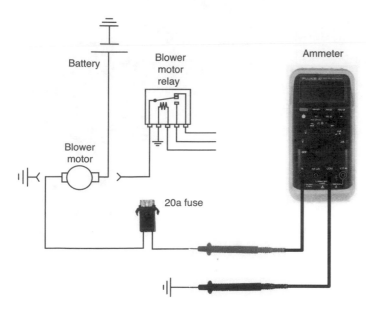

8. The motor does not operate when connected as shown in the illustration above. The fuse is good and there is no reading on the ammeter. The MOST likely cause of this problem is that the:
- **A.** Windings are shorted
- **B.** Windings are open
- **C.** Brushes are defective
- **D.** Motor is seized

9. High-voltage "spikes" are eliminated when the clutch is engaged and disengaged with the use of a:
- **A.** Thermistor
- **B.** Resistor
- **C.** Transistor
- **D.** Diode

10. The outlet tube of an accumulator is cooler than the inlet tube.
Technician A says the accumulator may be clogged and is partially restricting the flow of refrigerant back to the compressor.
Technician B says the metering device is allowing too much refrigerant into the evaporator and is not evaporated until leaving the accumulator.
Who is right?
- **A.** Technician A
- **B.** Technician B
- **C.** Both A and B
- **D.** Neither A nor B

11. The fitting shown in the illustration to the right is used with a:
- **A.** Low-pressure hose
- **B.** Barrier hose
- **C.** Nonbarrier hose
- **D.** Vacuum hose

12. Refer to Figure 5-1, page 144, to determine which of the following metric fasteners may be used to replace an English $^5/_{16}$-24 capscrew.
- **A.** M8-20
- **B.** M8-25
- **C.** Either A or B
- **D.** Neither A nor B

13. The loss of a vacuum signal at the control will MOST likely cause the system to "fail safe" to the _____ mode.
- **A.** Heat
- **B.** Defrost
- **C.** Either A or B
- **D.** Neither A nor B

14. A "blend" refrigerant means that:
- **A.** It contains more than one component in its composition
- **B.** It may be mixed (blended with) another refrigerant
- **C.** Both A and B
- **D.** Neither A nor B

15. The screen in the fixed orifice tube is found to be clogged. The recommended repair is to determine and correct the problem that caused the clogging and to:
- **A.** Clean the screen
- **B.** Replace the screen
- **C.** Replace the fixed orifice tube
- **D.** Replace the fixed orifice tube and liquid line

16. The pressure of the refrigerant in the condenser _____ as it gives up its heat to the ambient air.
- **A.** Is increased
- **B.** Remains about the same
- **C.** Is reduced
- **D.** Any of the above, depending on its temperature

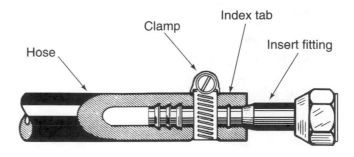

17. The clutch air gap may be measured using a:
    A. Dime
    B. Wire-type feeler gauge
    C. Stainless steel scale
    D. Nonmagnetic feeler gauge

18. The keyway shown in the illustration above should protrude about _____.
    A. 1/8 in. (3.18 mm)
    B. 3/16 in. (4.76 mm)
    C. 1/4 in. (6.35 mm)
    D. 9/32 in. (7.14 mm)

19. All of the following statements about a system with a variable displacement compressor are true, EXCEPT:
    A. There is no electromagnetic clutch.
    B. The cycling clutch is not used for temperature control.
    C. The swash plate angle determines the compressor displacement.
    D. When capacity demand is high the swash plate is at its greatest angle.

20. Refrigeration lubricant may be reused:
    A. If it has been cleaned by filtration
    B. If it has been dries by evacuation
    C. Both A and B
    D. Neither A nor B

21. All of the following are popular methods of leak detection, EXCEPT:
    A. Halogen
    B. Halide
    C. Dye
    D. Vacuum

22. When the refrigerant container is inverted, as shown in the illustration to the right:
    A. The air conditioning system must be charged through the high side with the compressor off.
    B. The air conditioning system must be charged through the high side with the compressor run-ning.
    C. The air conditioning system must be charged through the low side with the compressor off.
    D. The air conditioning system must be charged through the low side with the compressor run-ning.

23. Low voltage at the clutch coil may cause:
    A. The clutch to slip
    B. A noisy clutch
    C. Both A and B
    D. Neither A nor B

24. All except which of the following statements about a vacuum pump are true. A vacuum pump may be used to remove:
    A. Moisture from an air conditioning system
    B. Air from an air conditioning system
    C. Trace refrigerant from an air conditioning system
    D. Debris from an air conditioning system

25. The low-side gauge on a manifold and gauge set indicates a vacuum below 29 in. Hg while the vacuum pump is running. Five minutes after the pump is turned off, the gauge indicates 25 in. Hg.
    Technician A says the vacuum pump was not run long enough to remove residual refrigerant from the lubricant and the rise in pressure is caused by refrigerant out gassing.
    Technician B says the air conditioning system obviously has a leak and the rise in pressure is caused by the introduction of ambient air.
    Who is right?
    A. Technician A
    B. Technician B
    C. Both A and B
    D. Neither A nor B

26. All of the vacuum devices connected to the green hose are inoperative. All vacuum devices connected to the blue hose, however, operate properly.
Technician A says the green hose is defective, probably split or otherwise disconnected.
Technician B says the problem may be corrected by splicing the green hose to the blue hose.
Who is right?
   A. Technician A
   B. Technician B
   C. Both A and B
   D. Neither A nor B

27. The following statements are true about a check valve, EXCEPT:
   A. Air flow is blocked in one direction only.
   B. Vacuum flow is blocked in one direction only.
   C. Air or vacuum is permitted to flow in one direction.
   D. A check valve is omnidirectional when used in a vacuum system.

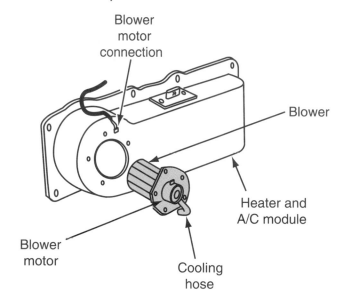

28. What is callout E in the illustration above?
   A. Wiring harness
   B. Cooling tube
   C. Ground wire
   D. Neither of the above

29. All of the following may result in inadequate air flow, EXCEPT:
   A. Duct or hose torn or disconnected
   B. Mode door binding, inoperative, or disconnected
   C. Defective or disconnected coolant flow control
   D. Outlet blocked or restricted

30. The most probable cause of windshield fogging is the:
   A. Heater coolant flow control valve is leaking
   B. Heater coolant flow control valve is out of adjustment
   C. Heater core is restricted
   D. Heater core is leaking

31. The high-side gauge needle is below 0 as shown in the illustration above. This is an indication that the:
   A. Gauge is hooked up to an air conditioning system that is under a vacuum
   B. Gauge is out of calibration and should be adjusted to zero before being used
   C. Both A and B
   D. Neither A nor B

32. The most important procedure before servicing an automotive air conditioning system is to:
   A. Ensure adequate ventilation
   B. Wear protective gear, such as goggles
   C. Disconnect the battery ground cable
   D. Hold a refrigerant purity test

33. To pass a purity test the refrigerant being tested must be at least _____ pure.
   A. 99 percent
   B. 98 percent
   C. 97 percent
   D. 96 percent

34. The receiver-drier or accumulator should be replaced during all of the following services, EXCEPT replacing:
   A. A defective service valve core
   B. A compressor
   C. A condenser
   D. An evaporator

35. All of the following information is required on a retrofit label EXCEPT:

- **A.** Date of retrofit
- **B.** Company or technician certificate number
- **C.** Type and amount of refrigerant
- **D.** Type and amount of lubricant

36. A thermistor's resistance is in proportion to:

- **A.** Its temperature
- **B.** The pressure applied to it
- **C.** The ambient light intensity
- **D.** The voltage applied to it

37. All of the following must be replaced when opened by an electrical overload, EXCEPT:

- **A.** Fusible link
- **B.** Panel-mounted fuse
- **C.** In-line fuse
- **D.** Circuit breaker

38. What instrument can be used to test a fuse or circuit breaker?

- **A.** Ohmmeter
- **B.** Voltmeter
- **C.** Both A and B
- **D.** Neither A nor B

39. The master control is turned to maximum cooling. The blower motor does not run. A jumper wire is connected to the blower motor case and to body metal. There is a slight spark when connected, and the motor runs.

Technician A says the problem is in the electrical control circuit.

Technician B says the problem is in the electrical ground circuit.

Who is right?

- **A.** A only
- **B.** B only
- **C.** Both A and B
- **D.** Neither A nor B

40. The blower motor in the schematic in the illustration to the right has _____ speeds.

- **A.** Two
- **B.** Three
- **C.** Four
- **D.** Variable

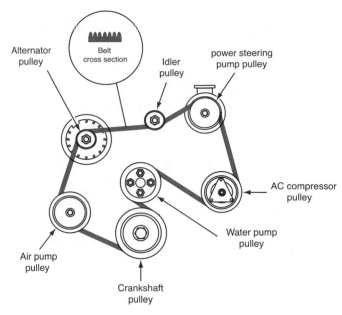

41. If the crankshaft pulley turns clockwise (cw), all of the following statements about the illustration above are true, EXCEPT:

- **A.** The compressor turns clockwise (cw).
- **B.** The power steering pulley turns clockwise (cw).
- **C.** The water pump turns clockwise (cw).
- **D.** The alternator pulley turns clockwise (cw).

42. At highway speeds, the coolant pump may turn at _____ rpm.

- **A.** 2,000
- **B.** 3,000
- **C.** 4,000
- **D.** 5,000

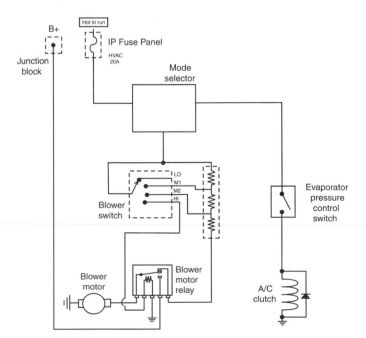

43. Which of the following is MOST likely to cause the belt to slip?
    A. Belt too loose
    B. Belt hardened and/or glazed
    C. Accessory bearing failure
    D. Coolant or oil on belt or pulley

44. The LEAST likely cause of engine overheating is a defective:
    A. Temperature sending unit
    B. Thermostat
    C. Radiator cap
    D. Water pump

45. Extended-life antifreeze has a useful life of up to _____ miles/kilometers.
    A. 50,000/80,450
    B. 100,000/160,900
    C. 150,000/241,350
    D. 200,000/321,800

46. The environmental and health problems associated with the venting of refrigerants are being discussed.
    Technician A says that some refrigerants vented near an open flame may produce a toxic vapor.
    Technician B says that one may be subject to heavy penalties for unlawfully venting refrigerant. Who is right?
    A. A only
    B. B only
    C. Both A and B
    D. Neither A nor B

47. What is the minimum number of manifold and gauge sets required to comply with federal regulations and to ensure against refrigerant contamination?
    A. Two
    B. Three
    C. Four
    D. Five

48. A refrigerant analyzer will not identify CFC-12 or HFC-134a refrigerant if it contains more than _____ percent impurities.
    A. Two
    B. Three
    C. Four
    D. Five

49. The information given on the label depicted in the above illustration is the vehicle's:
    A. Serial and model number
    B. Refrigerant and lubricant data
    C. Emissions control data
    D. Identification number

50. All of the following supply valuable service and technical information for the automotive air conditioning service technician, EXCEPT:
    A. MACS
    B. IATN
    C. Mitchell
    D. AAA

APPENDIX B

Metric Conversions

	to convert these	to these,	multiply by:
TEMPERATURE	Centigrade Degrees	Fahrenheit Degrees	1.8 then + 32
	Fahrenheit Degrees	Centigrade Degrees	0.556 after − 32
LENGTH	Millimeters	Inches	0.03937
	Inches	Millimeters	25.4
	Meters	Feet	3.28084
	Feet	Meters	0.3048
	Kilometers	Miles	0.62137
	Miles	Kilometers	1.60935
AREA	Square Centimeters	Square Inches	0.155
	Square Inches	Square Centimeters	6.45159
VOLUME	Cubic Centimeters	Cubic Inches	0.06103
	Cubic Inches	Cubic Centimeters	16.38703
	Cubic Centimeters	Liters	0.001
	Liters	Cubic Centimeters	1,000
	Liters	Cubic Inches	61.025
	Cubic Inches	Liters	0.01639
	Liters	Quarts	1.05672
	Quarts	Liters	0.94633
	Liters	Pints	2.11344
	Pints	Liters	0.47317
	Liters	Ounces	33.81497
	Ounces	Liters	0.02957
	Millileters	Ounces	0.3381497
	Ounces	Millileters	29.57
WEIGHT	Grams	Ounces	0.03527
	Ounces	Grams	28.34953
	Kilograms	Pounds	2.20462
	Pounds	Kilograms	0.45359
WORK	Centimeter Kilograms	Inch-Pounds	0.8676
	Inch-Pounds	Centimeter-Kilograms	1.15262
	Meter Kilograms	Foot-Pounds	7.23301
	Foot-Pounds	Newton-Meters	1.3558
PRESSURE	Kilograms/Square Centimeter	Pounds/Square Inch	14.22334
	Pounds/Square Inch	Kilograms/Square Centimeter	0.07031
	Bar	Pounds/Square Inch	14.504
	Pounds/Square Inch	Bar	0.0689
	Pounds/Square Inch	Kilopascals	6.895
	Kilopascals	Pounds/Square Inch	0.145

APPENDIX C

Automotive Heating and Air Conditioning Special Tool Suppliers.

Bright Solutions, Inc.
Troy, MI

Carrier Corporation
Syracuse, NY

Clardy Manufacturing Corporation
Fort Worth, TX

Classic Tool Design, Inc.
New Windsor, NY

Component Assemblies, Inc.
Bryan, OH

Corrosion Consultants, Inc.
Roseville, MI

CPS Products, Inc.
Hialeah, FL

Envirotech Systems, Inc.
Niles, MI

FJC, Inc.
Davidson, NC

Floro Tech, Inc.
Pitman, NJ

Four Seasons
Division of Standard Motor Products, Inc.
Lewisville, TX 75057

Interdynamics, Inc.
Brooklyn, NY

K. D. Binnie Engineering Pty. Ltd.
Kirrawee, Australia NSW

KD Tools
Lancaster, PA

Kent Moore Division
SPX Corporation
Warren, MI

Lincor Distributors
N. Hollywood, CA

MAC Tools
Washington Courthouse, OH

Mastercool, Inc.
Rockaway, NJ

Neutronics, Inc.
Exton, PA

OTC Division
SPX Corporation
Owatonna, MN

Owens Research, Inc./Tubes 'N Hoses
Dallas, TX

P & F Technologies Ltd.
Mississauga, ONT, Canada

Ritchie Engineering Company Inc.
Garrett, IN

Robinair Division
SPX Corporation
Montpelier, OH

RTI Technologies, Inc.
York, PA

The S. A. Day Manufacturing Company, Inc.
Buffalo, NY

Snap-On Tools Corporation
Kenosha, WI

Superior Manufacturing Company
Morrow, GA

Technical Chemicals Company
Dallas, TX

Thermolab, Inc.
Farmersville, TX

TIF Instruments, Inc.
Miami, FL

Tracer Products Division
Spectronics Corporation
Westbury, NY

Uniweld Products
Ft. Lauderdale, FL

Uview Ultraviolet
Mississauga, ONT, Canada

Varian Vacuum Technologies
Lexington, MA

Viper/T-Tech Division
Century Manufacturing Company
Minneapolis, MN

Yokagawa Corporation of America
Newnan, GA

Any technician opening the refrigeration circuit of an automotive air conditioning system must be certified in refrigerant recovery and recycling procedures to be in compliance with Section 609 of the Clean Air Act (CAA) Amendments of 1990.

Three major automotive organizations administer a certification test at this time: International Mobile Air Conditioning Association (IMACA), Mobile Air Conditioning Society (MACS) Worldwide, and National Institute for Automotive Service Excellence (ASE).

The following is a copy of the MACS 40-page Certification Training Manual, reprinted with permission, as a study guide for the convenience of any who wish to take the MACS-administered certification exam.

More than a half-million technicians have enrolled in the MACS certification program, which has the approval of the Environmental Protection Agency (EPA). This program is designed to help the technician comply with the law, protect the environment, and conserve the dwindling supply of CFC-12 refrigerant through on-site recovery and recycling procedures. This program encourages proper equipment use as well as safe and proper service procedures. It also includes information on HFC-134a refrigerant recovery and recycling procedures, so that all technicians certified for servicing CFC-12 systems are also considered certified for servicing HFC-134a systems.

The test is not included in this appendix. It may be ordered from MACS online, by phone, by fax, or by mail. A small testing fee is required for each technician taking the test. For further information, contact MACS at:

ONLINE: http://www.info@macsw.org
PHONE: (215) 679-2220
FAX: (215) 541-4635
MAIL: P.O. Box 100, E. Greenville, PA 18041

If ordering by mail, phone, or fax, you will receive a copy of the study guide. If ordering online, your payment will be acknowledged by e-mail, and you may then download all additional information.

Your test must be completed and returned to the designated independent test scoring company within 90 days. You will then be advised of your score. If a passing score is attained, certification credentials, including a wall certificate and laminated wallet card, will be mailed to you. If the test was taken online, a temporary certificate will be made available for you to print out.

Those not attaining a passing score of 84 percent may take the exam a second time without additional cost. To pass the test one must correctly answer 21 of the 25 questions.

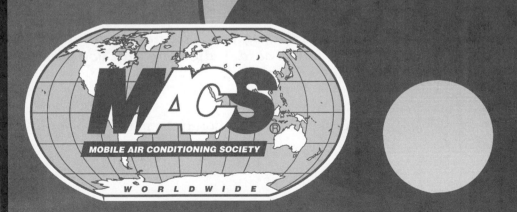

Certification Training Manual

APPROVED BY THE U.S. EPA FOR TECHNICIAN TRAINING
REQUIREMENTS UNDER SECTION 609 OF THE CLEAN AIR ACT

REVISED 6/98

REFRIGERANT RECYCLING AND SERVICE PROCEDURES

For Automotive Air Conditioning Technicians

MACS
MOBILE AIR CONDITIONING SOCIETY
WORLDWIDE

This technician certification program is not intended to test the technical skills of technicians regarding the diagnosis and repair of motor vehicle air conditioners. The basic goal of the technician training and certification program is to teach technicians how to properly recover and recycle refrigerant, and why it must be done to protect the stratospheric ozone layer. In addition. it provides information for servicing and retrofitting air conditioning systems with non-ozone-depleting alternate refrigerant.

Important Notes!!!

■ Do not mix up tests or exchange tests with other individuals at your place of business. All tests are coded with names matching assigned numbers.

■ Please review your test upon completion. Any questions marked with more than one answer will be scored incorrect.

■ Completely fill in the block (■) to the left of the correct answer.
  Do not mark with a check (✓) or an x (X).

General Information and Instructions

You have registered for MACS certification in REFRIGERANT RECYCLING & SERVICE PROCEDURES FOR AUTOMOTIVE AIR CONDITIONING TECHNICIANS. Following are the steps* necessary for you to take to complete the prescribed training:

* The following instructions apply to those technicians taking the MACS training course by mail. Those participating in a classroom program should follow the instructions of their trainer/proctor. Tests given in a classroom setting must be closed-book tests. The required score for passing these closed-book tests is less than that required for passing the open-book test.

1 • Read the instruction manual that came with your recycling machine (and review the training video, if provided). Then, read this manual cover to cover. Reread as necessary to gain full comprehension of the material presented.

2 • Take the enclosed test. The test is an untimed, "open-book" test, so you may refer to the training manual as often as necessary to research answers to the questions posed. (Note, however, that you must correctly answer a minimum of 21 of the 25 questions to earn certification.) **You must complete the test by yourself, without assistance from anyone, and submit it for scoring. (See 4 below.)**

3 • Fill out and sign the "Identification and Statement of Testing Conditions" block on the back of your test.

4 • Mail your test in the stamped, addressed envelope provided to: NSC, Inc., P.O. Box 149, East Greenville, PA 18041-9979.

5 • NSC will advise MACS of your test results.

6 • MACS will advise you of your score and, providing that you have attained a passing score, will issue a certificate and a wallet-sized I.D. card, indicating that you have successfully completed this MACS certification training program.

Enclosed:
• Test with identification material, job experience, and declaration to be mailed to scoring facility
• Stamped return envelope (NSC)

Important • Please note: Tests must be returned for scoring within 90 days of the date they are issued. MACS assumes no responsibility for tests submitted for scoring after this 90-day period. MACS will charge an additional fee for re-issuing tests which are lost, misplaced or destroyed.

Introduction

There is worldwide consensus that chlorofluorocarbons (such as the CFC-12 used as a refrigerant in mobile air conditioning systems) destroy the stratospheric ozone layer. This industry has moved with all possible speed to implement refrigerant containment and recycling of refrigerants in automotive air conditioning systems, and to develop systems that use a non-ozone-depleting refrigerant, HFC-134a.

Now it's all up to you!

If you fail to operate and maintain your refrigerant recycling equipment as required by federal law, and as recommended by the equipment manufacturer, the development of recycling technology to reduce the release of refrigerant to the atmosphere will have been in vain.

You are key to the success of the national refrigerant recovery/recycling program, and a lot is riding on the success of the technology. Please take the time to read this manual completely and do your part to make the recovery/recycling program work.

Table of Contents

A note on refrigerant terminology...

This manual makes repeated reference to two refrigerants: CFC-12 and HFC-134a.

CFC-12 (also known as R12) is a chlorofluorocarbon, and is composed of the elements chlorine, fluorine and carbon. Its specific nomenclature is dichlorodifluoromethane.

HFC-134a (also known as R134a) is a hydrofluorocarbon, and is composed of the elements hydrogen, fluorine and carbon. Its specific nomenclature is 1,1,1,2-tetrafluoroethane.

Stratospheric Ozone Depletion and CFCs

The Issue

Common practice in the service of mobile air conditioning systems has been to add refrigerant to a leaking system and then vent the charge when any other service was required. That practice was acceptable because refrigerant was relatively inexpensive in the past and thought to be environmentally benign. But, knowing what we do today about the role of CFC-12 in the degradation of the earth's protective ozone layer, the service practice of venting CFC-12 is irresponsible and is no longer allowed. Under the Clean Air Act, such service practice is illegal. In other rulings, the EPA prohibits the venting of other substitute refrigerants that are used in mobile air conditioning systems.

Overview

Stratospheric ozone acts as a shield against harmful solar ultraviolet (UV) radiation. A significant reduction in ozone in the upper atmosphere could result in long-term increases in skin cancer and cataracts. Also it may cause damage to the human immune system. Reductions in the abundance of stratospheric ozone also may reduce crop yields and alter terrestrial and aquatic ecosystems.

A worldwide consensus has emerged that substances, including chlorine from synthetic chemicals called chlorofluorocarbons (CFCs) and bromine from chemicals called halons, react in a way which depletes ozone in the stratosphere. CFCs have been used as blowing agents in plastic foam products (cushioning, insulation and packaging), as refrigerants, as solvents, as sterilants, and in aerosol applications. Additionally, halons are used as fire extinguishing agents. To protect the ozone layer, the United States and over 160 other nations have now ratified the 1987 Montreal Protocol on Substances which Deplete the Ozone Layer. This landmark international agreement is designed to control the production and consumption of certain chlorofluorocarbon and halon compounds.

How Stratospheric Ozone Protects Us

Ozone is a pungent, slightly blue gas that absorbs certain wavelengths of the sun's radiation. Ozone is concentrated in a part of the atmosphere called the stratosphere, between 6 and 30 miles above the earth's surface. Stratospheric ozone should not be confused with ground level ozone, commonly referred to as smog.

Ozone depletion results in increased levels of UVB (ultraviolet beta radiation) reaching the earth and related health and environmental effects. Ozone normally absorbs incident UVB; decreasing the amount of ozone results in higher UVB levels.

The Theory

Concern about possible depletion of the ozone layer from CFCs was first raised in 1974 with publication of research which theorized that chlorine released from CFCs could migrate to the stratosphere and destroy ozone molecules (Molina and Rowland, 1974). Some CFCs have an atmospheric lifetime of more than 120 years (i.e., they do not break down in the lower atmosphere). As a result, they migrate slowly to the stratosphere where higher energy radiation strikes them, releasing chlorine. Once freed, the chlorine acts as a catalyst, repeatedly combining with, and breaking apart ozone molecules. **It is believed that one CFC molecule can destroy as many as 100,000 ozone molecules.**

When ozone depletion occurs, more UV radiation penetrates to the earth's surface. Moreover, because of the long atmospheric lifetimes of CFCs, it will take many decades for the ozone layer to return to its former concentration and safety when CFCs are no longer used.

Ozone Depletion

The link between CFCs and ozone depletion is supported by scientific evidence.

Chlorofluorocarbons (CFCs) have been widely used, and they migrate into the upper atmosphere after use. Because CFCs are very stable, and they are heavier than air, they do not break down until they are carried by wind currents into the stratosphere, a process that can take as long as 5 to 10 years.

In the stratosphere, these chemicals absorb UV radiation, break apart, and react with ozone, taking one oxygen atom away and forming highly reactive chlorine monoxide. Chlorine monoxide in turns breaks down ozone again by pulling away a single oxygen atom, creating two oxygen molecules, and allowing the chlorine to move freely to another ozone molecule. (See illustration, page 462.)

With the increased release of CFCs to the atmosphere, a British researcher found that vortex winds prevented the mixing of ozone-rich air over the Antarctic, producing the ozone hole.

When scientists began studying ozone depletion in the early 1970s, they investigated several natural phenomena, such as volcanoes and evaporation of seawater.

Volcanoes can produce large quantities of hydrochloric acid. However, most volcanic discharges are not powerful enough to reach the stratosphere. Chlorine evaporation from seawater is dissolved in rain and does not reach the stratosphere.

Chlorine produced by volcanoes or oceans does not leave the troposphere and poses no threat to the ozone layer. However, CFCs, being extremely stable, do not release chlorine until they reach the stratosphere.

In December 1994, NASA announced that three years of satellite data confirmed that CFCs are the primary source of stratospheric chlorine.

Scientists predict that CFC levels should peak by the year 2000 and return to 1979 levels between the years 2020 and 2050. As the CFC levels are reduced, the natural atmospheric process will rebuild the ozone level. Until that time, increased UV levels can lead to a greater chance of overexposure to UV radiation and the consequent health effects.

Health & Environmental Effects

Shielding the earth from much of the damaging part of the sun's radiation, the ozone layer is a critical resource which safeguards life on this planet. Should the ozone layer be depleted, more of the sun's damaging rays would penetrate to the earth's surface. Each 1% depletion, it is estimated, would increase exposure to ultraviolet radiation by 1.5–2%.

The Environmental Protection Agency's (EPA) assessment of the risks from ozone depletion focused on the following areas:
• Increases in skin cancers
• Damage to the human immune response system
• Increases in cataracts
• Damage to crops
• Damage to aquatic organisms
• Increases in ground level ozone
• Increased global warming

Human Health Effects

Skin cancer is already a serious problem in the United States and will only increase with further depletion of the ozone layer.

If the ozone layer continues to be depleted, three distinct types of skin cancer will increase.

Basal and squamous cell skin cancers are the two most common types. They now affect about 500,000 people annually in the United States. If detected early, these cancers are treatable. Even so, approximately 1% of all cases result in premature deaths.

Malignant melanoma is far less common but substantially more harmful. About 25,000 cases now occur annually, resulting in 5,000 deaths each year. In 1996, the lat-

HOW CFCs DEPLETE THE OZONE LAYER

CFCs drift high up into the stratosphere where the sun's rays break them apart, starting a chain reaction in which chlorine destroys ozone. As the level of protective ozone diminishes, larger amounts of ultraviolet (UV) radiation reach the earth's surface. For people, overexposure to UV rays can lead to skin cancer, eye cataracts, and can weaken the immune system.

est year for which information is available, about 38,000 cases of melanoma occurred, with 7,300 fatalities. Six out of seven skin cancer fatalities are due to melanoma. While the relationship between exposure to UV radiation and melanoma is complex, existing studies provide a basis for estimating fu-

ture risks associated with ozone depletion.

Cataracts cloud the lens of the eye, thus limiting vision. Although cataracts develop for a variety of reasons, scientific evidence supports the conclusion that increased exposure to UV radiation from ozone depletion would increase the number of people experiencing this particular eye disorder.

Based on epidemiological studies, if current trends in the use of ozone-depleting gases continue, the number of cataract cases would increase by 18 million (for the population alive today and the people born before 2075). Actions required by the Montreal Protocol and the U.S. Clean Air Act to limit the use of these chemicals would reduce the number of additional cases by 92% during this time period.

Damage to the immune system is another possible threat to human health resulting from ozone depletion. Research to date suggests that exposure to UV radiation weakens the immune system's ability to fend off certain diseases (i.e., herpes simplex and leishmaniasis, a parasitic disease common in the tropics). However, more needs to be known about the exact way the immune system is affected and the implication UV radiation exposure has for a wide variety of other diseases.

Plant & Marine Effects

Crops and other terrestrial ecosystems also could be adversely affected by increased exposure to UV radiation. In greenhouse studies, approximately two-thirds of the crops exposed to elevated levels of UV radiation proved sensitive. Field studies of soybeans have shown that ozone depletion of up to 25% could decrease yield by over 20%, with substantially greater losses in years when climatic stresses are also a factor.

Certain marine organisms, particularly phytoplankton and the larvae of many species, also may be sensitive to increased exposure to UV radiation because they spend much of their existence near the surface of the water. Although it is difficult to design experiments replicating aquatic environments, research to date suggests that adverse effects on productivity and species diversity are related to increased exposure to UV radiation.

Other Impacts

Ground Level Ozone. Ozone depletion would increase the rate of formation of ground level (tropospheric) ozone, a major component of what is commonly called smog.
Degradation of Polymers. Ozone depletion would accelerate the weathering (i.e., chalking, yellowing, and cracking) of plastics used in outdoor applications.
Climate Change. CFCs are greenhouse gases (i.e., they have similar properties to carbon dioxide) and thus may contribute to global warming and rising sea levels.

Global Problem

Unlike many other environmental issues, stratospheric ozone protection is truly a global problem. CFCs and halons have been used in many nations, and, given their long atmospheric lifetimes, they become widely dispersed over time. As a result, the release of these chemicals in one country could adversely affect the stratosphere above, and therefore the health and welfare of other countries. Many

developed countries produced CFCs and halons. Most consumed these chemicals in a variety of different products. The United States, for example, has been one of the largest consumers of the world's CFCs. Other nations also have been significant users.

Therefore, to protect the ozone layer from the damage that may be caused by the venting of CFCs and halons, an international solution was critical.

Montreal Protocol

Recognizing the global nature of the problem, on September 16, 1987, in Montreal, Canada, 24 nations and the European Economic Community (EEC) signed the Montreal Protocol on Substances which Deplete the Ozone Layer. Currently more than 160 nations, representing more than 95 percent of the world's consumption of CFCs, are parties to the Protocol.

The Chemicals

Listed below are the five chemicals controlled by the Montreal Protocol and the associated **ODP** (Ozone Depletion Potential) value, a measure of a chemical's relative ability to destroy ozone molecules in the stratosphere:

•Fully-Halogenated Chlorofluorocarbons (Grp. 1)	ODP
CFC-11	1.0
CFC-12	1.0
CFC-113	0.8
CFC-114	1.0
CFC-115	0.6

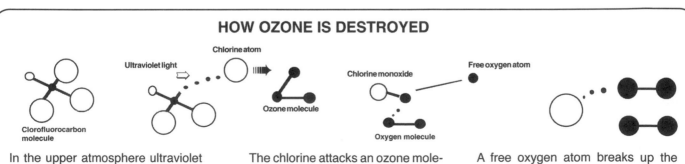

HOW OZONE IS DESTROYED

In the upper atmosphere ultraviolet light breaks off a chlorine atom from a chlorofluorocarbon molecule.

The chlorine attacks an ozone molecule breaking it apart. An ordinary oxygen molecule and a molecule of chlorine monoxide are formed.

A free oxygen atom breaks up the chlorine monoxide. The chlorine is then free to repeat the process.

The higher the ODP value, the greater the potential to destroy ozone in the stratosphere.

Focus on CFC-12

Mobile air conditioners, which cool the passenger compartments of automobiles, trucks, and buses, have been the largest single users of CFCs in the United States. CFC-12 refrigerant was used for this purpose. A/C systems used an estimated 54,000 metric tons of CFC-12 in 1985. Out of 15.3 million motor vehicles manufactured or imported in that year, over 85% had factory-installed air conditioners.

Until the fleet of mobile A/C systems produced prior to the mid-1990s is retired or retrofitted, CFCs used in these systems represent a large percentage of the total ozone-depleting potential of CFCs in the United States.

It is therefore important to reduce the release of CFC-12 by identifying, repairing and/or replacing A/C system parts which leak or cause leaks to occur.

Further, government studies show that many A/C-equipped vehicles sent to salvage still contain some refrigerant. Under Section 608 of the federal law, this refrigerant can not be released, and must be recovered before vehicle disposal.

U.S. Joins in Worldwide Action

The U.S. and other countries have signed the Protocol, agreeing to phase out production of ozone-depleting substances. The 1990 Clean Air Act Amendments incorporated the Protocol's original phase-out date: the year 2000. In 1992, President Bush pledged to halt almost* all U.S. production of CFCs by the end of 1995. (*The President's declaration left open the possibility of allowing some minimum production beyond this date for "essential uses." Automotive A/C systems are not considered "essential uses.") Millions of vehicles will still require CFCs to maintain A/C system operation or they will have to be retrofitted.

CFC-12 Supply

With the U.S. Dec. 31, 1995 phase-out date for the production of new CFC-12, the supply of both new and reclaimed CFC-12 can be expected to diminish as stockpiles are used up. Vehicle OEMs were fully aware of this situation, and worked both to conserve the available supply of CFC-12 and to efficiently retrofit existing vehicles to HFC-134a refrigerant when appropriate.

In addition, the EPA requested SAE and the industry to develop a retrofit refrigerant program for replacement of CFC-12 systems.

Sources of Refrigerant Venting

Motor vehicle air conditioning systems can release refrigerant into the atmosphere in a number of ways:

• There are system leaks from part failures and venting which occurs when a leak is not repaired and the system is recharged, or "topped off," and returned to the customer.

• Refrigerant can be vented during various service procedures.

• It can be vented from manifold gauges and equipment hoses.

• And when using small cans, refrigerant remaining in the container is likely to ultimately be vented. The refrigerant remaining in larger, 20- and 30-pound one-way containers is also sometimes vented when they are scrapped.

Important Dates

Jan. 1, 1992: Since this date, containment and recycling of CFC-12 (and HCFC refrigerants, none of which are approved by A/C systems manufacturers to date) has been required.

Nov. 15, 1992: Since this date, sales of containers of CFCs under 20 pounds to anyone other than certified Section 609 technicians has been prohibited.

Nov. 14, 1994: Since this date, the sale of ozone-depleting refrigerants in any size container is restricted to certified technicians (Ref.: Sect. 608, CAA).

July 1995: Since this date any CFC-12 mobile air conditioning system that is converted (retrofitted) to use an acceptable alternate refrigerant must have the appropriate unique service fittings and label for that refrigerant.

Nov. 15, 1995: Since this date, recovery and recycling of any substitute substance for CFC-12, such as HFC-134a, used in a motor vehicle air conditioner has been required.

The above requirements apply to both new businesses and those which change ownership.

EPA Q & A

Q: How was HFC-134a selected as a replacement refrigerant for CFC-12 in automobile air conditioning systems?

A: Engineers for automobile manufacturers conducted research and testing on many potential substitutes for CFC-12 before selecting HFC-134a. As part of this research and testing, they reviewed the potential health effects, toxicity, flammability, and corrosivity of each potential substitute, evaluated the effect of each compound on the life and performance of the air-conditioning components in the various models made by each manufacturer, and also determined the effect of each compound on the system's cooling capacity. It was determined that HFC-134a was the most suitable alternative.

Federal Regulations Affecting Motor Vehicle Air Conditioning & Repair

Summary of Federal Mobile A/C Service Requirements

In 1993, the EPA under Section 608 changed the rule so that farm, construction and off-road vehicles must comply with training and service requirements under either Section 608 or Section 609. Since these types of vehicles generally use automotive type air conditioning systems, certification for compliance can be made under Section 609.

New rules issued by the EPA were effective Jan. 29, 1998. These rules will affect the daily operation of the mobile A/C service industry.

Additional information can be obtained from the EPA Hotline at (800) 296-1996 or the EPA Web Site: http://www.epa.gov/ozone/title6/609/.

In summary, current federal requirements for mobile A/C systems are:

1. Any refrigerant including R12, R134a and other alternates, removed during service from a mobile air conditioning system must be recovered and recycled, not vented.

2. All R12 and R134a recovery-only or recovery/recycling equipment must be certified to meet EPA (SAE) standards and certified by UL or ETL.

3. Recovery-only equipment for each SNAP-listed alternate refrigerant must meet EPA equipment standards.

4. All automotive service technicians must be certified to handle non-ozone-depleting refrigerants including R134a. Certification is also required for alternate refrigerants listed under the SNAP rule for automotive use. Any technician that had been certified for CFC-12 does not require a new certification. Every technician that opens the refrigerant circuit must be certified to work on the mobile A/C system. This includes any service facilities whose technicians enter the refrigerant circuit, including those that only add refrigerant to ("top off") the system. Facilities that typically only change or add fluids (lube-oil-filter operations) also require certified technicians and equipment. The rule also requires that the refrigerant (R12 and R134a) removed from the vehicle must be recycled before it can be recharged, even if it has been removed from the same vehicle. This does not include alternate blend refrigerants, since currently it is illegal to recycle blends on site.

5. Under federal law it is legal (except as required under more stringent requirements by certain state and local laws) to add refrigerant to a pre-existing leaking system.

6. It is not required under federal regulations to remove refrigerant from a leaking system. (This action may be required under state and local laws.) Service facilities may adopt a policy not to add refrigerant to a leaking system; however, the policy should be explained to the customer in advance, including the fact that the policy is not a federal requirement.

7. If the customer arrives with some unknown amount of refrigerant in the system, and it is removed by the technician, and the system is not repaired, the technician must return to the system any refrigerant which was in the system when it arrived, unless the customer agrees to its removal.

It should be noted that some state and local laws are more stringent regarding servicing of mobile air conditioning systems. Determine if your area of employment requires compliance with both federal and state or local servicing requirements.

Remember that it is mandatory to be a certified technician and to use recovery/recycling equipment whenever you are doing work that might allow refrigerant to escape.

Offer to fix leaks in the air conditioning system. It helps to protect the environment and conserves refrigerant supplies. It is not correct, however, to state or imply that the leak repair is required under federal law. Doing so would constitute consumer fraud. It is important that you determine if local or state laws require that a leaking A/C system be repaired before adding refrigerant.

Salvage and/or Disposal Facilities

1. Salvage and/or disposal facilities must remove R12 and R12 substitutes, including R134a, from salvaged or scrapped mobile A/C systems.

2. Certified 609 A/C technicians can remove R12 and R134a from mobile A/C systems at salvage and/or disposal facilities. The refrigerant they recover can be moved to their facility for recycling and reuse in other vehicles. (Alternate blends cannot be recycled onsite for reuse.)

3. Salvage and/or disposal facilities that have purchased certified recovery equipment can recover refrigerant at their facility and also move the equipment to another salvage and/or disposal facility and recover refrigerant. This refrigerant can be sold to technicians certified under section 609. The salvage employee does not have to be certified for this salvage operation. However, this uncertified person cannot charge this or any other refrigerant into a mobile A/C system without being certified under section 609 and the refrigerant has to be properly processed prior to reuse.

The fact that refrigerant removed at these salvage and/or disposal facilities can be contaminated when removed from vehicles, or be mixed with contaminated refrigerant in recovery equipment, can be a problem. Since there is no requirement to identify the purity of

the refrigerant being removed from these vehicles, or label their containers, there is concern about the possibility of contaminated supplies coming from these facilities.
4. There are record-keeping requirements for salvage and/or disposal facilities to verify that the refrigerant was removed by another, and to record the sale of all ozone-depleting refrigerant. In addition, they must certify to EPA that they have certified recovery equipment.

Off-Site Service

1. The rule allows transportation of certified recovery/recycling equipment to another site to perform mobile A/C system refrigerant service. This allows the use of the certified equipment at body shops, used car dealerships, farms, construction sites, mines and other remote sites.
2. Service of the refrigerant circuit requires certification under section 609 for anyone performing "service for consideration." (See EPA Note on page 467.) Persons working on their own equipment, such as do-it-yourselfers (DIYers) and farmers are not covered under this rule and can add refrigerant without being certified. However, there are two important factors that cover anyone servicing a mobile A/C system. Anyone attempting to remove refrigerant from the system legally cannot vent it and can be fined if they do so. This means that recovery equipment and section 609 certification to operate it is required. So the end result is, working on the A/C system refrigerant circuit at some point requires equipment and refrigerant, both of which require 609 certification.

Section 609, 1990 Clean Air Act Amendments

Section 609 of the Act gives the EPA authority to establish requirements to prevent the release of refrigerants during the servicing of motor vehicle air conditioners. On July 14, 1992, the EPA published the final rules (regulation) implementing this section of the Act (40 CFR Part 82). Requirements outlined in the regulation include the following:

Please Note: Some state and local laws are more stringent than, or have requirements in addition to the federal requirements discussed here. Shop owners/technicians should determine if their local area or state has additional requirements.

Equipment Use

Since Jan. 1, 1992, for R12 (and January 29, 1998 for R134a and other alternates), any technician servicing, repairing or opening a motor vehicle air conditioning system "for consideration"—anything other than free service—must use either refrigerant recovery/recycling or recovery-only equipment approved by EPA.

Technician Training/Certification

Technicians using approved equipment must be trained and certified by an EPA-approved organization. To be certified, technicians must pass a test demonstrating their knowledge in the use of recycling equipment in compliance with SAE Standard J1989, the regulatory requirements, the importance of refrigerant containment and the effects of ozone depletion.

Any person that opens the refrigerant circuit for service must recover the refrigerant and therefore must be certified.

Equipment Certification

It is the responsibility of the equipment owner or another responsible officer to certify (report) to EPA that they own approved equipment. This requirement became effective Jan. 1, 1993. To certify equipment, the following information must be mailed to the EPA at this address: MVAC Recycling Program Manager, Stratospheric Ozone Protection Branch (6205J), 401 M Street, S.W., Washington, DC 20460. **Information to be provided:**

•**Name, address and telephone number** of the establishment where the recovery/recycling

Additional information is available through EPA's Stratospheric Ozone Information Hotline. This toll free public service is available Monday through Friday, 10 a.m. to 4 p.m. (Eastern), except on federal holidays.
The Hotline number is (800) 296-1996.

equipment is located;
•**Name brand, model number, year and serial number(s)** of the equipment acquired for use at the above establishment.
•**The certification statement must include the above information and can be on a plain sheet of paper and, must be signed** by the person who has acquired the equipment (the person may be the owner of the establishment or another responsible officer). The person who signs is certifying that they have acquired the equipment, that each individual authorized to use the equipment is properly trained and certified, and that the information provided is true and correct. Regardless of the refrigerant type, the service shop certification statement must be sent to the EPA. The repair facility is required to file this certification only one time.
NOTE: A copy of the equipment certification should be kept on file by the shop owner.

EPA Q & A

Q: Will shops that sent in their recover/recycling equipment certification forms receive an EPA number for purchasing small cans of refrigerant?

A: No, the equipment certification form is filed and used with inspection information in enforcement actions. The shop will not receive verification of receipt from EPA. Technician certification cards issued under section 609 are the only identification accepted for the purchase of small containers of refrigerant.

Record-Keeping Requirements

Any person who owns approved refrigerant recovery or recovery/recycling equipment must maintain records of the name and address of any facility to which refrigerant is sent.

Records of the amount of refrigerant recovered during service are not required. If refrigerant is sent off-site for reclamation, only the address of the reclamation facility used is required to be kept on record.

Any person who owns approved refrigerant recycling equipment must have records demonstrating that all persons authorized to operate the equipment are currently certified.

If the purchaser is buying small cans for resale, the seller must obtain a written statement from the purchaser that the containers are for resale only and indicate the purchaser's name and business address, Records must be retained for a period of three years.

Any person who sells a Class I or Class II substance for use as a refrigerant in a motor vehicle air conditioner must prominently display a sign which states: "It is a violation of federal law to sell containers of Class I and Class II refrigerant to any person who is not properly trained and certified to operate approved refrigerant recycling equipment."

All service facilities which service motor vehicle air conditioners for consideration must allow an authorized representative of the Administrator entry onto their premises, upon presentation of his or her credentials, and give the authorized representative access to all records required to be maintained.

Some local and state governments have additional requirements.

Overlap Between Section 608 and Section 609

Section 608 of the Act directs EPA to establish requirements to prevent the release of ozone-depleting substances during the servicing, repair, or disposal of appliances and industrial process refrigeration. Section 609 of the Act establishes standards specifically for the service of motor vehicle air conditioners (MVACs). MVACs are included in the definition of appliances under the definition put forth in section 608; however, since their service and repair are regulated under section 609 they are not subject to the servicing requirements put forth in section 608. Procedures involving MVACs that are not covered by section 609, such as the disposal of MVACs, are covered by section 608. Below is information concerning specific areas where the overlap between these two sets of regulations may require clarification.

Technician Certification

Both regulations require that technicians become certified. Technicians who repair or service MVACs must be trained and certified by an EPA-approved section 609 program. These programs are specifically designed to cover MVAC recycling equipment in accordance with SAE Standards and section 609 regulatory requirements. After completing a required training program MVAC technicians must pass a test to become certified. These tests are different from the section 608 certification tests.

Under section 608 EPA has established four types of certification for technicians that service and repair appliances other than MVACs. These technicians must be certified by passing a test in the appropriate area. All training and review classes for section 608 are voluntary; only passing the test is mandatory. The four categories of certification are:

Type I—small appliances

Type II—high-pressure appliances, except small appliances & MVACs

Type III—low-pressure appliances

Type IV (Universal)—all appliances except MVACs

In addition, people who service or repair MVAC-like appliances (e.g. farm equipment and other non-road vehicles) can choose to be certified by either the section 609 program or under section 608 Type II. Due to the similarities between MVAC and MVAC-like appliances, EPA recommends that technicians servicing MVAC-like appliances consider certification under section 609.

Please note: While buses using CFC-12 are MVACs, buses using HCFC-22 are not MVACs or MVAC-like appliances, but rather high-pressure equipment covered under Type II of the section 608 test. This also applies to cargo refrigeration equipment.

Refrigerant Sales Restriction

The sale of small cans of CFC-12 will always be limited to Section 609 technicians. After Nov. 14, 1994, under EPA regulations, only certified technicians can purchase CFC or HCFC refrigerants. However, the Clean Air Act itself further restricts the sale of the small containers of CFC-12.

Under the Clean Air Act, only section 609 technicians can purchase small cans (less than 20 pounds) of CFC-12. Traditionally small cans of CFC-12 have been used for recharging MVAC and MVAC-like appliances. The sales restriction provision in the Act was intended to discourage "do-it-yourselfers" who may release refrigerant because they lack access to recovery/recycling equipment. This restriction did not change after Nov. 14, 1994.

Record-Keeping

Section 608 requires that all persons who sell CFC and HCFC refrigerants retain invoices that indicate name of the purchaser, the date of the sale, and quantity of refrigerant purchased. These requirements are for all sales affected by section 608.

However, since the sale of small containers of CFC-12 is restricted to technicians certified under section 609, these record-keeping requirements do not apply to the sale of small containers of CFC-12.

Therefore, while records must be maintained for the sale of all other CFC and HCFC refrigerants in any size container, and for the sale of CFC-12 in containers of 20 pounds or more, it is not necessary to maintain records for the sale of small containers of CFC-12.

EPA's Significant New Alternatives Policy (SNAP) Rule

Under authority of Section 612 of the Clean Air Act, regulations promulgated on March 18, 1994, effective April 18, 1994, establish a program in which EPA will evaluate applications for use of substitute chemicals and technology to replace ozone depleters in specific uses. The Agency does not evaluate the performance or compatibility of substitute chemicals in an automotive air conditioning system.

SNAP requires the manufacturer or importer of a proposed substitute for an ozone-depleting chemical to provide EPA notification 90 days before introducing the substitute into interstate commerce. During the 90-day period, the Agency will evaluate company studies and other information and decide whether the substitute is either acceptable or unacceptable for a specific use, based on whether the substance may have adverse effects on human health or the environment. Some of the criteria EPA will consider in the risk screening include flammability, chemical toxicity, global warming potential and exposure of workers, consumers, the general population and aquatic life.

If EPA places a substance on the unacceptable list, it becomes unlawful to use it as a substitute for an ozone depleter.

Tax on CFC-12

On January 1, 1998 the tax on CFC-12 refrigerant increased to $6.70 per pound. Each year the floor tax increases 45 cents on each pound of refrigerant in stock.

On Jan. 1 of each year, shops with an inventory, or floor stock, of 400 pounds of CFC-12 or more, are required to report their inventory and pay the difference between the prior year tax rate per pound of refrigerant in stock. (If a shop's inventory is 399 pounds or less, no tax payment is required. If inventory is 400 pounds or more, tax is required on all of the refrigerant — the first 399 pounds is not exempted.)

The floor stocks tax on ozone depleting chemicals is due and payable without assessment or notice on or before June 30. The tax must be deposited, together with Form 8109, Federal Tax Coupon, at an authorized depository of the Federal Reserve Bank serving the taxpayer's area.

Every person liable for the floor stocks tax must file a return of tax on Form 720, Quarterly Federal Excise Tax Return, to which Form 6627, Environmental Taxes, is attached, by August 31.

Note: Consult your tax advisor for additional information before filing.

Refrigerant recycled on-site from mobile A/C systems is not taxable.

Imported Used & Recycled Refrigerant

Since January 1, 1996 new CFC refrigerant cannot be imported. However, used or recycled refrigerant can be imported from overseas. There is no federal requirement that containers of used or recycled refrigerant be labeled to identify the content or its purity.

EPA Tip

Handle refrigerants with care to prevent mixing. It is critical that supplies of CFC-12 and R134a are kept free of contamination.

EPA Q & A

Q: Are off-road vehicles, such as agricultural or construction equipment, covered by the section 609 regulation?

A: No, they are not covered by the section 609 regulation directly, but they are covered by the section 608 regulation. This rule was published on May 14, 1993, and requires those servicing MVAC-like appliances, as this type of air-conditioning equipment is classified, to use approved equipment. Also, technicians must be certified. The air conditioning equipment found on construction and agricultural equipment is similar to the motor vehicle equipment and, as a result, the section 608 rule allows technicians to use the equipment and technician certification programs approved under section 609.

EPA Note:

Service for consideration includes persons who are paid to perform service on motor vehicle air conditioners, thus subjecting to regulation all service except that done for free. Fleets of vehicles, whether private, or federal, state or local government owned, are covered because the technicians doing the service are paid. Other examples of establishments covered by the regulations include, but are not limited to: independent repair shops, service stations, fleet shops, body shops, chain or franchised repair shops, new and used car and truck dealers, rental establishments, radiator repair shops, mobile repair operations, vocational technical schools (because instructors are paid), farm equipment dealerships and fleets of vehicles at airports.

Refrigerant Recycling: An Introduction

Why Recycle?

It is important that a supply of CFC-12 be available to ensure that vehicles built to use CFC-12 can be serviced. Without a supply of CFC-12 for service use during ownership, conversion or obsolescence of vehicles could result in additional cost to the owners.

CFC-12 systems were not designed to use any other refrigerant, and A/C system manufacturers recommend that CFC-12 continue to be used for those systems as long as it is available.

Although the mobile air conditioning industry has phased in HFC-134a systems, vehicles with CFC-12 systems will still be in use beyond the year 2000. With a 5-to-10 year vehicle life expectancy, CFC-12 will be required for future service. If CFC-12 is not available for service, the consumer may have to choose among retrofitting to an alternate refrigerant (HFC-134a), purchasing a new vehicle with an HFC-134a system, or doing without air conditioning.

Controlled sale of CFCs, proper repair of leaking systems and recycling of existing CFCs are required to assure consumers the use of their CFC-12 automotive air conditioning systems.

With effective control of CFC supplies and mandatory recycling at all servicing levels, the automotive service industry has effected a major reduction of new CFC-12 requirements.

Field Study

Due to the severity of the ozone depletion issue, industry efforts were immediately directed toward determining if CFCs used in the mobile air conditioning service industry could be recycled.

During the summer of 1988, the EPA, with the support of the Mobile Air Conditioning Society Worldwide (MACS) initiated a sampling program of used refrigerant from 227 vehicles located in four regions of the country. These vehicles included properly operating systems, failed compressors, low-mileage vehicles and vehicles with over 100,000 miles. The chemical analysis of the removed refrigerant indicated a low amount of contamination. From the field study results, the task force established specifications for recycled refrigerant and requested the world's auto manufacturers to determine and approve a level of recycled refrigerant purity in December, 1988. Vehicle and A/C system manufacturers have accepted recycled CFC-12 and HFC-134a meeting the appropriate SAE standard for service and warranty repairs.

System Contaminants

Data obtained from the field study of CFC-12 sampling taken from mobile air conditioning systems identified moisture, refrigerant oil and non-condensable gases (air) as contaminants in used refrigerant which could affect system performance and life.

Standards Developed

Based on the field study, the SAE Interior Climate Control Standards Committee established documents to cover the automotive air conditioning industry handling and use of refrigerants. The documents include:

CFC-12 SAE Documents
- SAE J1989: Service Procedures
- SAE J1990: Specifications for Recycling Equipment
- SAE J1991: Standard of Purity
- SAE J2209: CFC-12 Extraction Equipment

HFC-134a SAE Documents
- SAE J2211: Service Procedures
- SAE J2210: Specifications for Recycling Equipment
- SAE J2099: Standard of Purity
- SAE J1732: HFC-134a Extraction Equipment

SAE documents have been developed for HFC-134a to assure the same level of service integrity as CFC-12 systems, and to protect the present and future environment by preventing the release of the refrigerant into the atmosphere during service operations.

SAE J Documents: CFC-12 and HFC-134a

Refrigerant Purity Standards

The intent of the applicable SAE J standards is to ensure that recycled refrigerant used in servicing mobile A/C systems provides a level of purity which will not affect the performance or warranty of the system.

The SAE, in conjunction with the mobile A/C industry, has developed standards of purity which allow reuse of refrigerant. In the SAE standard of purity, the document states: "Recycling equipment developed under SAE standards is for the purpose of cleaning the refrigerant that has been directly removed from, and intended to be returned to, a mobile air-conditioning system."

Also: "Purity specification of recycled refrigerants supplied in containers from other (non-automotive) recycle sources, for service in mobile automotive air-conditioning systems, shall meet the appropriate ARI 700 standard."

These standards of purity are designated SAE J1991 for CFC-12 and SAE J2099 for HFC-134a which has been recycled onsite. All refrigerants sent offsite for processing and/or from other sources, must meet the specific ARI 700 standard to assure that the refrigerant is not contaminated and to be in compliance with federal law.

The recycling requirements are referred to in section 609 of the 1990 Clean Air Act and also in other state and local laws.

Under the law, recycling of CFC-12 has been in effect since January 1, 1992, and recycling of HFC-134a has been in effect since November 15, 1995.

Equipment Certification

The standard for equipment certified to be used for refrigerant recovery/recycling is established by SAE. This equipment must also be certified by an appropriate EPA-approved testing laboratory (e.g., Underwriters Laboratories, ETL) to meet the required purity specifications. This level of purity for recycled refrigerant is recognized by the auto industry for warranty service applications.

Compliance of Recovery/ Recycling Equipment

To comply with section 609 of the Clean Air Act, recovery/recycling equipment must be certified to SAE specifications. Recovery/recycling equipment used for commercial refrigeration not certified to the SAE standards, does not meet the federal compliance requirements and cannot be used. To prevent refrigerant contamination, recovery/ recycling equipment must only be used with one refrigerant.

Dual Refrigerant Equipment

There are two major designs for single cabinet, dual refrigerant recovery/recycling equipment. Seperate refrigerant circuit equipment mounted on the same cabinet for R12 and R134a must have a labelling that it meets SAE J1991 (R12) or J2099 (R134a) to be in compliance with Section 609. Recovery/recycling equipment having a common refrigerant circuit for R12 and R134a in the same cabinet must be certified to SAE J1770 to meet federal compliance.

Such equipment contains special features to prevent cross-contamination in the refrigeration circuit. The technician must carefully follow the required procedures to switch from one refrigerant to another to prevent cross-contamination.

SAE J1989 & SAE J2211 Service Procedures

The SAE documents J1989 for CFC-12 and J2211 for HFC-134a provide guidelines for containment and assurance that all refrigerant has been removed from a system.

Improperly Recycled Refrigerant

If recycled refrigerant contains non-condensable gases (air) in excess of the allowable amount, high system operating pressure will occur. This will result in loss of air conditioning performance and possible system damage.

Properly operating recycling equipment will remove excess air, provide the maximum level of allowable air in recycled refrigerant, and also provide recycled refrigerant ready for use.

Verification for excess noncondensable contents in auxiliary portable containers of recycled refrigerant is important. Proper procedure to assure correct non-condensable levels is outlined in section 5 of SAE J1989 for CFC-12 (tables on page 473), and SAE J2211 for HFC-134a (tables on page 475).

When determining the pressure/ temperature of refrigerant containers, the location can become critical. If the container is located in a garage, the floor temperature can effect the temperature of the refrigerant. Attaching a temperature measuring device to the lower one half of the refrigerant container surface can provide a more accurate reading. Using only the air temperature surrounding the refrigerant container can result in incorrect refrigerant temperature information.

Only DOT CFR Title 49 or UL-approved storage containers for recycled refrigerant (containers marked DOT 4BW or 4BA) must be used.

Contaminated Refrigerant

Refrigerant recovery/recycling equipment will not separate or clean contaminated refrigerants.

If either CFC-12 or HFC-134a refrigerant has been contaminated with another refrigerant, or with each other, the refrigerant tank pressure will be higher than that noted in the SAE tables. Refrigerant contamination can also occur from excessive air in the recycled refrigerant. This high NCG level can be caused by improperly operating manual or automatic equipment air purge cycle.

If the pressure is 5% or more higher than the pressure indicated in the SAE tables for either refrigerant, it should be assumed that contamination has occurred. (See example below.) Automotive recycling equipment will not remove this contamination. The tank should be sent off-site for reclamation.

For example, using the SAE tables for each refrigerant at 80°F, your results would be as shown in Chart 1 below.

Chart 1.

For example, using the SAE tables for each refrigerant at 80°F, your results would be as shown in the chart below:

CFC-12	96 psig	(SAE Ref. Chart Pressure)
	x 1.05	(Multiplication Factor)
	= 100.8 psig	(Contaminated Ref. Press.)
HFC-134a	91 psig	(SAE Ref. Chart Pressure)
	x 1.05	(Multiplication Factor)
	= 95.5 psig	(Contaminated Ref. Press.)

Using a pressure gauge for A/C system/container readings will only identify possible refrigerant contamination and will not identify the refrigerant type. Certified SAE refrigerant identification equipment (SAE J1771) will help determine type.

Note: A/C system refrigerant contamination, by air or other refrigerants, in excess of 3% by weight can cause system operating problems.

Identifying Refrigerants

EPA requires that when any vehicle is retrofitted from R12, a label identifying the new refrigerant in the system must be placed under the hood, and new fittings that are unique to that refrigerant must be attached to the high- and low-side service ports of the A/C system. These EPA requirements obviously don't solve the entire refrigerant identification problem. Your shop could encounter a vehicle that has been retrofitted to another refrigerant but has not been properly labeled, or a vehicle that has the right label, but highly contaminated refrigerant.

Refrigerant Identifiers

Purchasing a refrigerant identifier unit can help pinpoint many refrigerant identification problems, and EPA strongly recommends (but does not require) that technicians obtain this equipment. You can use the identifier to confirm that the refrigerant your supplier is sending you is exactly what he says it

is—pure and uncontaminated. The equipment you choose will depend on what you plan to do once you discover that refrigerant in a vehicle is not pure R12 or R134a. If, for example, you decide to turn the customer with a contaminated system away, then a less-expensive identifier that simply tells you whether the refrigerant is pure R12 of R134a (go/no-go) may be sufficient for you.

Keep in mind that even the most sophisticated diagnostic units on the market today cannot properly identify all combinations of chemicals used in blend refrigerants.

Whether you are interested in purchasing a "go/no-go" unit or a diagnostic unit, check that the unit meets the SAE J1771 standard, which is an indication that the unit correctly identifies refrigerants. When claiming to meet this standard, manufacturers of identifier equipment are required to label the unit stating its level of accuracy,

Recovering Contaminated Refrigerant

As a first step, the contaminated or unfamiliar refrigerant must be recovered. EPA prohibits venting any automotive refrigerants (including "unacceptable" refrigerants), no matter what combination of chemicals is in the refrigerant. The best way today that a technician can recover contaminated or unfamiliar refrigerant is to dedicate a recovery-only unit to anything that is not pure R12 or pure R134a.

Some equipment manufacturers may also be marketing types of recovery-only stations specifically designed to remove these refrigerants.

If the refrigerant you extract into a recovery unit contains a high level of flammable substances such as propane and butane, a fire hazard may result if the refrigerant comes into contact with an ignition source within the equipment. Make sure you determine what features have been incorporated into the equipment to guard against risks of ignition.

Contaminated Refrigerant Storage and Disposal

Once the refrigerant has been recovered, if you can't recycle it, what do you do with it? The answer is that it depends.

If the refrigerant in your "junk" tank contains significant amounts of flammable substances, it may be considered hazardous and you should make sure you follow any local, state or federal requirements that govern the storage of combustible mixtures.

If the refrigerant in your "junk" tank is a chemical "soup"—either contaminated R12 and R134a, or a mixture of those contaminated refrigerants and some blend refrigerants that you are unfamiliar with—then the contents should be reclaimed or destroyed. You should investigate all your options and pick the one that makes the most economic sense for you.

EPA Q & A

Q: Does the EPA require that all leaks in motor vehicle air conditioners be repaired?

A: The EPA does not require that leaks be repaired, although it recommends that vehicle owners consider repairing leaks to reduce emissions and extend the useful life of their air conditioner. Repair of leaking systems will help vehicle owners avoid the need to continue to refill systems with high-priced refrigerant. EPA recognizes that good service practices include recovering and recycling refrigerant and performing leak detection. If a leak is identified, the customer should be presented with all the options for service, including

repair. If leak repair is not chosen, the technician may refill the system if requested to do so by the customer (unless a state or local leak repair requirement exists).

Q: Is a technician required to recover and recycle any refrigerant added to a system for the purpose of leak detection?

A: If a technician adds refrigerant to a system for the purpose of leak detection and if the refrigerant is then removed, it must be recovered and recycled and not released to the environment. The leak detection charge may be left in the system at the request of the customer.

SAE J1991: Purity of Recycled CFC-12

The SAE J1991 standard of purity for recycled CFC-12 refrigerant for use in mobile A/C systems, which has been directly removed from automotive A/C systems, shall not exceed the following levels of contaminants:

- **Moisture: 15 PPM (parts per million) by weight**
- **Refrigerant Oil: 4000 PPM by weight**
- **Non-condensable Gases (air): 330 PPM by weight**

Certified recycling equipment is required by the Clean Air Act to meet the SAE J1990 standard.

Equipment which has safety certification, such as Underwriters Laboratories "UL," does not mean it is in compliance with J1990 and J1991.

The equipment also must have a label which states: "Design certified for compliance with SAE J1991," to comply with the Clean Air Act.

SAE J2099: Purity of Recycled HFC-134a

The SAE J2099 standard of purity for recycled HFC-134a refrigerant for use in mobile A/C systems, which has been directly removed from automotive A/C systems, shall not exceed the following levels of contaminants:

- **Moisture: 50 PPM (parts per million) by weight**
- **Refrigerant Oil: 500 PPM by weight**
- **Non-condensable Gases (air): 150 PPM by weight**

Certified recycling equipment is required by the Clean Air Act to meet the SAE J2210 standard.

Equipment which has safety certification, such as Underwriters Laboratories "UL," does not mean it is in compliance with J2099 and J2210.

The equipment must also have a label which states: "Design certified for compliance with SAE J2210," to comply with the Clean Air Act.

SAE J2209: CFC-12 Extraction Equipment

SAE J2209 establishes certification requirements for CFC-12 extraction equipment. Extraction equipment that meets SAE J2209 is designed for the purpose of removing CFC-12 from an automotive air conditioning system. This extraction-only equipment does not recycle the refrigerant for reuse. Refrigerant which is taken from a mobile A/C system by extraction equipment must be recycled (or reprocessed to the appropriate ARI 700 specification) before it can be reused in an automotive A/C system.

The operation of extraction equipment is similar to the recovery portion of recovery/recycling equipment, but it will not clean the removed refrigerant.

Refrigerant removed with extraction equipment shall not be directly reused without being recycled or reprocessed to the appropriate ARI 700 specification.

Extraction equipment is also equipped with a device to indicate the amount of lubricant taken out during the removal process. (The procedure for lubricants is found in the "System Lubricant" section on page 486 of this manual.)

The equipment and refrigerant tanks have SAE 3/8-in., high-side service fittings to prevent possible direct use of the dirty refrigerant into an automotive air conditioning system.

Do not use adaptor fittings. Use of adaptor fittings may result in possible contamination of clean CFC-12 supplies and mobile systems.

The refrigerant tanks not only have the approved SAE fitting, but are gray in color, with a yellow top and identification label marking: "DIRTY CFC-12 • DO NOT USE: MUST BE REPROCESSED."

This is intended to prevent possible misuse.

To comply with the regulations of the Clean Air Act, refrigerant removed with extraction equipment must be sent off-site to be reprocessed to the appropriate specifications. Records must be maintained for three years identifying where the refrigerant was sent.

The federal Clean Air Act allows only one exception in the use of extraction equipment: That is, if the owner of the extraction equipment also owns certified recovery/recycling equipment and can assure direct recycling of refrigerant from motor vehicles for re-use in motor vehicles serviced at his facilities.

SAE J1732: HFC-134a Extraction Equipment

Extraction equipment which meets SAE J1732 is designed for the purpose of removing HFC-134a from an automotive air conditioning system. Extraction-only equipment does not recycle the refrigerant for re-use. The operation of extraction equipment is similar to the recovery portion of recovery/recycling equipment, but it will not clean the removed refrigerant.

Refrigerant removed with extraction equipment shall not be directly reused without being reprocessed to the appropriate ARI 700 specification.

Extraction equipment is also equipped with a device to indicate the amount of lubricant taken out during the removal process. (The procedure for lubricants is found in the "System Lubricant" section on page 486 of this manual.)

The equipment and refrigerant tanks have a 1/2-in. Acme thread service fitting.

The refrigerant tanks not only have the approved SAE fitting, but are gray in color, with a yellow top and identification label marking: "DIRTY HFC-134a • DO NOT USE: MUST BE REPROCESSED."

This is intended to prevent possible misuse.

To comply with the regulations of the Clean Air Act, refrigerant removed with extraction equipment must be sent offsite to be reprocessed to the appropriate specifications. Records must be maintained for three years identifying where the refrigerant was sent.

The federal Clean Air Act allows only one exception in the use of extraction equipment. That is, if the owner of the extraction equipment also owns certified recovery/recycling equipment and can assure direct recycling of refrigerant from motor vehicles for re-use in motor vehicles serviced at his facilities.

SAE J1989 • Recommended Service Procedure

SAE J1989 • Issued October 1989
Recommended Service Procedure for the Containment of CFC-12
© Society of Automotive Engineers, Inc., 1989

1. SCOPE:
During service of mobile air conditioning systems, containment of the refrigerant is important. This procedure provides service guidelines for technicians when repairing vehicles and operating equipment defined in SAE J1990.

2. REFERENCES:
SAE J1990, Extraction and Recycle Equipment for Mobile Automotive Air Conditioning Systems

3. REFRIGERANT RECOVERY PROCEDURE:

3.1 Connect the recovery unit service hoses, which shall have shutoff valves within 12 in. (30 cm) of the service ends, to the vehicle air conditioning system service ports.

3.2 Operate the recovery equipment as covered by the equipment manufacturer's recommended procedure.

3.2.1 Start the recovery process and remove the refrigerant from the vehicle A/C system. Operate the recovery unit until the vehicle system has been reduced from a pressure to a vacuum. With the recovery unit shut off for at least 5 minutes, determine that there is no refrigerant remaining in the vehicle A/C system. If the vehicle system has pressure, additional recovery operation is required to remove the remaining refrigerant. Repeat the operation until the vehicle A/C system vacuum level remains stable for 2 minutes.

3.3 Close the valves in the service lines and then remove the service lines from the vehicle system. Proceed with the repair/service. If the recovery equipment has automatic closing valves, be sure they are operating properly.

4. SERVICE WITH MANIFOLD GAUGE SET:

4.1 Service hoses must have shutoff valves in the high-side, low-side and center service hoses within 12 in. (30 cm) of the service ends. Valves must be closed prior to hose removal from the air conditioning system. This will reduce the volume of refrigerant contained in the service hose which would otherwise be vented to atmosphere.

4.2 During all service operations, the valves should be closed until connected to the vehicle air conditioning system or the charging source to avoid introduction of air, and to contain the refrigerant rather than vent to the open atmosphere.

4.3. When the manifold gauge set is disconnected from the air conditioning system, or when the center hose is moved to another device which cannot accept refrigerant pressure, the gauge set hoses should first be attached to the reclaim equipment to recover the refrigerant from the hoses.

5. RECYCLED REFRIGERANT CHECKING PROCEDURE FOR STORED PORTABLE AUXILIARY CONTAINER:

5.1 To determine if the recycled refrigerant container has excess non-condensable gases (air), the container must be stored at a temperature of 65°F (18.3°C) or above, for a period of 12 hours, protected from direct sun.

5.2 Install a calibrated pressure gauge, with 1 psig divisions (0.07 kg), to the container and determine the container pressure.

5.3 With a calibrated thermometer, measure the air temperature within 4 inches (10 cm) of the container surface.

5.4 Compare the observed container pressure and air temperature to determine if the container exceeds the pressure limits found in Table 1 (opposite page), e.g.: at air temperature of 70°F (21°C) pressure must not exceed 80 psig (5.62 kg/cm²).

5.5 If the container pressure is less than the Table 1 values, and has been recycled, limits of non-condensable gases (air) have not been exceeded and the refrigerant may be used.

5.6 If the pressure is greater than the range, and the container contains recycled material, slowly vent from the top of the container a small amount of vapor into the recycle equipment, until the pressure is less than the pressure shown on Table 1.

5.7 If the container still exceeds the pressure shown in Table 1, the entire contents of the container shall be recycled.

MAXIMUM ALLOWABLE CONTAINER PRESSURE—RECYCLED CFC-12

TABLE 1 (Standard)

TEMP°F	PSIG	TEMP°F	PSIG	TEMP°F	PSIG	TEMP°F	PSIG	TEMP°F	PSIG
65	74	75	87	85	102	95	118	105	136
66	75	76	88	86	103	96	120	106	138
67	76	77	90	87	105	97	122	107	140
68	78	78	92	88	107	98	124	108	142
69	79	79	94	89	108	99	125	109	144
70	80	80	96	90	110	100	127	110	146
71	82	81	98	91	111	101	129	111	148
72	83	82	99	92	113	102	130	112	150
73	84	83	100	93	115	103	132	113	152
74	86	84	101	94	116	104	134	114	154

TABLE 1 (Metric)

TEMP°C	PRES	TEMP°C	PRES	TEMP°C	PRES	TEMP°C	PRES	TEMP°C	PRES
18.3	5.20	23.9	6.11	29.4	7.17	35.0	8.29	40.5	9.56
18.8	5.27	24.4	6.18	30.0	7.24	35.5	8.43	41.1	9.70
19.4	5.34	25.0	6.32	30.5	7.38	36.1	8.57	41.6	9.84
20.0	5.48	25.5	6.46	31.1	7.52	36.6	8.71	42.2	9.98
20.5	5.55	26.1	6.60	31.6	7.59	37.2	8.78	42.7	10.12
21.1	5.62	26.6	6.74	32.2	7.73	37.7	8.92	43.3	10.26
21.6	5.76	27.2	6.88	32.7	7.80	38.3	9.06	43.9	10.40
22.2	5.83	27.7	6.95	33.3	7.94	38.8	9.13	44.4	10.54
22.7	5.90	28.3	7.03	33.9	8.08	39.4	9.27	45.0	10.68
23.3	6.04	28.9	7.10	34.4	8.15	40.0	9.42	45.5	10.82

PRES kg/sq cm

6. CONTAINERS FOR STORAGE OF RECYCLED REFRIGERANT:

6.1 Recycled refrigerant should not be salvaged or stored in disposable refrigerant containers. This is the type of container in which virgin refrigerant is sold. Use only DOT CFR Title 49 (marked DOT 4BA or DOT 4BW) or UL-approved storage containers for recycled refrigerant.

6.2 Any container of recycled refrigerant which has been stored or transferred must be checked prior to use as defined in Section 5.

7. TRANSFER OF RECYCLED REFRIGERANT:

7.1 When external portable containers are used for transfer, the container must be evacuated to at least 27 inches of vacuum (75 mm Hg absolute pressure) prior to transfer of the recycled refrigerant. External portable containers must meet DOT and UL standards.

7.2 To prevent on-site overfilling when transferring to external containers, the safe filling level must be controlled by weight and must not exceed 60% of container gross weight rating.

8. DISPOSAL OF EMPTY/NEAR EMPTY CONTAINERS:

8.1 Since all the refrigerant may not be removed from disposable refrigerant containers during normal system charging procedures, empty or near empty container contents should be reclaimed prior to disposal of the container.

8.2 Attach the container to the recovery unit and remove the remaining refrigerant. When the container has been reduced from a pressure to a vacuum, the container valve can be closed. The container should be marked empty and is then ready for disposal.

SAE J2211 • Issued December 1991
Recommended Service Procedure for the Containment of HFC-134a
© Society of Automotive Engineers, Inc., 1991

1. SCOPE:

Refrigerant containment is an important part of servicing mobile air conditioning systems. This procedure provides service guidelines for technicians when repairing vehicles and operating equipment designed for HFC-134a (described in SAE J2210).

2. REFERENCES:

2.1 Applicable Documents—The following publications form a part of this specification to the extent specified herein. The latest issue of SAE publications shall apply.

2.1.1 SAE Publications—Available from SAE, 400 Commonwealth Drive, Warrendale, Pa. 15096-0001.
SAE J2196—Service Hoses for Mobile Air Conditioning Systems
SAE J2197—Service Hose Fittings for Automotive Air-Conditioning
SAE J2210—Refrigerant Recycling Equipment for HFC-134a Mobile Air Conditioning Systems
SAE J2219—Concerns to the Mobile Air Conditioning Industry

2.2 DEFINITIONS

2.2.1 Recovery/Recycling (R/R) Unit—Refers to a single piece of equipment which performs both functions of recovery and recycling of refrigerants per SAE J2210.

2.2.2 Recovery—Refers to that portion of the R/R unit operation which removes the refrigerant from the mobile air conditioning system and places it in the R/R unit storage container.

2.2.3 Recycling—Refers to that portion of the R/R unit operation which processes the refrigerant for reuse on the same job site to the purity specifications of SAE J2099.

3. SERVICE PROCEDURE:

3.1 Connect the recycling unit service hoses, which shall have shutoff devices (e.g., valves) within 30 cm (12 in.) of the service ends, to the vehicle air-conditioning (A/C) service ports. Hoses shall conform to SAE J2196 and fittings shall conform to SAE J2197.

3.2 Operate the recycling equipment per the equipment manufacturer's recommended procedure.

3.2.1 Verify that the vehicle A/C system has refrigerant pressure. Do not attempt to recycle refrigerant from a discharged system as this will introduce air (noncondensable gas) into the recycling equipment which must later be removed by purging.

3.2.2 Begin the recycling process by removing the refrigerant from the vehicle A/C system. Continue the process until the system pressure has been reduced to a minimum of 102 mm (4 in.) of mercury below atmospheric pressure (vacuum). If A/C components show evidence of icing, the component can be gently heated to facilitate refrigerant removal. With the recycling unit shut off for at least 5 minutes, check A/C system pressure. If this pressure has risen above vacuum (0 psig), additional recycler operation is required to remove the remaining refrigerant. Repeat the operation until the system pressure remains stable at vacuum for 2 minutes.

3.3 Close the valves in the service lines and then remove the service lines from the vehicle system. If the recovery equipment has automatic closing valves, be sure they are operating properly. Proceed with the repair/service.

3.4 Upon completion of refrigerant removal from the A/C system, determine the amount of lubricant removed during the process and replenish the system with new lubricant, which is identified on the A/C system label. Used lubricant should be discarded per applicable federal, state, and local requirements.

4. SERVICE WITH A MANIFOLD GAUGE SET:

4.1 High-side, low-side, and center service hoses must have shutoff devices (e.g. valves) within 30 cm (12 in.) of the service ends. Valves must be closed prior to hose removal from the A/C system to prevent refrigerant loss to the atmosphere.

4.2 During all service operations, service hose valves should be closed until connected either to the vehicle A/C system or the charging source so as to exclude air and/or contain the refrigerant.

4.3 When the manifold gauge set is disconnected from the A/C system, or when the center hose is moved to another device which cannot accept refrigerant pressure, the gauge set hoses should be attached to the recycling equipment to recover refrigerant from the hoses.

5. SUPPLEMENTAL REFRIGERANT CHECKING PROCEDURE FOR STORED PORTABLE CONTAINERS:

5.1 Certified recycling equipment and the accompanying recycling procedure, when properly followed, will deliver use-ready refrigerant. In the event that the full recycling procedure was not followed, or the technician is unsure about the noncondensable gas content of a given tank of refrigerant, this procedure can be used to determine whether the recycled refrigerant container meets the specification for noncondensable gases (air).

Note: The use of refrigerant with excess air will result in higher system operating pressures and may cause A/C system damage.

5.2 The container must be stored at a temperature of 18.3°C (65°F) or above for at least 12 hours, protected from direct sunlight.

5.3 Install a calibrated pressure gauge, with 6.9 kPa (1 psig) divisions, to the container and determine the container pressure.

5.4 With a calibrated thermometer, measure the air temperature within 10cm (4 in.) of the container surface.

5.5 Compare the observed container pressure and air temperature to the values given in Table 2 (below) to determine whether the container pressure is below the pressure limit given in the Table. For example, at an air temperature of 21°C (70°F), the container pressure must not exceed 524 kPa (76 psig).

5.6 If the refrigerant in the container has been recycled and the container pressure is less than the limit in Tables 1 and 2, the refrigerant may be used.

5.7 If the refrigerant in the container has been recycled and the container pressure exceeds the limit in Tables 1 and 2, slowly vent, from the top of the container, a small amount of vapor into the recycle equipment until the pressure is less than the pressure shown in Tables 1 and 2.

5.8 If, after shaking the container and letting it stand for a few minutes, the container pressure still exceeds the pressure limit shown in Tables 1 and 2, the entire contents of the container shall be recycled.

6. CONTAINERS FOR STORAGE OF RECYCLED REFRIGERANT:

6.1 Recycled refrigerant should not be salvaged or stored in disposable containers (this is one common type of container in which new refrigerant is sold). Use only DOT CFR Title 49 or UL-approved storage containers, specifically marked for HFC-134a, for recycled refrigerant.

6.2 Any container of recycled refrigerant which has been stored or transferred must be checked prior to use as defined in Section 5.

6.3 Evacuate new tanks to at least 635 mm Hg (25 in. Hg) below atmospheric pressure (vacuum) prior to first use.

7. TRANSFER OF RECYCLED REFRIGERANT:

7.1 When external portable containers are used for transfer, the

---

MAXIMUM ALLOWABLE CONTAINER PRESSURE—RECYCLED HFC-134a
TABLE 2 (Metric)

TEMP°C(F)	kPa	TEMP°C(F)	kPa	TEMP°C(F)	kPa	TEMP°C(F)	kPa
18 (65)	476	26 (79)	621	34 (93)	793	42(108)	1007
19 (66)	483	27 (81)	642	35 (95)	814	43(109)	1027
20 (68)	503	28 (82)	655	36 (97)	841	44(111)	1055
21 (70)	524	29 (84)	676	37 (99)	876	45(113)	1089
22 (72)	545	30 (86)	703	38(100)	889	46(115)	1124
23 (73)	552	31 (88)	724	39(102)	917	47(117)	1158
24 (75)	572	32 (90)	752	40(104)	945	48(118)	1179
27 (77)	593	33 (91)	765	41(106)	979	49(120)	1214

TABLE 2 (English)

TEMP°F	PSIG	TEMP°F	PSIG	TEMP°F	PSIG	TEMP°F	PSIG
65	69	79	90	93	115	107	144
66	70	80	91	94	117	108	146
67	71	81	93	95	118	109	149
68	73	82	95	96	120	110	151
69	74	83	96	97	122	111	153
70	76	84	98	98	125	112	156
71	77	85	100	99	127	113	158
72	79	86	102	100	129	114	160
73	80	87	103	101	131	115	163
74	82	88	105	102	133	116	165
75	83	89	107	103	135	117	168
76	85	90	109	104	137	118	171
77	86	91	111	105	139	119	173
78	88	92	113	106	142	120	176

container must be evacuated to at least 635 mm (25 in Hg) below atmospheric pressure (vacuum) prior to transfer of the recycled refrigerant to the container. External portable containers must meet DOT and UL standards.

7.2 To prevent onsite overfilling when transferring to external containers, the safe filling level must be controlled by weight and must not exceed 60% of the container gross weight rating.

8. SAFETY NOTE FOR HFC-134a:

8.1 HFC-134a has been shown to be nonflammable at ambient temperature and atmospheric pressure. However, recent tests under controlled conditions have indicated that, at pressures above atmospheric, and with air concentrations greater than 60% by volume, HFC-134a can form combustible mixtures. While it is recognized that an ignition source is also required for combustion to occur, the presence of combustible mixtures is a potentially dangerous situation and should be avoided.

8.2 Under NO CIRCUMSTANCE should any equipment be pressure tested or leak tested with air/HFC-134a mixtures. Do not use compressed air (shop air) for leak detection in HFC-134a systems.

9. DISPOSAL OF EMPTY OR NEAR EMPTY CONTAINERS:

9.1 Since all refrigerant may not have been removed from disposable refrigerant containers during normal system charging procedures, empty or near empty container contents should be recycled prior to disposal of the container.

9.2 Attach the container to the recycling unit and remove the remaining refrigerant. When the container has been reduced from a pressure to a vacuum, the container valve can be closed and the container removed from the unit. The container should be marked "Empty," after which it is ready for disposal.

Recovery & Recycling

Recycling versus reclaiming refrigerant... There is a difference!

When this text refers to **recycled** refrigerant, it is referring to refrigerant that has been recycled on-site at a service facility with automotive recycling equipment certified to the appropriate SAE J standard.

When this text refers to **reclaimed** refrigerant, it is referring to refrigerant that has been sent to an EPA-listed reclamation facility where it processed and returned to the appropriate ARI 700 specification.

The standards of purity for reclaimed refrigerant are much higher than the standards of purity for recycled refrigerant.

Please Note!

Recovery/recycling equipment is not designed to recycle or separate contaminated refrigerants. Contaminated or unknown refrigerant must be removed, using separate recovery equipment, from the mobile A/C system or equipment and properly disposed. Contaminated refrigerant containing CFCs, HCFCs and HFC-134a, under federal law, cannot be vented.

Check Equipment

Recycling equipment should be checked on a monthly basis to ensure that no leakage occurs. Establish records for maintenance and service equipment filter changes as recommended by the equipment manufacturer. These procedures will ensure that the SAE J standards of purity for recycled refrigerant are maintained. Check equipment manual for filter location.

Recover All Refrigerant

All cylinders which contain any unused refrigerant should be connected to the proper recovery/recycling machine and brought to a vacuum before they are discarded.

Always Follow Equipment Manufacturer's Instructions.... when using refrigerant recovery/recycling equipment, it is important to observe the general service procedures in this manual as well as the operating instructions provided by the equipment manufacturer.

Figure 1.

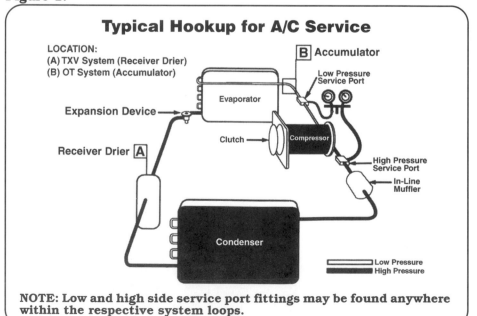

Typical Hookup for A/C Service

LOCATION:
(A) TXV System (Receiver Drier)
(B) OT System (Accumulator)

B Accumulator

Low Pressure Service Port

Expansion Device

Evaporator

Clutch

Compressor

Receiver Drier A

High Pressure Service Port

In-Line Muffler

Condenser

Low Pressure
High Pressure

NOTE: Low and high side service port fittings may be found anywhere within the respective system loops.

Types of Recovery/Recycling Equipment

Shown here are two types of refrigerant recovery/recycling systems: single-pass and multi-pass. Both systems remove the refrigerant from the vehicle and provide a process for cleaning and storing recycled refrigerant. The single-pass system makes recycled refrigerant available for re-use immediately. The multi-pass system does not.

Single-Pass System

In single-pass systems (Figure 2), refrigerant drawn from the vehicle A/C system passes through an oil separator. This removes any oil. The filter/drier assembly removes moisture and particle contamination. After a single cycle, the contaminant-free recycled refrigerant is then sent to a storage tank.

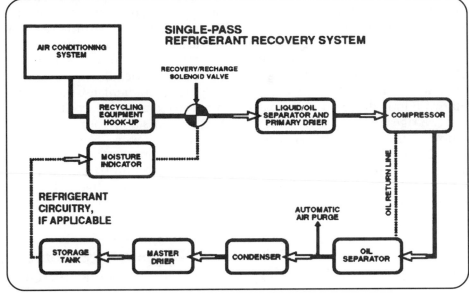

Figure 2.

Multi-Pass System

In multi-pass systems (Figure 3), refrigerant is drawn from the vehicle, passed through an oil separator which removes any oil, and a filter/drier assembly which removes moisture and particle contamination, and then is stored in a tank. (**NOTE:** Arrangement of system components differs among equipment manufacturers, as shown by diagrams A and B below.)

When recycling is desired, the recycle solenoid valve is opened, allowing a continuous loop-filtering process in which the refrigerant passes through a desiccant (drier) cartridge several times, until the moisture is fully removed. The station has an indicator to alert the service technician—or will automatically vent the recovery tank—to remove air. A moisture indicator will show when the refrigerant is ready for use.

Figure 3.

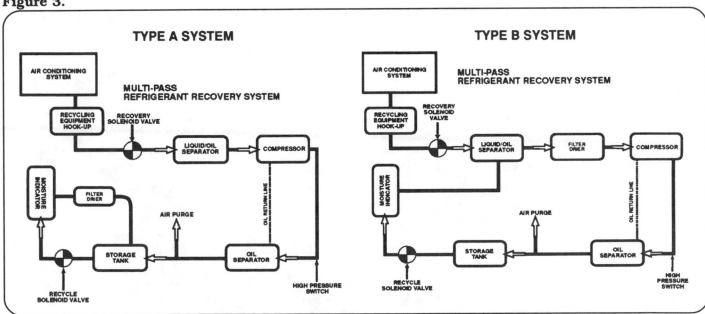

Servicing Alternate Refrigerants

With the transition from CFC-12, alternate refrigerants have entered the service industry. Under the federal SNAP rule, many alternate refrigerants have been listed for use in automotive A/C systems. HFC-134a is a single chemical refrigerant and it is the refrigerant of choice of the automobile industry. The other alternate refrigerants are blends containing from two to as many as four different chemicals.

All refrigerants listed by EPA for automotive use must be recovered, not vented. This includes R134a and blends containing R22, R124 and R142b. Alternate refrigerants require separate service equip-

ment, and separate hoses and gauge manifolds for each type of refrigerant. Refrigerant and lubricant residue in the hoses are common sources of system contamination; that's why you should never switch service equipment back and forth between different refrigerants. To comply with federal law, every service facility must have service equipment for the type of refrigerant being serviced.

Using equipment for more than one refrigerant will contaminate A/C systems and your refrigerant supply. Recovery and recycling equipment that's SAE certified for use with R12 and R134a isn't certified for use with alternate refrigerants.

Since no certification standards exist for blend refrigerant recovery/recycling equipment, it is illegal to recover/recycle on site for reuse, any blend refrigerant listed under SNAP for automotive use. The use of recovery only equipment will not process the refrigerant for reuse. The refrigerant can be removed from the vehicle using recovery equipment and sent off-site for processing.

As this manual goes to press, the EPA is working to develop standards for equipment and regulations that would allow a blend to be recovered from a vehicle, processed to remove lubricant and moisture, and returned to the same vehicle from which it was recov-

Chart 2.

Pressure/Temperature Comparison
FreeZone, Freeze 12, GHG-X4, GHG, ICOR, FRIGC, CFC-12 and HFC-134a.

Deg.	FreeZone psig (b) (d)	Freeze 12 psig Avg.	(R-414A)/GHG-X4 psig (b) (d)	R406A/GHG psig (b) (d)	(R-414B)/IKOR psig (b) (d)	FRIGC psig (b) (d)	R-12 psig	R-134a psig
20	15.7-15.8	18.1	25.4-13.3	39.9-28.4	26.9-15.8	15.9-14.2	21.1	18.4
25	19.0-19.2	23.3	29.5-16.5	43.9-31.5	31.1-19.1	19.3-17.5	24.6	22.1
30	22.6-22.8	26.5	33.8-19.8	48.3-34.9	35.6-23.1	22.9-20.9	28.5	26.1
35	26.5-26.7	30.2	38.5-23.5	52.9-38.5	40.5-26.7	26.8-24.6	32.6	30.4
40	30.7-30.9	34.0	43.5-27.5	57.9-42.5	45.6-30.9	31.1-28.7	37.0	35.0
45	35.2-35.4	39.5	48.8-31.7	63.2-46.8	51.2-35.5	35.7-33.0	41.7	40.1
50	40.0-40.3	43.1	54.5-36.3	68.8-51.3	57.0-40.3	40.6-37.6	46.7	45.4
55	45.2-45.5	48.3	60.5-41.3	74.8-56.3	63.6-45.6	45.9-42.6	52.1	51.2
60	50.8-51.1	55.5	67.0-45.5	81.3-61.5	70.0-51.2	51.5-45.7	57.7	57.4
65	56.8-57.7	60.0	73.9-52.2	88.1-67.2	77.8-57.2	57.5-53.7	63.8	64.0
70	63.7-63.5	67.4	81.2-58.3	95.3-73.2	84.6-63.6	64.0-59.8	70.2	71.1
75	69.9-70.4	75.0	88.9-64.7	103.-79.6	92.7-70.5	70.9-66.3	77.0	78.6
80	77.2-77.6	80.6	97.1-71.6	111.1-86.5	101.1-77.8	78.7-73.3	84.2	86.7
85	84.8-85.3	89.1	105.7-79.0	119.7-93.8	110.0-85.6	86.0-79.8	91.8	95.2
90	93.0-93.5	95.4	114.8-86.8	128.8-101.6	119.5-93.8	94.3-88.6	99.8	104.3
95	101.7-102.2	102.2	124.5-95.1	138.3-109.9	129.4-102.5	103.0-97.1	108.3	113.9
100	110.8-111.4	116.7	134.6-103.9	148.4-118.6	139.-111.8	112.3-106.0	117.2	124.1
105	120.5-121.7	125.0	145.3-113.2	159.0-127.8	150.9-121.6	122.1-115.4	126.6	134.9
110	130.7-131.3	135.1	156.5-123.0	170.1-137.6	162.5-132.0	132.5-125.5	136.4	146.4
115	141.5-142.1	145.6	168.3-133.5	181.8-148.0	174.6-142.9	143.4-136.1	146.8	158.4
120	152.9-153.4	157.0	180.6-144.5	194.0-158.9	187.4-154.5	154.9-147.3	157.7	171.1
125	164.8-165.4	165.1	193.6-156.1	206.9-170.4	200.7-166.7	167.1-159.1	169.1	184.5
130	177.4-177.9	183.3	207.1-168.3	220.3-182.6	214.7-179.5	179.8-171.6	181.0	198.7
140	204.3-204.9	211.8	236.1-194.8	249.1-208.8	244.6-207.2	207.2-198.5	206.5	229.3
150	234.0-234.5	243.2	267.7-224.1	280.4-237.9		237.3-228.1	234.5	263.0

ered. Note that this is not the same as recycling, where a refrigerant must be returned to a minimum standard of purity. There is no assurance that the A/C system will provide the same level of performance with the processed blend refrigerant as it did with new refrigerant containing the proper blend formulation.

Systems charged with blends may provide a high level of performance when initially charged. However, each refrigerant in these blends has a different pressure/temperature relationship, and different leakage rates through flexible hose. These blends can separate while in use, and a leak in the system can allow just one component of the blend to escape from the system. This partial leakage can change the entire refrigerant mixture and cause system operating problems.

Chart 3. This is a list of MVAC substitutes for CFC-12, which have been reviewed under EPA's SNAP Program. The most current version of this list may be obtained by calling the EPA's Stratospheric Ozone Hotline at 1-800-296-1997, or through the EPA website, located at http://www.epa.gov/ozone/title6/609/609.html.

Motor Vehicle Air Conditioning Substitutes for CFC-12 Reviewed Under EPA's SNAP Program as of June 3, 1997									
				Components / Reasons Unacceptable					
Name (1)	Status (2)	Date	Manfacturer	HCFC-22	HCFC-124	HCFC-142b	HFC-134a	Butane (R-600) (3)	Isobutane (R-600a) (3)	HFC-227ea
HFC-134a	ASU	3/18/94	Several				100			
FRIGC FR-12	ASU	6/13/95	Intermagnetics General 800-555-1442		39		59	2		
Free Zone/ RB-276 (4)	ASU	5/22/96	Freezone 888-373-3066			19	79			
Ikon-12	ASU	5/22/96	Ikon Corp. 601-868-0755	Composition claimed as confidential business information						
R-406A/ GHG/McCool (5)	ASU	10/16/96	People's Welding 800-382-9006	55		41			4	
GHG-X4/ Autofrost/Chill-It (5)	ASU	10/16/96	People's Welding 800-382-9006	51	28.5	16.5			4	
Hot Shot/Kar Kool (5)	ASU	10/16/96	ICOR 800-357-4062	50	39	9.5			1.5	
GHG-HP (5)	ASU	10/16/96	People's Welding 800-382-9006	65		31			4	
FREEZE 12	ASU	10/16/96	Technical Chemical 800-527-0885			20	80			
GHG-X5	ASU	6/3/97	People's Welding 800-382-9006	41		15			4	40
OZ-12	UNA	3/18/94	OZ Technology	Flammable blend of hydrocarbons; insufficient data to demonstrate safety						
R-176	UNA	3/18/94	Arctic Chill	Contains CFC-12, which is inappropriate in a CFC-12 substitute						
HC-12a	UNA	6/13/95	OZ Technology	Flammable blend of hydrocarbons; insufficient data to demonstrate safety						
Duracool 12a	UNA	6/13/95	Duracool Limited	This blend is identical to HC-12a ®						
R-405A	UNA	6/13/95	Greencool	Perfluorocarbon component; extremely high global warning potential and lifetime						

1 — R-401A (made by DuPont), R-401B (DuPont), R-409A (Elf Atochem), Care 30 (Calor Gas), Adak-29/Adak-12 (TACIP Int'l), MT-31 (Millenia Tech), and ES-12R (Intervest) have not been submitted for review in motor vehicle air conditioning, and it is therefore illegal to use these refrigerants in such systems.

2 — See text for details on legality of use according to status.
ASU = acceptable subject to fittings, labeling, no drop-in, and compressor shutoff switch use conditions
UNA = unacceptable; illegal for use as a CFC-12 substitute in motor vehicle air conditioners

3 — Although some blends contain hydrocarbons, all blends that are ASU are nonflammable as blended.

4 — Freezone contains 2% of a lubricant.

5 — HCFC-22 content results in an additional use condition: must be used with barrier hoses.

Blend Refrigerants

Blend refrigerants contain more than one refrigerant, and have a bubble and dew pressure value that affects A/C controls. This bubble and dew characteristic of blends results in a temperature difference across the evaporator and condenser known as "glide." The auto industry doesn't design automotive A/C systems to use blend refrigerants. Installing blend refrigerants with the original A/C system refrigerant controls (expansion valves, pressure controls) could cause system performance problems. The information in Chart 2 compares the different refrigerant pressures.

The terms *bubble* and *dew* refer to the different condensing and vaporization characteristics of a given blend refrigerant. This may seem confusing since the automotive industry has always used single composition refrigerants such as R12 and R134a. Typically, with single composition refrigerants, by controlling the refrigerant pressure, the vaporization and condensing points are at the same temperature at a given pressure. Keep in mind that in order for the refrigerant to change states from a liquid to a vapor (vaporization) or from a vapor to liquid (condensation), it must first transfer a considerable amount of latent heat which is typically measured in BTUs.

With blend refrigerants, there's a variation between the actual vaporization and condensing temperature. The difference is referred to as glide and the greater the difference, the higher the glide. In other words, blend refrigerants do not have a specific pressure or temperature that allows them to be in either a vapor or liquid state, depending on the amount of heat contained.

Chart 4.

Container Fittings

Refrigerant	30# Container			Small Cans		
	Diameter (Inches)	Thread (Pitch/inch)	Thread (Direction)	Diameter (Inches)	Thread (Pitch/inch)	Thread (Direction)
R12	7/16	20	Right			
R134a	8/16	16 Acme	Right	8/16	16 Acme	Right
Freeze 12	8/16	18	Right	6/16	24	Right
Free Zone/RB-276	9/16	18	Right	6/16	24	Left
Hot Shot/Kar Kool	10/16	18	Right			
GHG-X4/Autofrost/Chill-it	.368	26	Right	14 mm	1.25 mm spacing	Left
R-406A/GHG/McCool	.368	26	Left	8/16	20	Left
FRIGC/FR-12	8/16	20	Left	7/16	20	Left

Based on EPA information dated October 1996

Mobile A/C Service Fittings

Refrigerant	High Side Service Port			Low Side Service Port		
	Diameter Inches	Thread Pitch/inch	Thread Direction	Diameter Inches	Thread Pitch/inch	Thread Direction
R12	3/8	24	Right	7/16	20	Right
R134a	Quick	Couple	16 mm	Quick	Couple	13 mm
Freeze 12	7/16	14	Left	8/16	18	Right
Free Zone/RB-276	8/16	13	Right	9/16	18	Right
Hot Shot/Kar Kool	10/16	18	Left	10/16	18	Right
GHG-X4/Autofrost/Chill-it	.305	32	Right	.368	26	Right
R-406A/GHG/McCool	.305	32	Left	.368	26	Left
FRIGC/FR-12	Quick Couple	Different than R134a		Quick Couple	Different than 134a	

Based on EPA information dated October 1996

Retrofitted Vehicles

CFC-12 vehicles that have been retrofitted to use an alternate refrigerant must comply to federal law. Retrofitted vehicles using an alternate refrigerant must have unique service fittings and a new system identification label and compressor high pressure cut-off switch if not already installed.

If a technician is servicing a system with nonbarrier hose, and rertofits it to a refrigerant that contains HCFC-22, the technician must install barrier hose.

Chart 3 identifies the SNAP listed alternate refrigerants for automotive A/C systems.

Charts 4 and 5 identify the refrigerant container fittings, service fittings and labels.

System Service

The design of A/C systems affects the amount of time required to extract all of the refrigerant prior to opening the system for repair.

Systems using an accumulator require special attention. Refrigerant removal from accumulator systems requires additional time and precautions. When refrigerant is removed during extraction from an accumulator system, low system pressure results in the accumulator becoming very cold, with external frost sometimes in evidence.

Since the accumulator contains both lubricant and refrigerant, a large quantity of refrigerant will remain in the system until the system has equalized. Also, until the accumulator achieves the temperature of the surrounding area, it will continue to outgas refrigerant.

Because both the lubricant and refrigerant are potentially at this outgas condition, venting and safety are of concern. If the liquid refrigerant has not been completely removed and the refrigerant lines are opened, as the accumulator warms, a sudden release of the mixture can occur.

Use of external heating sources, such as hair dryers and electric heating pads, will raise the pressure in the accumulator and reduce the extraction time.

At no time should an open flame torch be used.

All the refrigerant must be removed before opening any of the system's refrigerant connections.

SAE Standards J1989 and J2211 provide procedures to assure that the refrigerant has been extracted.

Chart 5.

Suppliers, Composition, Labels
EPA "SNAP" Listed Refrigerants for Automotive Use

Refrigerant	Supplier	Refrigerant Chemicals	Recommended Desiccant	Label Background Color	Label Foreground Color
R-12	Many	100% CFC-12	XH-5, XH-7, XH-9		
R-134a	Many	100 % HFC-134a	XH-7, XH-9	Sky Blue	Black
Freeze 12	Technical Chemical 800-527-0885	80% HFC-134a 20% HCFC-142b	XH-7, XH-9	Yellow	Black
Free Zone RB-276	Refrigerant Gases 888-373-3066	79% HFC-134a 19% HCFC-142b	XH-7, XH-9	Light Green	White
R-414B Hot Shot Kar Kool	ICOR 800-357-4062	50% HCFC-22 39% HCFC-124 9.5% HCFC-142b 1.5% R600a	XH-9	Medium Blue	Black
R-414A GHG-X4 Autofrost Chill-it	People's Welding 800-382-9006	51% HCFC-22 28.5% HCFC-124 16.5% HCFC-142b 4% R600a	XH-9	Red	White
R-406A GHG McCool	People's Welding 800-382-9006	55% HCFC-22 41% HCFC-142b 4% R600a	XH-9		
FRIGC FR-12	Intercool 800-55-1445	59% HFC-134a 39% HCFC-124 2% R600a	XH-7, XH-9	Grey	Black

ASHRAE Chemical spec: Non-flammable +/- 2% Flammable < 1%

Current Systems Using HFC-134a Refrigerant

Starting with some 1992 models with completion by the 1995 model year, HFC-134a replaced CFC-12, which has been used in mobile A/C systems for many years.

Design changes in A/C systems have resulted in the adoption of improved hose and seal materials which reduce system leakage.

The development of recovery/recycling equipment has also resulted in reduced consumption of new refrigerant during normal service operations. Onsite recycling is both good for the environment and required by law.

Industry and government cooperation with SAE has resulted in the development of additional SAE documents which provide service equipment guidelines and procedures for mobile A/C system service.

Service hoses and fittings for refrigerants are covered in J2196 and J2197. This includes requirements for lower refrigerant emission leakage rates for service hoses; and to prevent system cross contamination, 14-mm service hose and gauge manifold connections are specified for HFC-134a.

Containers for HFC-134a use a 1/2-in. Acme thread for hose connection and are light blue in color.

It is important to remember that CFC-12 and HFC-134a refrigerant are not directly interchangeable between systems. Should they become mixed, refrigerant contamination will occur, resulting in higher system operating pressures.

There is no "drop-in" replacement refrigerant for CFC-12 automotive A/C systems. **Some system modification will be required with the use of any alternate refrigerant.**

Use of service parts for CFC-12 systems, such as accumulators or receiver dryers, compressors, seals and hoses may cause problems if installed on an HFC-134a system. It is important to use only parts which meet the OEM specifications to assure compatibility with the refrigerant.

HFC-134a does not provide the same system lubricating circulation as CFC-12, and it is important that the proper lubricant be used in an HFC-134a system to ensure adequate lubricant flow to the compressor.

SAE J639 requires that the type of lubricant be identified on the system label. To ensure that the proper lubricant is used in an HFC-134a system, confirm the system requirements. The industry is using many different formulations of PAG lubricants (polyalkylene glycol) with various additives to provide compressor lubrication. It is important that the proper lubricant be used and it is recommended that PAG lubricants not be mixed.

PAG lubricant will absorb moisture; it is very important when working on HFC-134a systems, that the system, hoses and containers of lubricant be kept tightly closed to prevent moisture entry. Protective impervious gloves are required to prevent lubricant contact with the skin.

System Identification
Only two refrigerants, CFC-12 and HFC-134a, are recognized and approved by the OE A/C system manufacturers for use in mobile A/C systems. As the availability of CFC-12 is exhausted, CFC-12 systems will require retrofitting.

SAE/industry guidelines are established for retrofitting CFC-12 systems. Use of alternate refrigerants in CFC-12 systems which are not approved and do not meet the guidelines of the industry, could cause problems in the servicing of mobile A/C systems.

Servicing of mobile A/C systems with an alternate refrigerant which has not been approved for retrofit could also contaminate both the CFC-12 and HFC-134a recycled refrigerant supply.

SAE J639 establishes service fittings for both CFC-12 and HFC-134a systems.

CFC-12 systems use screw thread service fittings, with the high side smaller than the low side. HFC-134a systems use a quick-couple fitting with the high side larger than the low side fitting.

In addition, the system service label will identify the specific system refrigerant type, the amount and the lubricant type.

SAE J1660 identifies the requirements for service fittings and labels when retrofitting CFC-12 systems to HFC-134a. Federal law requires that vehicles that are retrofitted shall have unique service fittings and proper identification label for the refrigerant installed to prevent contamination of refrigerant supplies.

System Changes
Changes were required for HFC-134a systems to ensure performance equal to systems using CFC-12.

Changes include new hose and seal materials which are compatible with the new refrigerant and lubricant. This includes new hose construction to reduce hose leakage, and a new desiccant material in the accumulator or receiver/dryer for reduction of moisture level in the system.

The most noticeable change, however, is the increased condenser capacity, or increased air flow, to reduce system pressures at low speed operation and city traffic conditions. In general, condenser performance has been increased by approximately 30%, which results in comparable performance for HFC-134a systems as experienced in CFC-12 systems.

One major difference is that on some systems, the sight glass, which in the past has been used to determine system refrigerant charge, has been eliminated. The sight glass may not be reliable for determining system charge with HFC-134a since layering conditions can exist with the lubricant and HFC-134a. A misreading will result in possible improper servicing of the system refrigerant charge.

It is advisable to charge mobile air conditioning systems only with known charge amounts. This is due to reduced system charge capacities, which, when overcharged can result in high system pressures. The customary "top-off" method of refrigerant charging is no longer recommended for servicing any mobile air conditioning system.

It is important to remember that the mineral oil lubricant used with CFC-12 systems, and the PAG or ester lubricant used with HFC-134a systems, are different. Mixing of these lubricants in systems which have not been retrofitted may cause problems. It is advisable to follow the A/C system manufacturer's recommendations concerning lubricants.

In order to assure that refrigerant supplies are not contaminated, use of separate service equipment, including recovery/recycling equipment, gauge manifold and hoses for each refrigerant is required. Lubricant and refrigerant left in hoses and equipment is a major source of contamination which will cause problems when servicing systems.

Use separate equipment to keep your refrigerant supply pure.

With the concern about the CFC effects on the environment, and the cost of refrigerant, it is important that the service technician, who is the key person involved in servicing the mobile A/C system, maintains the highest level of professional service.

Safety Precautions & Warnings

1. Failure to follow instructions provided by recycling equipment manufacturers could result in personal injury or equipment damage.

2. Always wear safety goggles when servicing an air conditioning system or when handling refrigerant.

REFER TO MATERIAL SAFETY DATA SHEET PROVIDED BY YOUR REFRIGERANT SUPPLIER FOR INFORMATION REGARDING THE PROPER HANDLING OF REFRIGERANT.

3. NEVER perform service on the recycling equipment (other than routine maintenance) without first consulting authorized service personnel. The removal of internal fittings and filters can cause refrigerant under pressure to be expelled. Use care and always wear safety glasses.

4. NEVER perform maintenance or service on the recycling equipment with the unit plugged into electric power unless directed otherwise.

5. NEVER transfer refrigerants to a cylinder or tank unless it is Department of Transportation approved for refilling. DOT approval is indicated by "DOT 4BW" or "DOT 4BA."

6. Recycling equipment incorporates parts, such as snap switches. These tend to produce arcs or sparks. Therefore, when located in a repair shop, recycling equipment should be used in locations with mechanical ventilation.

7. Avoid the use of an extension cord with recycling equipment to assure safe and proper operation. Under extreme situations where extension cords may be required, use minimum length 3-wire (No. 14 AWG minimum) with a ground circuit. To prevent shock hazard and reduce the risk of fire, make sure the extension cord is in good condition (not worn or frayed) with the ground circuit intact.

EPA Q & A
Q: How do I know if a vehicle uses CFC-12 (R12, also known by the trade name Freon)?
A: You can check under the hood for a label that identifies the refrigerant used in the vehicle's A/C system. The change to R134a, a non-ozone-depleting refrigerant, began in 1992 and was completed in 1995.

MACS Recommended Service Procedures

Ensure System Integrity

As a first step in service, always perform a visual inspection to spot obvious problems. If the system does not have refrigerant, installing 10% to 15% of the total system charge is sufficient for leak testing using a certified SAE J1627 electronic leak detector.

Be sure to check service valve fittings and O-ring integrity (the seals) in dust caps. Missing caps and seals at system valve ports are major leak sources. Caps and seals must be properly inspected and installed after service to minimize refrigerant loss.

Service Procedures

To provide containment and reduce unnecessary venting of refrigerant, proper service procedures must be followed.

1. Refrigerant in mobile A/C systems shall not be vented to the atmosphere during service or repair operations.
2. Refrigerant introduced into a system for the purpose of leak detection must be recovered and not vented.
3. The leak should be identified and repaired.

When a customer arrives at your service facility, it is necessary to determine if you are about to work on a contaminated system.

Questions such as when, who, and what was charged into the system the last time it was serviced might keep you out of a problem. Any past history available from the customer also may help keep you from working on a known-to-be contaminated system.

Service Guidelines

When the customer's car has poor cooling, it may be caused by low refrigerant charge, which in most cases may be caused by a leak. Most mobile A/C systems have controls which shut off a low charge system. In this case, the refrigerant remaining in the system could be in the 1/2-pound range or more with no way to identify the actual amount of refrigerant remaining in the system. **Installing gauges on the system and reading pressure will not identify the amount of refrigerant in the system.**

Checking System For Leaks

MACS recommends using the SAE J1628 technician service procedure when checking a mobile A/C system for a leak. It requires the use of an SAE J1627 certified electronic leak detector.

Leak Detection

To ensure that serviced systems are returned to their original design-intent leakage specification, leak detection devices should be used. Proper use of leak detection equipment is important since leaks may occur in locations not directly visible to the technician.

SAE J1628 provides guidelines for the use of leak detection devices.

This document follows the guidelines of the A/C system manufacturers in providing technicians with the proper procedures to identify system leaks. The vehicle engine should not be operating during the leak checking procedure. All fittings and components should be checked on all surfaces. Refrigerant leaks are under pressure and can be present at any point, at the top or bottom of the part being checked.

To prevent system chemical contamination, it is recommended that leak detection be done only with the refrigerant which is specified for the system.

Do not use shop air for this procedure. Use of shop air for leak detection may introduce both air and moisture into the system. Use of other gasses having higher pressures, such as nitrogen, can result in damage to the A/C system (e.g., evaporator failure).

The SAE J1628 procedure **does** not require a fully charged A/C system to identify if the system has a leak. If the system has only a few ounces of refrigerant and at least 50 psig system pressure (at 59°F ambient), that is sufficient to check for a leak. This pressure will be higher with the same amount of refrigerant when the service procedure is performed at warmer temperatures.

1. The system should be inspected for leaks by identification of oil on refrigeration parts or broken parts.
2. If the system has to be charged with refrigerant, only a few ounces are required to obtain a minimum system pressure (about 10–15% of the total system charge). The system is then checked with the electronic leak detector to identify any leaks. Use of soap bubbles will only identify leaks that are in excess of 40 oz. per year, as compared to electronic identification of less than one ounce per year.
3. With this limited refrigerant amount you can identify if the system has a leak, however, you can not determine if the refrigerant system will provide cooling. To operate the compressor the system must have additional refrigerant, approximately 1–2 pounds.

*CAUTION: Do not operate the compressor without the full charge of refrigerant as specified by the OE manufacturer.

Under the Clean Air Act, refrigerant added for leak detection must be recovered and not vented.

After the system problem has been identified, and the necessary repair completed, the system should again be thoroughly leak tested.

Some states and local laws have more stringent requirements.

4. After adding refrigerant to the system you can identify which items will require replacement, such as a failed or leaking part/parts.

The EPA, MACS and many facilities recommend that leaking syslems should be repaired for environmental reasons and to save R12 for future use.

Electronic Detectors

Some electronic leak detectors will only indicate CFC-12 leakage and will not indicate an HFC-134a leak. Newer design electronic detectors will provide leakage identification of both refrigerant types.

SAE J1627 equipment manufacturers provide rating information on the leak detecting capability of electronic leak detectors. Detectors which have the most sensitive detection levels will help identify systems which have smaller leaks.

EPA Q & A

Q: I understand that pr o-duction of CFC-12 is being banned because it depletes the ozone layer . What does this mean for me? How do I keep down the cost of ser vicing vehicle A/C systems?

A: The continued use of CFC-12 is not banned. Even though *production* of CFC-12 ended on December 31, 1995, *use* of CFC-12 will still be permitted, so you can continue to use the CFC-12 that is in the vehicle now, and can continue to put it in vehicles, as long as supplies are available. CFC-12 used today is constantly being recovered and recycled, and some CFC-12 produced in 1994 and 1995 has been placed into inventory, so that there is still refrigerant available for sale after the 1995 deadline, although the price will most likely increase.

In order to minimize paying increasingly higher prices to replace CFC-12 that has leaked out of the A/C system, you should suggest preventive maintenance by checking A/C systems for leaks once a year, and you should suggest that leaks be fixed. Keep in mind that fixing leaks is not an EPA requirement, although some state and local regulations may require it.

Manufacturers' service and maintenance procedures should be followed to ensure proper operation of the equipment.

Trace Dyes

The chemical composition and amount of trace (leak) dyes when injected in mobile A/C systems may cause problems. Trace dye material can affect the A/C system compressor lubricant. Caution should be taken when using trace dyes since they may not be compatible for use with the different refrigerants. R12 dye material may not be compatible when used in an R134a system and may result in a compressor lubrication problem.

Leak dyes should not be added to any automotive A/C system unless the specific product has been approved by the original air conditioning system manufacturer.

If trace dyes are used, dyes meeting SAE J2297 for R134a refrigerant and/or A/C system manufacturer requirements should be considered to maintain system chemical stability.

Proper System Processing

Experimentation has shown that, even when your gauge set reads 28 or 29 in. of mercury, the inside of the system is at a higher value. The gauge indicates vacuum in the hose, not the system. Consider that the small opening in the service valve core fitting is a major restriction. It is very difficult to reduce the actual system to this level and even with a very good vacuum pump, can take a long time. Therefore, manufacturers suggest evacuation times of 30–45 minutes to ensure that a vacuum has been attained.

When it comes to removing moisture, the entire system volume, not just the service gauge reading, must be below 29 inches of mercury. Actual moisture removal does not start until the system, not what you read on a questionable service gauge, has been reduced to at least 29.25 inches vacuum of mercury. Water boils at 212°F at 1 atmo-

sphere of pressure. By reducing the actual system pressure to 27 inches of mercury to accomplish moisture removal, the entire system would have to be raised to a temperature of 115°F. This is not very practical.

So the bottom line is that your service equipment may not remove moisture by evacuation. The best assurance for control of excess moisture in the A/C system is to install a new receiver/drier or accumulator with fresh desiccant.

It is important that processing the system by pulling a vacuum to remove air and refrigerant be done prior to charging additional refrigerant.

Recovery/recycling equipment which meets SAE requirements only has to reduce the system to a minimal level of vacuum to remove refrigerant.

Use of equipment which has a limited vacuum pump capacity may not reduce the system to complete refrigerant outgassing and removal of air (noncondensable gasses), prior to adding the new refrigerant charge.

To ensure that you are processing the system to the lowest obtainable vacuum levels, use a calibrated vacuum gauge and check the system vacuum level obtained with a good vacuum pump and compare that value with the recovery equipment.

Some equipment manufacturers have either a separate vacuum pump or a procedure to vent the vacuum pump to atmosphere, allowing the equipment to pull 29 in. of mercury.

Be sure that you have pulled at least 29 in. of vacuum on the system before charging the system with refrigerant. (**NOTE:** At facilities located at higher elevations, such as Denver, the gauge reading will be less.)

Desiccant Failure

The desiccant, which removes system moisture, is located in the receiver/drier or accumulator. Desiccant may break down when exposed to an incompatible refrigerant and cause system plugging. It is important that when servicing the A/C system the correct desiccant is used for the type of refrigerant used in the system.

Desiccant identified as XH5 has been used in CFC-12 systems, and HFC-134a systems require XH7 desiccant. It is advisable that when a receiver/drier or accumulator is being replaced, XH7 should be used since it is compatible with both R12 and R134a refrigerants. Blend refrigerants that contain R22 require XH9 desiccant that is generally not available to the service industry. If the A/C system has contained R22, it may result in damage to the XH5 and XH7 desiccants.

System Lubricant

It is important that a correct system lubricant charge be maintained to ensure proper system operation. Component replacement has general guidelines, supplied by the manufacturer, for lubricant addition during system service.

In general, recycling equipment will remove very little, if any, lubricant from the A/C system during the extraction operation. Design of recycling equipment requires that the amount of lubricant removed during refrigerant extraction be identified.

REMOVED LUBRICANT MUST NOT BE REUSED IN THE SYSTEM.

When a large quantity of lubricant is removed during extraction, the A/C system may have a lubricant overcharge.

If the indicated sample, removed during the recycling operation, contains refrigerant dissolved in the lubricant, use of this indicated amount may result in replacement of excess new lubricant and cause system damage.

Use only new lubricant, as identified on the A/C system label, to replace the amount removed during the recycling process. Used lubricant should be disposed of in accordance with federal, state and local requirements.

Recovery only and recovery/recycle equipment which meets SAE requirements will separate the lubricant during the extraction process, so that the recycled refrigerant will not contain sufficient lubricant to cause a problem. This is important since HFC-134a systems use a number of different lubricant types. The recovery/recycling equipment must be serviced and maintained to assure its proper operation.

Lubricant Mixing

Do not mix lubricants in systems. Use only the type of lubricant specified by the system manufacturer.

CFC-12 systems use mineral-based lubricants. New HFC-134a systems use several types of PAG lubricants. The proper type of lubricant and amount used is extremely important.

The A/C system label will identify the correct type of lubricant required. Mixing of PAG lubricants may also cause system problems. Use only the lubricant specified by either the A/C system manufacturer or the vehicle maker. Ester lubricant is not generally used by the A/C system manufacuters.

If systems have an overcharge of lubricant, the lubricant may collect in the evaporator and result in warmer outlet air temperatures. Proper system lubricant charge is important. Systems retrofitted without the removal of the mineral oil may have excess lubricant return to the compressor, if systems are overcharged with refrigerant. Follow the system manufac-turer's recommendations.

HFC-134a Lubricants

Caution must be taken when handling both PAG and ester lubricants. Protective impervious gloves are required to prevent lubricant contact with the skin. If lubricant does come into contact with the skin, wash the material off skin with plenty of soap and water. Skin irritation may occur with repeated and/or prolonged contact. Additional health and safety information may be obtained from lubricant manufacturers.

Coating of O-rings and seals prior to installation on A/C parts for HFC-134a systems should only be done with mineral oil, rather than with HFC-134a lubricants, to prevent skin contact. Also, since the PAGs absorb moisture, the potential for connector corrosion at the O-rings is reduced by using mineral oil. Because of the part location, this small amount of mineral oil will not affect system operation.

In addition, care should be taken since damage may result when HFC-134a lubricants contact paint, plastic parts, engine drive belts and coolant hoses.

With the early phase-out of CFCs, the mobile A/C industry has had to consider retrofitting CFC-12 systems to HFC-134a. Certain compromises have to be considered in order to retrofit the CFC-12 fleet. These compromises have included the mixing of mineral oil and PAG system lubricants in retrofitting CFC-12 systems. It is important that when servicing any system, recommendations of the system manufacturer for the proper lubricant (mineral oil, PAG, or ester) should be strictly followed to ensure compressor lubrication.

Flushing of Systems

Open vent system flushing with CFCs is now not only illegal under the Clean Air Act, but has also become a system chemical stability concern. For many years, R11 and R113 have been used for open-vent flushing when cleaning mobile A/C systems. Technical information shows that even small amounts of R11 residue will cause problems in HFC-134a systems.

The practice of open vent flushing often will not remove failed compressor material from some condenser units. Some A/C system manufacturers recommend that flushing not be considered after mechanical failure. The use of an in-line filter is considered to be the more effective method of controlling particle residues.

Use of other flushing solvents are also areas of concern since, depending on their boiling point, the vacuum pump may not remove all the solvent. This will possibly affect the chemical stability of the refrigerant, seal and hose materials.

DO NOT use any cleaning solvents (e.g., solvent cleaners of any kind), since they will affect system seals and O-rings, and cannot be totally removed. This may result in future system failures.

If you are still not happy just using filters and want to power flush systems, flush each part with only the refrigerant type used in the system and, so you do not vent it, be sure to collect the discharge with recovery/recycling equipment.

To power-flush a separate part, or the complete system, the flushing equipment must be connected in series with the portion which is being flushed and filled so liquid refrigerant is used for the flushing action.

Attachment of the system at the gauge service ports, even with the valve cores removed, will not provide adequate system flushing. Use of this method will result in the flushing being confined to the system's lowest pressure circuit and may not result in the removal of material.

If another flushing solvent is used, determine if flushing material is classified as hazardous material. Dispose of it in accordance with local, state and federal regulations.

NEVER USE ANY CFC PRODUCT FOR FLUSHING HFC-134a SYSTEMS.

Refrigerant Identification

It is important that when using containers of refrigerant, you identify the container contents. Chemical companies have trademark names, and it is important to identify the refrigerant type to ensure proper system operation. Do not rely on a trade name alone to confirm that it is the correct refrigerant. Use only the specific refrigerant designated for the A/C system currently being serviced.

Some bus and truck air conditioning systems use R22, which is covered under section 608 of the Clean Air Act and requires compliance under that section.

It is essential that the service technician use only the OEM-recom-mended refrigerant and appropriate service equipment to ensure that refrigerant mixing does not occur.

Use of the wrong refrigerants during top-off service activity which is not a recommended service procedure, will not improve system performance and may cause system damage.

Purity of Refrigerant

The purity of reclaimed refrigerants supplied in containers from sources other than automotive air conditioners for service and reuse in automotive air-conditioning systems, must meet the appropriate ARI 700 standard.

Since there are many other non-automotive uses of both CFC-12 and HFC-134a, it is important that the source of the refrigerant be known.

Since CFC-12 and HFC-134a are also used in residential and commercial systems such as refrigerators, water chillers and central cooling systems, many different contaminants and acids may be present in used refrigerant from these sources.

Use of recycling equipment which meets SAE J1990 and J2210 requirements will not purify refrigerants from nonautomotive or other sources to the ARI 700 standard so as to meet automotive air-conditioning purity requirements, and federal requirements.

Refrigerant from any source, other than an automotive A/C system, should not be used unless it has been returned to a reclamation center which can return the refrigerant to the ARI 700 specifications.

Use of refrigerant from nonautomotive sources which contains acids and other contaminants, as well as a possible mixture of other refrigerants, will cause serious problems in automotive air conditioning systems.

There are no federal requirements that containers of reclaimed refrigerant must identify contents or purity.

Reclaimed refrigerant purchased for servicing mobile A/C systems must meet ARI specification of 0.5% purity to be in compliance with federal law.

Flammable Refrigerants

Replacement refrigerants which are or can become flammable have been sold for use in A/C systems. Caution should be taken before working on any system suspected of containing this type of refrigerant. Use of leak detection equipment and recovery/recycling service equipment for removal of this type of refrigerant may pose a safety concern.

Service Concerns

1. Some electronic or open flame leak detectors may become the source of ignition when used to attempt identification of leaks in a mobile air conditioning system containing flammable refrigerant.

2. Use of automotive recovery or recycling equipment that is operated by electricity, to remove flammable refrigerants that are not compatible with compressor motor-insulation systems, may pose electrical shock or fire hazards to the operator of such equipment.

3. Mobile air conditioning systems use service port fittings to gain access and refrigerants are released when connecting and removing them. Release of refrigerant that is flammable in a confined area containing an ignition source (such as a torch or pilot light on a gas appliance) could result in fire or explosion.

4. Flammable refrigerants may also contain CFCs or HCFCs. Flammable refrigerants have been installed in motor vehicles, therefore any facility servicing the vehicle may not be aware of a flammable refrigerant's existence in the system. Under federal law systems containing a mixture of these refrigerants can not be vented. Proper disposal is required.

WARNING: REMOVAL AND HANDLING OF FLAMMABLE REFRIGERANTS MAY BE DANGEROUS.

Consequences of Cross-Contamination

Mixing of Refrigerants

Mobile A/C system contamination will occur due to the installation of the wrong refrigerant without proper system identification.

Under no circumstances should refrigerant be mixed in a system, since any mixing will affect recycling programs and cause equipment and system problems.

Damage may include compressor failure, damage to recycling equipment and transfer of the mixed refrigerant to other vehicles, causing additional problems and reduced system performance.

If CFC-12 and HFC-134a are mixed in the same system, increased pressures will occur resulting in loss of performance, system damage (such as compressor failure), hose and seal leakage, and refrigerant contamination.

It is essential that the service technician use the designated separate refrigerant service equipment to make sure cross-contamination does not occur. Professional A/C service facilities should have separate recovery/recycling equipment for CFC-12 and HFC-134a. This also includes having the proper service equipment when servicing other alternate refrigerants. **Under federal law, designated service equipment is required for all SNAP listed refrigerants.**

Applicability to Manifold Gauges and Refrigerant Identifiers

Manifold gauges allow technicians to diagnose system problems and to charge, recover and/or recycle refrigerant. A standard fitting may be used at the end of the hoses attached to the manifold gauges, but unique fittings must be permanently attached at the ends of the hoses that attach to vehicle air conditioning systems and recovery or recycling equipment. Similarly, refrigerant identifiers may be used with multiple refrigerants. The connection between the identifier or similar service equipment and the service hose may be standardized and work with multiple hoses. For each refrigerant, however, the user must attach a hose to the identifier that has a fitting unique to that refrigerant permanently attached to the end going to the vehicle. Adapters for one refrigerant may not be attached to the end going to the vehicle and then be removed and replaced with the fitting for a different refrigerant. The guiding principle is that once attached to a hose, the fitting is permanent and is not to be removed.

CAUTION: This includes manifold gauge sets, hoses, charging and recycling equipment.

General Precautions

Contaminated refrigerant removed from a mobile A/C system can affect recovery/recycling equipment and servicing of other vehicles. When R12 and R134a refrigerant are mixed, it will have a different pressure than pure refrigerant. The indication of this pressure change is found when R12, or R134a have 2, 5, and 10 percent contamination (see Chart 6).

System contamination can occur when a refrigerant other than that specified for the system is added.

If 12 oz. of R22 were added to a typical CFC-12 system, discharge pressures would increase by approximately 30%.

CFC-12 Pressure

AMBIENT	PURE CFC-12	MAX. NCG/AIR	\%of HFC-134a MIXED IN CFC-12		
			2%	5%	10%
80° F	84	96	88	93	99
90° F	100	110	105	111	116
100° F	117	127	123	127	135

HFC-134a Pressure

AMBIENT	PURE HFC-134a	MAX. NCG/AIR	\%of CFC-12 MIXED IN HFC-134a		
			2%	5%	10%
80° F	86	91	92	95	98
90° F	104	109	109	112	116
100° F	124	129	130	133	136

Caution: You cannot rely on system gauge pressure readings to identify refrigerant contamination in an auto A/C system because all parts of the system may not be at equal temperature, which is required to have a stabilized pressure condition.

Chart 6.

Fuel System Fittings

Caution: The fitting on the fuel injection plumbing which contains combustible fluids is the same size as one of the CFC-12 A/C service fittings. Be careful not to attach A/C service lines to the vehicle fuel system.

Eye Protection

To prevent injury when working on refrigeration systems, eye protection must be worn.

Future of CFC-12 Systems

Blend Refrigerants/Retrofits

The only retrofit refrigerant that has been approved by the original mobile air conditioning system manufacturers for replacement in a CFC-12 system is HFC-134a.

Additional system retrofit modifications may include hoses, a high pressure cut-out device, seals, desiccant, lubricant, refrigerant control replacement, increased condenser capacity and other modifications as determined by the equipment manufacturer. Not following the OEM recommendation may result in system damage, loss of performance and affect warranty.

It should be noted that other refrigerants are not compatible with CFC-12 or HFC-134a and under federal law require separate service equipment to prevent cross-contamination.

Since January 1992, the Clean Air Act has required that any refrigerant containing HCFCs, such as blends, must comply with section 609. These cannot be vented and must be recycled.

Under federal law it is legal to store and use CFC-12 for servicing mobile A/C systems until it is no longer available. However, it can only be purchased, regardless of container size, by personnel certified under section 608 or 609 of the Clean Air Act.

The Federal Clean Air Act "SNAP" ruling has identified refrigerants that are considered acceptable for use in mobile A/C systems. However, simply because a refrigerant is considered acceptable under this ruling does not mean that refrigerant will provide satisfactory A/C system performance. The EPA only considers replacement refrigerants for environmental and safety reasons. The EPA does not test the refrigerants for performance or system durability.

Several refrigerants are considered acceptable under the "SNAP" rule. The only replacement refrigerant currently approved for use by the world automobile manufacturers and many A/C system replacement parts suppliers is HFC-134a. Use of other refrigerants for new vehicles or retrofitting CFC-12 systems may void warranty.

The Federal "SNAP" rule also requires that refrigerants used to service or retrofit CFC-12 mobile A/C systems must have unique service fittings and labels, as well as a high pressure compressor cut-off switch if not already installed.

It will be difficult to enforce requirements that fittings and labels be installed on a nationwide basis. It is most likely that these replacement refrigerants will be installed in some vehicle A/C systems without the required changes, resulting in contamination of refrigerant supplies and equipment.

The "SNAP" ruling has also identified OZ-12, HC-12a manufactured by OZ Technology and other refrigerants which contain hydrocarbons as unacceptable for use as replacements for CFC-12 in mobile A/C systems. Indications are that their use has been somewhat extensive, and there is no information on system material compatibility, and the possibility of early system failures.

These flammable refrigerants have been manufactured to duplicate CFC-12 and HFC-134a refrigerant pressure temperature curves. The pressure temperature relationships of hydrocarbon refrigerants in the field have not been consistent. Given the many possible ignition sources, from the vehicle, shop area and service equipment, flammable refrigerants pose safety concerns.

Recommendations for Retrofit

All mobile A/C system manufacturers encourage use of R12 refrigerant in A/C systems designed to operate with that refrigerant.

If R12 refrigerant is not available, then the more costly system retrofit must be considered.

Final Rule Conditions for Any EPA/SNAP R12 Substitute for Automotive A/C Use

1. No substitute refrigerant may be used to "top off" an automotive A/C system, unless the original refrigerant has first been extracted in accordance with EPA regulations.

2. Only substitute refrigerants listed as acceptable by EPA can be installed in automotive A/C systems in order to be in compliance with federal law. [To date, EPA has listed several refrigerants. To get an up-to-date SNAP list, call the EPA hotline at 1-800-296-1996.]

3. These refrigerants may only be used with a set of fittings unique to that refrigerant.

4. A unique label must be used to identify the refrigerant in the system for the purpose of proper future service. This label must comply with certain standards.

5. If the system does not already have a high pressure compressor clutch cut off switch, one must be installed.

6. Each replacement refrigerant cannot be vented and requires specific recovery equipment.

This means that any R12 system that is retrofitted, converted or altered to use a refrigerant other than R12 must use fittings and labels unique to the new refrigerant, and have a high pressure compressor cut-off switch to be in compliance.

Under the Significant New Alternative Policy (SNAP) program rules, the EPA has listed HFC-134a and several other alternatives as acceptable for use in retrofitting existing CFC-12 mobile A/C systems.

The world auto industry has identified HFC-134a as the refrigerant of choice for new and retrofitted mobile A/C systems.

Remember, there is no direct "drop in" replacement for R12 A/C systems.

The SAE, at the request of the U.S. Environmental Protection Agency and the air conditioning industry, has developed a set of retrofit guidelines. Conversion of a CFC-12 system to use HFC-134a is covered in SAE J1660 "Fittings and Labels for Retrofit of R12 Mobile Air Conditioning Systems to R134a," and SAE J1661 "Procedure for Retrofitting R12 Mobile Air Conditioning Systems to HFC-134a." Air conditioning system manufacturers' procedures follow these SAE requirements.

Conversion of a CFC-12 system not following these procedures, or conversion to another refrigerant may result in system problems.

Under the federal requirements, to prevent contamination of mobile A/C systems and refrigerant supplies, each system that uses a refrigerant listed acceptable must have the appropriate service fittings and label for that retrofit refrigerant. **The system must also have a high-pressure compressor cut-off switch to be in compliance.**

No CFC-12 automotive A/C system should have HFC-134a charged into that system until it has been retrofitted.

Mobile A/C Service Options for Leak Repair

For example, the A/C system has some pressure and an unknown amount of refrigerant in the system at first inspection. Additional refrigerant is also added to check operation of the system. The system has an identifiable leak and the customer declines to repair the leaking system.

The service facility's option under section 609 is to charge the customer for the inspection and added refrigerant and return the pre-existing leaking system to the customer. There is no EPA requirement under section 609 that refrigerant must be removed from a leaking system. If the leaking system is not repaired it must be returned to the customer with at least the same amount of refrigerant as was in the system when it arrived. Note: Some state and local laws have additional requirements regarding this aspect of A/C service.

What if the A/C system has no pressure at first inspection, so any refrigerant added to the system is owned by the service facility? If it is the policy of the facility not to charge a leaking system, and this policy is explained to the customer up front, and the customer declines to have the leaking system fixed, all of the refrigerant added can be removed. But remember, under section 609 regulations, the pre-existing leaking system can be charged. Note: Some state and local regulations are more stringent than section 609 on charging a leaking system.

EPA Q & A

Q: I've heard that I might have to convert vehicle A/C systems to use a different refrigerant. When will I need to do that?

A: You will need to allow the customer to decide whether to convert the vehicle's system to use an alternate refrigerant if the system becomes inoperative and requires a new refrigerant charge, and CFC-12 is no longer available. Although there is no way to predict with certainty when supplies of CFC-12 will be exhausted, the extensive recycling and banking of CFC-12 occurring now should make it available for several years. Depending on the age of the vehicle, it may well be the case that CFC-12 will be around for the remainder of its life.

It may also make sense to convert the system if major service is being performed on the A/C system (for example, a front-end collision or a compressor failure). In that event, the additional cost of doing the conversion over and above the cost of repair work may be minimal, because many steps in converting are also necessary in performing major repair.

Q: What will the cost be to convert vehicles to a different refrigerant?

A: EPA estimates that conversions will cost between $100 and $800 or more, depending on the make, model and age of the vehicle. Conversions at a minimum will require that the oil used to lubricate the A/C system, and the fittings, be changed. EPA estimates that this minimal conversion will add less than $200 to the cost of any repair work requested. Other components of the A/C system may have to be replaced, depending on whether the current system A/C components are compatible with the new refrigerant.

Q: If I decide to convert a vehicle, how do I know what changes are required?

A: EPA recommends that you consult vehicle manufacturers' guidelines. Manufacturers have available retrofit guidelines for vehicles manufactured after the late 1980s. The Ozone Protection Hotline will be able to tell you if the manufacturer of the automobile has established specific procedures for the conversion of the vehicle. When considering converting any vehicle, you should rely on the OEMs' retrofit guidelines.

Q: What new refrigerant should be put in my vehicle? Are there many substitute refrigerants that are OK?

A: Automakers are producing new vehicles with R134a, which does not deplete the ozone layer. EPA evaluates all substitutes for CFC-12 under its Significant New Alternatives Policy (SNAP) program in order to determine if they pose any risk to human health or to the environment. Currently, R134a is the only alternative listed as acceptable by EPA, which also has been fully tested and is specified by automakers in their guidelines.

Q: I've been seeing other substitutes for sale. If I find out that a particular alternate has not been reviewed by EPA, or that EPA has not finished its review of the product, can I legally buy the product? What happens if I buy it now, and EPA decides in the future that it is not acceptable?

A: While you can legally purchase the product if the Agency has not made a determination as to its acceptability under the SNAP program, you should keep in mind that such product has not been tested to determine whether it is safe to use. If EPA later declares the product unacceptable, you do not legally need to remove it from the vehicle's air-conditioning system, but you may choose to do so. You should be aware that it may be costly to convert the system back to an acceptable alternative, and that it's illegal to add any more refrigerant that EPA has declared unacceptable. The fine for sale of an unacceptable alternative is up to $25,000 per day and 5 years in jail.

Q: Will any alternative refrigerant listed by EPA as acceptable work in the vehicle?

A: Although EPA's SNAP program determines what risks an alternative poses to human health and the environment, the Agency does not determine whether the alternative will provide adequate performance or will be compatible with the components of the A/C system.

Keep in mind that using a refrigerant not yet reviewed and determined acceptable by EPA may result in damage to the A/C system components including the compressor, and may limit the ability to have the vehicle's system serviced in the future.

Q: I have heard that R134a does not cool nearly as well as CFC-12. Is this true?

A: Vehicle manufacturers have designed air conditioning systems for new vehicles that use R134a while maintaining reliability and cooling performance. Conversion specifications for A/C units using R134a are also designed to maintain performance, but this may vary depending on the condition of the unit prior to the conversion, and on other factors.

Be prepared to provide consumers with up-to-date information about the use of CFC-12 and substitute refrigerants. Service shops should be able to offer information as well as respond to questions. Having brochures, fact sheets, posters, and/or videos on hand will help educate consumers about their options.

HFC-134a Cautionary Statements

Safety Issues

There has been considerable activity by promoters of so-called drop-in refrigerants for mobile A/C systems and this activity has created a lot of misinformation regarding HFC-134a.

The toxicity data base for HFC-134a is even more extensive than that for CFC-12. HFC-134a was listed by the U.S. EPA as acceptable for use in mobile A/C systems in the April, 1994 final SNAP rule.

HFC-134a has been approved for use in metered dose medical inhalers for asthma suffers.

HFC-134a is at least as safe as CFC-12. Regarding reports of HFC-134a being associated with testicle tumors: rats were exposed to 50,000 ppm of HFC-134a for six hours per day, five days per week for two years. At the end of this period microscopic examination of the male rat testis, indicated an increased incidence of benign tumors. Such tumors are known to occur in rats, not in humans. At levels of 10,000 ppm, no life-threatening effects occurred during the two-year study.

These results are equivalent to the working lifetime of a human.

Illustrating the safety of R134a, a service technician would have to enclose himself in his garage, close all doors and windows, turn off all means of ventilation, and intentionally release 10 oz. of R134a directly into the garage air to create an exposure level of 1,000 ppm. The allowable occupational ex-posure limit tells us that a technician could do this for eight hours per day, five days per week, for a lifetime and suffer no adverse effects.

Material Safety Data Sheets for these refrigerants, from a major producer of R12, R22 and R134a, include some of the following statements: "Inhalation of high concentrations of vapor is harmful and may cause heart irregularities, unconsciousness, or death." "Vapor reduces oxygen available for breathing and is heavier than air."

In other words, acute risk from high short dosage will cause problems with any of these refrigerants. Also, "Material is stable. However, avoid open flames and high temperatures. Decomposition products are hazardous. Can be decomposed by high temperatures (open flames, glowing metal surfaces, etc.) forming hydrofluoric acids, and possibly carbonyl halides." All three refrigerants have a 1,000 ppm 8 hour applicable exposure limit.

Both R22 and R134a are not flammable at ambient temperatures and atmospheric pressure. However, R134a and R22 have been shown in tests to be combustible at certain pressures and ambient temperatures when mixed with air (when contained in a pipe or tank). Service equipment or vehicle A/C systems should not be pressure tested or leak tested with compressed air.

These mixtures may be potentially dangerous, causing injury or property damage. Additional health and safety information may be obtained from refrigerant and lubricant manufacturers.

Other Safety Considerations

CAUTION: Avoid breathing A/C refrigerant and lubricant vapor or mist. Exposure may irritate eyes, nose and throat. To remove HFC-134a from the A/C system, use service equipment certified to meet the requirements of SAE J2210 (HFC-134a recovery/recycling equipment). If accidental discharge occurs, ventilate the work area before resuming service. Additional health and safety information may be obtained from refrigerant and lubricant manufacturers.

Shut-Off Valves

Shut-off valves may be either manual or automatic. While SAE J standards say shut-off valves must be used within 12 in. (30 cm) of a

EPA Q & A

Q: I know that the old r e-frigerant, CFC-12, does not pose cancer risks when properly used. Is this tr ue of R134a?

A: R134a is regarded as one of the safest refrigerants yet introduced, based on current toxicity data. The chemical industry's Program for Alternative Fluorocarbon Toxicity Testing (PAFT), a series of protocols for testing fluorocarbons, determined that R134a does not cause cancer or birth defects.

Q: Is R134a flammable?

A: R134a is considered as safe or safer than CFC-12 in motor vehicle uses, including involvement in collisions. Like CFC-12, R134a is not flammable at ambient temperature and atmospheric pressures. However, R134a service equipment and vehicle A/C systems should not be pressure tested or leak tested with compressed air. Some mixtures of air and R134a have been shown to be combustible at elevated pressures. These mixtures may be potentially dangerous, causing injury or property damage.

connection, some equipment manufacturers use quick-couplers. These automatically shut off the flow of refrigerant when connections are broken. When quick-couplers are used, follow the equipment manufacturer's recommendations.

Proper hookup of refrigerant lines to the system must include shut-off valves on the end of every line. Shut-off valves should be no more than 12 in. (30 cm) from the port where the lines are connected.

Use of a shut-off valve at the hookup for refrigerant canisters also will help ensure minimum refrigerant loss.

Containers: Handle With Care

CAUTION: NEVER USE A STANDARD DISPOSABLE 30-LB. TANK (THE TYPE OF CONTAINER IN WHICH VIRGIN REFRIGERANT IS SOLD) TO RECLAIM REFRIGERANT CFC-12. USE ONLY DOT CFR TITLE 49 OR UL-APPROVED STORAGE CONTAINERS FOR RECYCLED REFRIGERANT (CONTAINERS MARKED DOT 4BW OR DOT 4BA).

Thermal Expansion

Safety codes recommend that closed tanks not be filled over 80% of the volume with liquid.* The remaining 20% is called head pressure room.

Refrigerant expands when it gets warm.

When refrigerant expands some of it boils, thus increasing the pressure.

Remaining liquid expands rapidly and may fill the container 100% full with liquid.

The fuller the tank the more liquid expansion takes place.

Pressure within the tank increases at a slower rate if there is room for the gases. The pressure increases according to the liquid saturation.

A tank filled with 80% liquid is relatively safe.* However, the tank filled with 90% liquid is in essence a time bomb. A 90% fill causes pressure within the tank to rise at a very rapid rate. At temperatures above 100°F, the tank may explode.

Cylinder Temp.	60° F	70° F	100° F	130° F	150° F
Starting with Cylinder 80% Full					
Space Occupied by Liquid	80% Full	81% Full	83% Full	90% Full	94% Full
Starting with Cylinder 90% Full					
Space Occupied by Liquid	91% Full	92% Full	96% Full	Tank is 100% Full Liquid @ 113° Pressurizes Very Rapidly	EXPLOSION

DO NOT OVERFILL REFRIGERANT TANKS

*SAE J1989, Section 7.2 states: "To prevent on-site over-filling when transferring to external containers, the safe filling level must be controlled by weight and must not exceed 60% of container gross weight rating."

Notes:

Notes:

Editor's Note:

Where possible, the information contained in this manual identified as "EPA Q & A" and "EPA Tips" have been excerpted from EPA fact sheets; however, this information is intended only as an overview, not a detailed accounting of the subject regulations. To learn more about the EPA stratospheric protection program or to order publications, call EPA's Hotline: 1-800-296-1996 or check EPA's website, located at:

http://www.epa.gov/ozone/title6/609/609.html.

Published by:
MOBILE AIR CONDITIONING SOCIETY WORLDWIDE

No part of this publication may be produced in any form, in an electronic retrieval system or otherwise, without the prior written permission of the publisher.

NATIONAL OFFICE
P.O. Box 100
East Greenville, PA 18041
Phone: (215) 679-2220
Fax: (215) 541-4635
Email: info@macsw.org
Website: http://www.macsw.org

SENIOR VICE PRESIDENT:
Elvis Hoffpauir

MACS TECHNICAL ADVISOR:
Ward Atkinson

GLOSSARY

Absolute Perfect in quality or nature, complete. Usually used in refrigeration context when referring to temperature or pressure.

Access valve See Service port and Service valve.
Valvula de acceso Ver Service port [Orificio de servicio] y Service value [Valvula de servicio].

Accumulator A tank located in the tailpipe to receive the refrigerant that leaves the evaporator. This device is constructed to ensure that no liquid refrigerant enters the compressor.
Acumulador Tanque ubicado en el tubo de escape para recibir el refrigerante que sale del evaporador. Dicho dispositivo esta disenado de modo que asegure que el refrigerante liquido no entre en el compresor.

Acme A type of fitting thread. The service hose connections to the R-134a manifold set have 1/2-16 acme threads.

Actuator A device that transfers a vacuum or electric signal to a mechanical motion. An actuator typically performs an on/off or open/close function.
Accionador Dispositivo que transfiere una serial de vacio o una senal electrica a un movimiento mecanico. Tipicamente un accionador lleva a cabo la funcion de modulacion de impulsos o la de abrir y cerrar.

Adapter A device or fitting that permits different size parts or components to be fastened or connected to each other.
Adaptador Dispositivo o ajuste que permite la sujeccion o connexion entre si de piezas de tamanos differentes.

Aftermarket A term generally given to a device or accessory that is added to a vehicle by the dealer after original manufacture, such as an air conditioning system.
Postmercado Termino dado generalmente a un dispositivo o accesorio que el distribuidor de automoviles agrega al automovil despues de la fabricacion original, como por ejemplo un sistema de acondicionamiento de aire.

Air gap The space between two components such as the rotor and armature of a clutch.
Espacio de aire El espacio entre dos componentes, como por ejemplo el rotor y la armadura de un embrague.

Ambient sensor A thermistor used in automatic temperature control units to sense ambient temperature. Also see Thermistor.
Sensor ambiente Termistor utilizado en unidades de regulacion automatica de temperatura para sentir la temperatura ambiente. Ver tambien Thermistor [Termistor].

Ambient temperature The temperature of the surrounding air.

Approved power source A power source that is consistent with the requirements of the equipment so far as voltage, frequency, and ampacity are concerned.
Fuente aprobada de potenica Fuente de potencia que cumple con los requisitos del equipo referente a la tension, frecuencia, y ampacidad.

Armature The part of the clutch that mounts onto the crankshaft and engages with the rotor when energized.
Armadura La parte del embrague que se fija al ciguenal y se engrans al exitarse el rotor.

Asbestos A silicate of calcium (Ca) and magnesium (Mg) mineral that does not burn or conduct heat. It has been determined that asbestos exposure is hazardous to health and must be avoided.
Asbesto Mineral de silicato de calcio (Ca) y magnesio (Mg) que no se quema ni conduce el calor. Se ha establecido que la exposicion al asbesto es nociva y debe evitarse.

Aspirator A device that uses a negative (suction) pressure to move air.

Atmospheric Pressure Air pressure at a given altitude. At sea level, atmospheric pressure is 14.696 psia (101.329 kPa absolute).
Presion atmosferica La presion del aire a una dada altitud. Al nivel del mar, la presion atmosferica es de 14,696 psia (101.329 kPa absoluto).

AUTO Abbreviation for automatic.
AUTO Abreviatura del automatico.

Back seat (service valve) Turning the valve stem to the left (ccw) as far as possible back seats the valve. The valve outlet to the system is open and the service port is closed.
Asentar a la izquierda (valvula de servicio) El girar el vastago de la valvula al jpunto mas a la izquerda posible asienta a la izquierda la valvula. La salida de la valvula al sistema esta abierta y el orificio de servicio esta cerrado.

Barb fitting A fitting that slips inside a hose and is held in place with a gear-type clamp. Ridges (barbs) on the fitting prevent the hose from slipping off.
Accesorio arponado Adjuste que se inserta dentro de una manguera y que se sujeta en su lugar con una abrazadera de tipo engranaje. Proyecciones (puas) en el adjuste impiden que se deslice la manguera.

Barrier hose A hose having an impervious lining to prevent refrigerant leakage through its wall. Air conditioning systems in vehicles have had barrier hoses since 1988.

BCM An abbreviation for Blower Control Module.
BCM Abreviatura de Modulo regulador del soplador.

Belt See V-belt and Serpentine belt.
Correa Ver V-belt [Correa en V], y Serpentine belt [Correa serpentina].

Belt tension Tightness of a belt or belts, usually measured in foot-pounds (ft-lb) or Newton-meters (N·m).
Tension de la correa Tension de una correa o correas, medida normalmente en libras-pies (ft-lb) o metros-Newton (N·m).

Blower See Squirrel-cage blower.
Soplador Ver Squirrel-cage blower [Soplador con jaula de ardilla].

Blower motor See Motor.
Motor de soplador Ver Motor.

Blower relay An electrical device used to control the function or speed of a blower motor.
Rele del soplador Dispositivo electrico utilizado para regular la funcion o velocidad de un motor de soplador.

Boiling point The temperature at which a liquid changes to a vapor.
Punto de ebullicion Temperatura a la que un liquido se convierte en vapor.

Break a vacuum The next step after evacuating a system. The vacuum should be broken with refrigerant or other suitable dry gas, not ambient air or oxygen.
Romper un vacio El paso que inmediantamente sigue la evacuacion de un sistema. El vacio debe de romperse con refrigerante u otro gas seco apropiado, y no con aire ambiente u oxigeno.

Breakout box A tool in which the probes of a digital volt-ohmmeter (DVOM) may be inserted to access various sensors and actuators through pin connectors to the computer.

Bypass An alternate passage that may be used instead of the main passage.

Desviacion Pasaje alternativo que puede utilizarse en vez del pasaje principal.

Bypass hose A hose that is generally small and is used as an alternate passage to bypass a component or device.

Manguera desviadora Manguera que generalmente es pequena y se utiliza como pasaje alternativo para desviar un componente o dispositivo.

CAA Clear Air Act.

CAA Ley para Aire Limpio.

Calibration To check, adjust, or determine the accuracy of an instrument used for measuring, for example, temperature or pressure.

Can tap A device used to pierce, dispense, and seal small cans of regrigerant.

Macho de roscar para latas Dispositivo utilizado para perforar, distribuir, y sellar pequenas latas de refrigerante.

Can tap valve A valve found on a can tap that is used to control the flow of refrigerant.

Valvula de macho de roscar para latas Valvula que se encuentra en un macho deroscar para latas utilizada para regular el flujo de refrigerante.

Cap A protective cover. Also used as an abbreviation for capillary (tube) or capacitor.

Tapadera Cubierta protectiva. Utilizada tambien como abreviatura del tubo capilar o capacitador.

Cap tube A tube with a calibrated inside diameter and length used to control the flow of refrigerant. In automotive air conditioning systems, the tube connecting the remote bulb to the expansion valve or to the thermostat is called the capillary tube.

Tubo capilar Tubo de diametro interior y longitud calibrados; se utiliza para regular el flujo de refrigerante. En sistemas automotrices para el acondicionamiento de aire el tubo que conecta la bombilla a distancia con la valvula de expansion o con el termostato se llama el tubo capilar.

Carbon monoxide (CO) A major air pollutant that is potentially lethal if inhaled, even in small amounts. An odorless gas composed of carbon (C) and hydrogen (H) formed by the incomplete combustion of any fuel containing carbon.

Carbon seal face A seal face made of a carbon composition rather than from another material such as steel or ceramic.

Frente de carbono de la junta hermetica Frente de la junta hermetica fabricada de un compuesto de carbono en vez de otro material, como por ejemplo el acero o material ceramico.

Caution A notice to warn of potential personal injury situations and conditions.

Precaucion Aviso para advertir situaciones y condiciones que podrian causar heridas personales.

CCW Counterclockwise.

CCW Sentido inverso al de las agujas del reloj.

Celsius A metric temperature scale using zero as the freezing point of water. The boiling point of water is 100°C (212°F).

Celsio Escala de temperatura metrica en la que el cero se utilza como el punto de congelacion de agua. El punto de ebullicion de agua es 100°C (212°F).

Ceramic seal face A seal face made of a ceramic material instead of steel or carbon.

Frente ceramica de la junta hermetica Frente de la junta hermetica fabricada de un material ceramico en vez del acero o carbono.

Certified Having a certificate. A certificate is awarded or issued to those who have demonstrated appropriate competence through testing and/or practical experience.

Certificado El poseer un certificado. Se les otorga o emite un certificado a los que han demonstrado una cierta capacidad por medio de examenes y/o experiencia practica.

CFC-12 See Refrigerant-12.

CFC-12 Ver [Refrigerante-12].

Charge A specific amount of refrigerant or oil by volume or weight.

Carga Cantidad especifca de refrigerante o de aceite por volumen o peso.

Check valve A device located in the liquid line or inlet to the drier. The valve prevents liquid refrigerant from flowing the opposite way when the unit is shut off.

Valvula de retencion Dispositivo ubicado en la linea de liquido o en la entrada al secador. Al cerrarse la unidad, la valvula impide que el refrigerante liquido fluya en el sentido contrario.

Clean Air Act A Title IV amendment signed into law in 1990 that established national policy relative to the reduction and elimination of ozone-depleting substances.

Ley para Aire Limpio Enmienda Titulo IV firmado y aprobado en 1990 que establecio la politica nacional relacionada con la reducion y eliminacion de sustancias que agotan el ozono.

Clockwise A term referring to a clockwise (cw), or left-to-right rotation or motion.

Sentido de las agujas del reloj Termino que se refiere a un movimiento en el sentido correcto de las agujas del reloj (cw por sus siglas en ingles), es decir, rotacion o movimiento desde la izquierda hacia la derecha.

Clutch An electromechanical device mounted on the air conditioning compressor used to start and stop compressor action, thereby controlling refrigerant circulating through the system.

Embrague Dispositivo electromecanico montado en el compresor del acondicionador de aire y utilizado para arrancar y detener la accion del compresor, regulando asi la circulacion del refrigerante a traves del sistema.

Clutch coil The electrical part of a clutch assembly. When electrical power is applied to the clutch coil, the clutch is engaged to start and stop compressor action.

Bobina del embrague La parte electrica del conjunto del embrague. Cuando se aplica una potencia electrica a la bobina del embrague, este se engrana para arrancar y detener la accion del compresor.

Clutch pulley A term often used for "clutch rotor"; that portion of the clutch in which the belt rides.

Compound gauge A gauge that registers both pressure and vacuum (above and below atmospheric pressure); used on the low side of the systems.

Manometro compuesto Calibrador que registra tanto la presion como el vacio (a un nivel superior e inferior a la presion atmosferica); utilizado en el lado de baja presion de los sistemas.

Compression fitting A type of fitting used to connect two or more tubes of the same or different diameter together to form a leak-proof joint.

Adjuste de compresion Tipo de ajuste utilizado para sujetar dos o mas tubos del mismo tamano o de un tamano diferente para formar una junta hermetica contra fugas.

Compression nut A nut-like device used to seat the compression ring into the compression fitting to ensure a leak-proof joint.

Tuerca de compresion Dispositivo parecido a una tuerca utilizado para asentar el anillo de compresion dentro del ajuste de compresion para asegurar una junta hermetica contra fugas.

Compression ring A ring-like part of a compression fitting used for a seal between the tube and fitting.

Aro de compresion Pieza parecida a un anillo del adjuste de compresion utilizada como una junta hermetica entre el tubo y el adjuste.

Compressor shaft seal An assembly consisting of springs, snap rings, O-rings, shaft seal, seal sets, and gasket. The shaft to be turned without a loss of refrigerant or oil.

Junta hermetica del arbol del compresor Conjunto que consiste de muelles, anillos de muelles, juntas toricas, una junta hermetica del arbol, conjuntos de juntas hermeticas, y una guarnicion. La junta hermetica del arbol esta montada en el ciguerial del compresor y permite que el arhol se gire sin una perdida de refrigerante o aceite.

Constant-tension hose clamp A hose clamp, often referred to as a "spring clamp": so designed that it is under constant tension.

Contaminated A term generally used when referring to a refrigerant cylinder or a system that is known to contain foreign substances such as other incompatible or hazardous refrigerants.

Contaminado Temino generalmente utilizado al referirse a un cilindro para refrigerante o a un sistema que es reconocido contener sustancias extranas, como por ejemplo otros refrigerantes incompatibles o peligrosos.

Contaminated refrigerant Any refrigerant that is not at least 98 percent pure. Refrigerant may be considered to be contaminated if it contains excess air or another type refrigerant.

Counterclockwise (ccw) A direction, right to left, opposite that which a clock runs.

Sentido contrario al de las agujas del reloj Direccion de la derecha hacia la izquierda contraria a la correcta de las agujas del reloj.

Cracked position A mid-seated or open position.

Posicion parcialmente asentada Posicion abierta o media asentada.

Customer The vehicle owner or a person that orders and/or pays for goods or services.

CW Abbreviation for clockwise. Also cw.

CW Abreviatura del sentido de las agujas del reloj. Tambien cw.

Cycle clutch time (total) Time from the moment the clutch engages until it disengages, then reengages. Total time is equal to on time plus off time for one cycle.

Duracion del ciclo del embrague (total) Espacio de tiempo medido desde el momento en que se engrana el embrague hasta que se desengrane y se engrane de nuevo. El tiempo total es equivalente al trabajo efectivo mas el trabajo no efectivo por un ciclo.

Cycling clutch pressure switch A pressure-actuated electrical switch used to cycle the compressor at a predetermined pressure.

Automata manometrico del embrague con funcionamiento ciclico Interruptor electrico accionado a presion utilizado para ciclar el compresor a una presion predeterminada.

Cycling clutch system An air conditioning system in which the air temperature is controlled by starting and stopping the compressor with a thermostat or pressure control.

Sistema de embrague con funcionamiento ciclico Sistema de acondicionamiento de aire en el cual la temperatura del aire se regula al arrancarse y detenerse el compresor con un termostato o regulador de presion.

Cycling time A term often used for "cycling clutch time." The total time from when the clutch engages until it disengages and again engages; equal to one on time plus one off time for one cycle.

Debris Foreign matter such as the remains of something broken or deteriorated.

Decal A label that is designed to stick fast when transferred. A decal affixed under the hood of a vehicle is used to identify the type of refrigerant used in a system.

Calcomania Etiqueta disenada para pegarse fuertemente al ser transferido. Una calcomania pegada debajo de la capota se utiliza para identificar el tipo de refrigerante utilizado en un sistema.

Department of Transportation The United States Department of Transportation is a federal agency charged with regulation and control of the shipment of all hazardous materials.

Departamento de Transportes El Departamento de Transportes de los Estados Unidos de America es una agencia federal que tiene a su cargo la regulacion y control del transporte de todos los materiales peligrosos.

Dependability Reliability; trustworthiness.

Caracter responsible Digno de confianza; integridad.

Depressing pin A pin located in the end of a service hose to press (open) a Schrader-type valve.

Pasador depresor Pasador ubicado en el extremo de una manguera de servicio para forzar que se abra una valvula de tipo Schrader.

Diagnosis The procedure followed to locate the cause of a malfunction.

Diagnosis Procedimiento que se sigue para localizar la causa de una disfuncion.

Disarm To turn off; to disable a device or circuit.

Disinfectant A cleansing agent that destroys bacteria and other microorganisms.

Dissipate To reduce, weaken, or use up; to become thin or weak.

Desarmar Apagar; incapacitar un dispositivo o circuito.

Dry nitrogen The element nitrogen (N) that has been processed to ensure that it is free of moisture.

Nitrogeno seco El elemento nitrogeno (N) que ha sido procesado para asegurar que este libre de humedad.

Dual Two

Doble Dos.

Dual system Two systems; usually refers to two evaporators in an air conditioning system, one in the front and one in the rear of the vehicle, driven off a single compressor and condenser system.

Sistema doble Dos sistemas; se refiere normalmente a dos evaporadores en un sistema de acondicionamiento de aire; uno en la parte delantera y el otro en la parte trasera del vehiculo; los dos son accionados por un solo sistema compresor condensador.

Duct A tube or passage used to provide a means to transfer air or liquid from one point or place to another.

Conducto Tubo o pasaje utilizado para proveer un medio para transferir aire o liquido desde un punto o lugar a otro.

EATC Electronic automatic temperature control.

EATC Regulador automatico y electronico de temperatura.

ECC Electronic climate control.

ECC Regulador electrnico de clima.

Electronic charging meter A term often used for "electronic scale," a device used to accurately dispense and/or monitor the amount of refrigerant being charged into an air conditioning system.

English fastener Any type fastener with English size designations, numbers, decimals, or fractions of an inch.

Asegurador ingles Cualquier tipo de asegurador provisto de indicaciones, numeros, decimales, o fraciones de una pulgada del sistema ingles.

Environmental Protection Agency (EPA) An agency of the U.S. government that is charged with the responsibility of protecting the environment and enforcing the Clean Air Act (CAA) of 1990.

Agencia para la Proteccion del Medio Ambiente (EPA) Agencia del gobierno estadounidense que tiene a su cargo la responsabilidad de proteger el medio ambiente y ejecutar la Ley para Aire Limpio (CAA por sus singlas en ingles) de 1990.

EPA Environmental Protection Agency.

EPA Agencia para la Proteccion del Medio Amblente.

Etch An intentional or unintentional erosion of a metal surface generally caused by an acid.

Atacar con acido Desgaste previsto o imprevisto de una superficie metalico, ocacionado generalmente por un acido.

Etching See Etch.

Ataque con acido Ver Etch [Atacar con acido].

Evacuate To create a vacuum within a system to remove all traces of air and moisture.

Evacuar El dejar un vacio dentro de un sistema para remover completamente todo aire y humedad.

Evacuation See Evacuate.

Evacuacion Ver Evacuate [Evacuar].

Evaporator core The tube and fin assembly located inside the evaporator housing. The refrigerant fluid picks up head in the evaporator core when it changes into a vapor.

Nucleo del evaporador El conjunto de tubo y aletas ubicado dentro del alojamiento de evaporador. El refrigerante acumula calor en el nucleo del evaporador cuando se conviente en vapor.

Expansion tank An auxiliary tank that is usually connected to the inlet tank or a radiator and that provides additional storage space for heated coolant. Often called a coolant recovery tank.

Tanque de expansion Tanque auxiliar que normalmente se conecta al tanque de entrada o a un radiador y que provee almacenaje adicional del enfriante calentado. Llamado con frecuencia tanque para la recuperacion del enfriante.

External On the outside.

Externo Al exterior.

External snap ring A snap ring found on the outside of a part such as a shaft.

Anillo de muelle exterior Anillo de muelle que se encuentra en el exterior de una pieza, como por ejemplo un arbol.

Facilities Something created and equipped to serve a particular function, such as a specialty garage used to service motor vehicles.

Fan relay A relay for the cooling and/or auxiliary fan motors.

Rele del ventilador Rele para los motores de enfriamiento y/o los auxiliares.

Federal Clean Air Act See Clean Air Act.

Ley Federal para Aire Limpio Ver Clean Air Act [Ley para Aire Limpio].

Fill neck The part of the radiator on which the pressure cap is attached. Most radiators, however, are filled via the recovery tank.

Cuello de relleno La parte del radiador a la que se fija la tapadera de presion. Sin embargo, la mayoria de radiadores se llena por medio del tanque de recuperacion.

Filter A device used with the drier or as a separate unit to remove foreign material from the refrigerant.

Filtro Dispositivo utilizado con el secador o como unidad separada para extraer material extrano del refrigerante.

Filter drier A device that has a filter to remove foreign material from the refrigerant and a desiccant to remove moisture from the refrigerant.

Secador del filtro Dispositivo provisto de un filtro para remover el material extrano del refrigerante y un desecante para remover la humedad del refrigerante.

Flammable refrigerant Any refrigerant that contains a flammable material and is not approved for use. Any refrigerant, however, may be considered flammable under certain abnormal operating conditions.

Flange A projecting rim, collar, or edge on an object used to keep the object in place or to secure it to another object.

Brida Cerco, collar, o extremo proyectante ubicado sobre un objeto utilizado para mantener un objeto en su lugar o para fijarlo a otro objeto.

Flare A flange or cone-shaped end applied to a piece of tubing to provide a means of fastening to a fitting.

Abocinado Brida o extremo en forma conica aplicado a una pieza de tuberia para proveer un medio de asegurarse a un ajuste.

Fluorescent tracer dye A dye solution introduced into the air conditioning system for leak testing procedures. An ultraviolet (UV) lamp is used to detect the site of the leak.

Forced air Air that is moved mechanically such as by a fan or blower.

Aire forzado Aire que se mueve mecanicamente, como por ejemplo por un ventilador o soplador.

Fringe benefits The extra benefits aside from a salary that an employee may expect, such as vacation, sick leave, insurance, or employee discounts.

Front seat Closing of the line leaving the compressor open to the service port fitting. This allows service to the compressor without purging the entire system. Never operate the system with the valves front seated.

Asentar a la derecha El cerrar la linea dejando abierto el compresor al ajuste del orificio de servicio, lo cual permite prestar servicio al compresor sin purgar todo el sistema. Nunca haga funcionar el sistema con las valvulas asentadas a la derecha.

Functional test See Performance Test.

Prueba funcional Ver Performance Test [Prueba de rendimiento].

Fungi Plural of "fungus," an organism, such as mold, that grows in the damp atmosphere inside an evaporator plenum, often producing an undesirable odor.

Fusible link A type of fuse made of a special wire that melts to open a circuit when current draw is excessive.

Cartucho de fusible Tipo de fusible fabricado de un alambre especial que se funde para abrir un circuito cuando ocurre una sobrecarga del circuito.

Gasket A thin layer of material or composition that is placed between two machined surfaces to provide a leakproof seal between them.

Guarnicion Capa delgada de material o compuesto que se coloca entre dos superficies maquinadas para proveer una junta hermetica para evitar fugas entre ellas.

Gauge A tool of a known calibration used to measure components. For example, a feeler gauge is used to measure the air gap between a clutch rotor and armature.

Calibrador Herramienta de una calibracion conocida utilizada para la medicion de componentes. Por ejemplo, un calibrador de espesores se utiliza para medir el espacio de aire entre el rotor del embrague y la armadura.

Graduated container A measure such as a beaker or measuring cup that has a graduated scale for the measure of a liquid.

Recipiente graduado Una medida, como por ejemplo un cubilete o una taza de medir, provista de una escala graduada para la medicion de un liquido.

Gross weight The weight of a substance or matter that includes the weight of its container.

Peso bruto Peso de una sustancia o materia que incluye el peso de su recipiente.

Ground A general term give to the negative (–) side of an electrical system.

Tierra Termino general para indicar el lado negativo (–) de un sistema elcctrico.

Grounded An intentional or unintentional connection of a wire, positive (+) or negative (–), to the ground. A shortcircuit is said to be grounded.

Puesto a tierra Una conexion prevista o imprevista de un alambre, positiva (+) o negativa (–), a la tierra. Se dice que un cortocircuito es puesto a tierra.

Hazard A possible source of danger that may cause damage to a structure or equipment or that may cause personal injury.

HCFC Hydrochlorofluorocarbon refrigerant.

HCFC Refrigerante de hidroclorofluorocarbono.

HFC-134a A hydrofluorocarbon refrigerant gas used as a refrigerant. The refrigerant of choice to replace CFC-12 is automotive air conditioning systems. Often referred to as R-134a, this refrigerant is not harmful to the ozone.

Header tanks The top and bottom tanks (downflow) or side tanks (crossflow) of a radiator. The tanks in which coolant is accumulated or received.

Tanques para alimentacion por gravedad Los tanques superiores e inferiores (flujo descendente) o los tanques laterales (flujo transversal) de un radiador. Tanques en los cuales el enfriador se acumula o se recibe.

Heater That part of the climate control comfort system consisting of the heater core, hoses, coolant flow control valve, and related controls used to provide air to the vehicle interior.

Heater core A radiator-like heat exchanger located in the case/duct system through which coolant flows to provide heat to the vehicle interior.

Nucleo del calentador Intercamiador de calor parecido a un radiador y ubicado en el sistema de caja/conducto a traves del cual fluye el enfriador para proveer calor al interior del vehiculo.

Heat exchanger An apparatus in which heat is transferred from one medium to another on the principle that heat moves to an object with less heat.

Intercambiador de calor Aparato en el que se transfiere el calor de un medio a otro, lo cual se basa en el principio que el calor se atrae a un objeto que tiene menos calor.

HI The designation for high as in blower speed or system mode.

HI Indicacion para indicar marcha rapida, como jpor ejemplo la velocidad de un soplador o el modo de un sistema.

High-side gauge The right-side gauge on the manifold used to read refrigerant pressure in the high side of the system.

Calibrador del lado de alta presion El calibrador del lado derecho del multiple utilizado para medir la presion del refrigerante en el lado de alta presion del sistema.

High-side hand valve The high-side valve on the manifold set used to control flow between the high side and service ports.

Valvula de mano del lado de alta presion Valvula del lado de alta presion que se encuentra en el conjunto del multiple, utilizada para regular el flujo entre el lado de alta presion y los orificios de servicio.

High-side service valve A device located on the discharge side of the compressor; this valve permits the service technician to check the high-side pressures and perform other necessary operations.

Valvula de servicio del lado de alta presion Dispositivo ubicado en el lado de descarga del compresor; dicha valvula permite que el mecanico verifique las presiones en lado de alta presion y lleve a cabo otras fuciones necesarias.

High-side switch See Pressure switch.

Automata manometrico del lado de alta presion Ver Pressure switch [Automata manometrico].

High-torque clutch A heavy-duty clutch assembly used on some vehicles known to operate with higher-than-average head pressure.

Embrague de alto par de torsion Conjunto de embrague para servicio pesado utilizado en algunos vehiculos que fncionan con una altura piezometrica mas alta que la normal.

Hot A term given the positive (+) side of an energized electrical system. Also refers to an object that is heated.

Cargado/caliente Termino utilizado para referirse al lado positivo (+) de un sistema electrico excitado. Se refiere tambien a un objeto que es calentado.

Hot knife A knife-like tool that has a heated blade. Used for separating objects, e.g., evaporator cases.

Cuchillo en caliente Herramienta parecida a un cuchillo provista de una hoja calentada. Utilizada para separar objetos; p.e. las cajas de evaporadores.

Housekeeping A system of keeping the shop floors clean, lighting proper, tools in proper repair and operating order, and storing materials properly.

Hub The central part of a wheel-like device such as a clutch armature.

Cubo Parte central de un dispositivo parecido a una rueda, como por ejemplo la armadura del embrague.

Hygiene A system of rules and principles intended to promote and preserve health.

Higiene Sistema de normas y principios cuyo proposito es promover y preservar la salud.

Hygroscopic Readily absorbing and retaining moisture.

Higroscopico Lo que absorbe y retiene facilmente la humedad.

Idler A pulley device that keeps the belt whip out of the drive belt of an automotive air conditioner. The idler is used as a means of tightening the belt.

Polea loca Polea que maintiene la vibracion de la correa fuera de la correa de transmision de un acondicionador de aire automortiz. Se utiliza la polea loca para proveerle tension a la correa.

Idler pulley A pulled used to tension or torque the belt(s).

Polea tensora Polea utilizada para proveer tension o par de torsion a la(s) correa(s).

Idle speed The speed (rpm) at which the engine runs while at rest (idle).

Marcha minima Velocidad (rpm) a la que no hay ninguna carga en el motor (marcha minima).

In-car temperature sensor A thermistor used in automatic temperature control units for sensing the in-car temperature. Also see Thermistor.

Sensor de temperatura del interior del vehiculo Termistor utilizado en unidades de regulacion automatica de temperatura para sentir la temperatura del interior del vehiculo. Ver tabien Thermistor [Termistor].

Indexing tab A mark or protrusion on mating components to ensure that they will be assembled in their proper position.

Insert fitting A fitting that is designed to fit inside, such as a barb fitting that fits inside a hose.

Ajuste inserto Ajuste disenado para insertarse dentro de un objecto, como por ejemplo un ajuste arponado que se inserta dentro de una manguera.

Internal Inside; within.

Interno Al interior, dentro de una cosa.

Internal snap ring A snap ring used to hold a component or part inside a cavity or case.

Anillo de muelle interno Anillo de muelle utilzado para sujetar un componente o una pieza dentro de una cavidad o caja.

Jumper A wire used to temporarily bypass a device or component for the purpose of testing.

Barreta Alambre utilizado para desviar un dispositivo o componente de manera temporal para llevar a cabo una prueba.

Kilogram A unit of measure in the metric system. One kilogram is equal to 2.2010-2-615 pounds in the English system.

Kilogramo Unidad de medida en el sistema metrico. Un kilogramo equivale a 2,205 libras en el sistema ingles.

Kilopascal A unit of measure in the metric system. One kilopascal (kPa) is equal to 0.145 pound per square inch (psi) in the English system.

Kilopascal Unidad de medida en el sistema metrico. Un kilopascal (kPa) equivale a 0,145 libras por pulgada cuadrada en el sistema ingles.

kPa Kilopascal.

kPa Kilopascal.

Liquid A state of matter; a column of fluid without solids or gas pockets.

Liquido Estado de materia; columna de fluido sin solidos ni bolsillos de gas.

Low-refrigerant switch A switch that senses low pressure due to a loss of refrigerant and stops compressor action. Some alert the operator and/or set a trouble code.

Interruptor para advertir un nivel bajo de refrigerante Interruptor que siente una presion baja debido a una perdida de refrigerante y que detiene la accion del compresor. Algunos interruptores advierten al operador y/o fijan un codigo indicador de fallas.

Low-side gauge The left-side gauge on the manifold used to read refrigerant pressure in the low side of the system.

Calibrador de lado de baja presion El calibrador en el lado izquierdo del multiple utilizado para medir la presion del refrigerante del lado de baja presion del sistema.

Low-side hand valve The manifold valve used to control flow between the low side and service ports of the manifold.

Valvula de mano del lado de baja presion Valvula de distribucion utilizada para regular el flujo enre el lado de baja presion y los orificios de servicio del colector.

Low-side service valve A device located on the suction side of the compressor that allows the service technician to check low-side pressures and perform other necessary service operations.

Valvula de servicio del lado de baja presion Dispositivo ubicado en el lado de succion del compresor; dicha valvula permite que el mecanico verifique las presiones del lado de baja presion y lleve a cabo otras funciones necesarias de servicio.

Manifold A device equipped with a hand shut-off valve. Gauges are connected to the manifold for use in system testing and servicing.

Multiple Dispositivo provisto de una valvula de cierre accionada a mano. Calibradores se conectan al multiple para ser utilizados para llevar a cabo pruebas del sistema y para servicio.

Manifold and gauge set A manifold complete with gauges and charging hoses.

Conjunto del multiple y calibrador Multiple provisto de calibradores y mangueras de carga.

Manifold hand valve Valves used to open and close passages through the manifold set.

Valvula de distribucion accionada a mano Valvulas utilizadas para abrir y cerrar conductos a traves del conjunto del multiple.

Manufacturer A person or company whose business is to produce a product or components for a product.

Fabricante Persona o empresa cuyo proposito es fabricar un producto o componentes para un producto.

Manufacturer's procedures Specific step-by-step instructions provided by the manufacturer for the assembly, disassembly, installation, replacement, and/or repair of a particular product manufactured by them.

Procedimientos del fabricante Instrucciones especificas a seguir paso por paso; dichas instrucciones son suministradas por el fabricante para montar, desmontar, instalar, reemplazar, y/o reparar un producto especifico fabricado por el.

MAX A mode, maximum, for heating or cooling. Selecting MAX generally overrides all other conditions that may have been programmed.

MAX (Maximo) Modo maximo para calentamiento o enfriamiento. El seleccionar MAX generalmente anula todas las otras condiciones que pueden haber sido programadas.

Metric fastener Any type fastener with metric size designations, numbers, or millimeters.

Asegurador metrico Cualquier asegurador provisto de indicaciones, numeros, o milimetros.

Mid-positioned The position of a stem-type service valve where all fluid passages are interconnected. Also referred to as "cracked."

Ubicacion-central Posicion de una valvula de servicio de tipo vastago donde todos los pasajes que conducen fluidos se interconectan. Llamado tambien parcialmente asentada.

Mildew A form of fungus formed under damp conditions.

Mold A fungus that causes disintegration of organic matter.

Motor An electrical device that produces a continuous turning motion. A motor is used to propel a fan blade or a blower wheel.

Motor Dispositivo electrico que produce un movimiento giratorio continuo. Se utiliza un motor para impoeler las aletas del ventilador o la rueda del soplador.

Mounting See Flange.

Brida de montaje Ver Flange [Brida].

Mounting boss See Flange.

Protuberancia de montaje Ver Flange [Brida].

MSDS Material Safety Data Sheet.

MSDS Hojas de informacion sobre la seguridad de un material.

Mushroomed A condition caused by pounding of a punch or a chisel, producing a mushroom-shaped end that should be gound off to ensure maximum safety.

Hinchado Condicion ocasionada por el golpeo de un punzon o cincel, lo cual hace que el extremo vuelva en forma de un hongo y que debe ser afilado para asegurar maxima seguridad.

Net weight The weight of a product only; container and packaging not included.

Peso neto Peso de solo el producto mismo; no incluye el recipiente y encajonamiento.

Neutral On neither side; the position of gears when force is not being transmitted.

Neutro Que no esta en ningun lado; posicion de los engranajes cuando no se transmite la potencia.

Noncycling cluch An electromechanical compressor clutch that does not cycle on and off as a means of temperature control; it is used to turn the system on when cooling is desired and off when cooling is not desired.

Embrague sin funcionamiento ciclico Embrague electromecanico del compresor que no se enciende y se apaga como medio de regular la temperatra; se utiliza para arrancar el sistema cuando se desea enfriamiento y para detener el sistema cuando no se desea enfriamiento.

Observe To see and note; to perceive; to notice.

Observar Ver y anotar; percibir; fijarse en algo.

OEM Original equipment manufacturer.

OEM Fabricante Original del Equipo.

Off-road A term often used for an "off-the-road" vehicle.

Off-the-road Generally refers to vehicles that are not licensed for road use, such as harvesters, bulldozers, and so on.

Fuera de carretera Generalmente se refiere a vehiculos que no son permitidos operar en la carretera, como por ejemplo cosechadoras, rasadoras, ecetera.

Ohmmeter An electrical instrument used to measure the resistance in ohms of a circuit or component.

Ohmiometro Instrumeno electrico utilizado para medir la resistencia en ohmios de un circuito o componente.

Open Not closed. An open switch, for example, breaks an electrical circuit.

Abierto No cerrado. Un interruptor abierto corta un circuito electrico, por ejemplo.

Orifice A small hole. A calibrated opening in a tube or pipe to regulate the flow of a fluid or liquid.

Orificio Agujero pequeno. Apertura calibrada en un tubo o caneria para regular el flujo de un fluido o de un liquido.

O-ring A synthetic rubber or plastic gasket with a round- or square-shaped cross-section.

Junta torica Guarnicion sintetica de caucho o de plastico provista de una seccion transversal en forma redonda o cuadrada.

OSHA Occupational Safety and Health Administration.

OSHA Direccion para la Seguridad y Salud Industrial.

Outside temperature sensor See Ambient sensor.

Sensor de la temperatura ambiente Ver Ambient sensor [Sensor ambiente].

Overcharge Indicates that too much refrigerant or refrigeration oil is added to the system.

Sobreccarga Indica que una cantidad excesiva de refrigerante o aceite de refrigeracion ha sido agregada al sistema.

Overload Anything in excess of the design criteria. An overload will generally cause the protective device such as a fuse or pressure relief to open.

Sobrecarga Cualquier cosa en exceso del criterio de diseno. Generalmente una sobrecarga causara que se abra el dispositivo de proteccion, como por ejemplo un fusible o alivio de presion.

Ozone friendly Any product that does not pose a hazard or danger to the ozone.

Sustancia no danina al ozono Cualquier producto que no es peligrosa o amenaza al ozono.

Park Generally refers to a component or mechanism that is at rest.

Reposo Generalmente se refiere a un componente o mecanismo que no esta functionado.

PCM Power control module.

PCM Modulo regulador del transmisor de potencia.

Performance test Readings of the temperature and pressure under controlled conditions to determine if an air conditioning system is operating at full efficiency.

Prueba de rendimiento Lecturas de la temperatura y presion bajo condiciones controladas para determinar si un sisteria de acondicionamiento de aire funciona a un rendimiento completo.

Piercing pin The part of a saddle valve that is used to pierce a hole in the tubing.

Pasador perforador Parte de la valvula de silleta utilizada para perforar un agujero en la tuberia.

Pin-type A single or multiple electrical connector that is round- or pin-shaped and fits inside a matcing connector.

Conectador de tipo pasador Conectador electrico unico o multiple en forma redonda o en forma de passador que se inserta dentro de un conectador emparajado.

Poly belt See Serpentine belt.

Correa poli Ver Serpentine belt [Correa serpentina].

Polyol ester (ESTER) A synthetic oil-like lubricant that is occasionally recommended for use in an HFC-134a system. This lubricant is compatible with both HFC-134a and CFC-12.

Poliolester Lubrificante sintetico parecido a aceite que se recomienda de vez en cuando para usar en un sistema HFC 134a. Dicho lubrificante es compatible tanto con HFC 134a como CFC 12.

Positive pressure Any pressure above atmospheric.

Presion positiva Cualquier presion sobre la de la atmosferica.

Pound A weight measure, 16 ounces. A term often used when referring to a small can of refrigerant, although the can does not necessarily contain 15 ounces.

Libra Medida de peso, 16 onzas. Termino utilizada con frecuencia al referirse a una lata pequena de refrigerante, aunque es posible que la lata contenga menos de 16 onzas.

"Pound" of refrigerant A term used by some technicians when referring to a small can of refrigerant that actually contains less than 16 ounces.

Libra de refrigerante Termino utilizado por algunos mecanicos al referirse a una laa pequena the refrigerante que en realidad contiene menos de 16 onzas.

Power module Contros the operation of the blower motor in an automatic tempeature control system.

Transmisor de potencia Regula el funcionamiento del motor del soplador en un sistema de control automatico de temperatura.

Predetermined A set of fixed values or parameters that have been programmed or otherwise fixed into an operating system.

Predeterminado Valores fojos o parametros que han sido programados o de otra manera fijados en un sistema de functrionamiento.

Pressure The application of a continuous force by one body onto another body. Force per unit area or force divided by area usually expressed in pounds per square inch (psi) or kilopascal (kPa).

Pressure gauge A calibrated instrument for measuring pressure.

Manometro Instrumento calibrado para medir la presion.

Pressure switch An electrical switch that is activated by a predetermined low or high pressure. A high-pressure switch is generally used for system protection; a low-pressure switch may be used for temperature control or system protection.

Automata manometrico Interruptor electrico accionado por una baja o alta presion predeterminada. Generalmente se utiliza un automata manometrico de alta presion para la proteccion del sistema; puede utilizarse uno de baja presion para la regulacion de temperatura o proteccion del sistema.

Propane A flammable gas used as a propellant for the halide leak detector.

Propano Gas inflamable utilizado como propulsor para el detector de fugas de halogenuro.

Psig Pounds per square inch gauge.

Psig Calibrador de libras por pulgada cuadrada.

Purge To remove moisture and/or air from a system or a component by flushing with a dry gas such as nitrogen (N) to removal all refrigerant from the system.

Purgar Remover humedad y/o aire de un sistema o un componente al descargarlo con un gas seco, como por ejemplo el nitrogeno (N), para remover todo el refrigerante del sistema.

Purity test A static test that may be performed to compare the suspect refrigerant pressure with an appropriate temperature chart to determine its purity.

Prueba de pureza Prueba estatica que puede llevarse a cabo para comparar la presion del refrigerante con un grafico de temperatura apropriado para determinar la pureza del mismo.

Radiation The transfer of heat without heating the medium through which it is transmitted.

Radiacion La transferncia de calor sin claentar el medio por el cual se transmite.

Ram air Air that is forced through the radiator and condenser coils by the movement of the vehicle or the action of the fan.

Air Admitido en sentido de la marcha. Air forzado a traves de las bobinas del radiador y del condensador por medio del movimiento del vehiculo o la accion del ventilador.

Rebuilt To build after having been disassembled, inspected, and worn and after damaged parts and components are replaced.

Reconstruido Fabricar despues de haber sido desmontado y revisado, y luego remplazar las piezas desgatadas y averiadas.

Receiver/drier A tank-like vessel having a desiccant and used for the storage of refrigerant.

Receptor/secador Recipiente parecido a un tanque provisto de un desecante y utilizado para el almacenaje de refrigerante.

RECIR An abbreviation for the recirculate mode, as with air.

RECIR Abreviatura del modo recirculatorio, como por ejemplo con aire.

Recover system A term often used to refer to the circuit inside the recovery unit used to recycle and/or transfer refrigeant from the air conditioning system to the recovery cylinder.

Sistema de recuperacion Termino utilizado con frecuencia para referirse al circuito dentro de la unidad de recuperacion interior utilizado para reciclar y/o transferir el refrigerante del sistema de acondicionamiento de aire al cilindro de recuperacion.

Recovery tank An auxiliary tank, usually connected to the inlet tank of a radiator, which provides additional storage space for heated coolant.

Tanque de recuperacion Tanque auxiliar que normalmente se conecta al tanque de entrada de un radiador, lo cual provee almacenaje adicional para el enfriante calentado.

Refrigerant A chemical compound, such as R-134a, used in an air conditioning system to achieve the desired refrigerating effect.

Refrigerant-12 The refrigerant used in automotive air conditioners, as well as other air conditioning and refrigeration systems. The chemical name of refrigerant-12 is dichlorodiluoromethane. The chemical symbol is $CC_{12}F_2$.

Refrigerante 12 Refrigerante utilizado tanto en acondicionadores de airc automotrices como en otros sistemas de acondicionamiento de aire y refrigeracion. Ele nombre quimico del refrigerante 12 es diclorodifloromentano, y el simbolo quimico es CC12F2.

Relay An electrical switch device that is activated by a low-current source and controls a high-current device.

Rele Interruptor electrico que es accionado por una fuente de corriente baja y regula un dispositivo de corriente alta.

Reserve tank A storage vessel for excess fluid. See Recovery tank; Receiver/dryer, and Accumulator.

Tanque de reserva Recipiente de almacenaje para un exceso de fluido. Ver Recovery tank [Tanque de recuperacion], Receiver/drier [Receptor/secador], y Accumulador [Acumulador].

Resistor A voltage-dropping device that is usually wire wound and that provides a means of controlling fan speeds.

Resistor Dispositivo de caida de tension que normalmente es devando con alambre y provee un medio de regular la velocidad del ventilador.

Respirator A mask or face shield worn in a hazardous environment to provide clean fresh air and/or oxygen.

Mascarilla Miscara o protector de cara que se lleva puesto en un amibiente peligroso para proveer air limpio y puro y/o oxigeno.

Responsibility Being reliable and trustworthy.

Responsibilidad Ser confiable y fidedigno.

Restricted Having limitations. Keeping within limits, confines, or boundaries.

Restringido Que tiene limitaciones. Mantenerse dentro de limites, confines, o fronteras.

Restrictor An insert fitting or device used to control the flow of refrigerant or refrigeration oil.

Limitador Pieza inserta o dispositivo utilizado para regular el flujo de refrigerante o aceite de refrigeracion.

Rotor The rotating or freewheeling portion of a clutch; the belt slides on the rotor.

Rotor Parte giratoria o con marcha a rueda libre de un embrague; la correa se desliza sobre el rotor.

RPM Revolutions per minute; also rpm or r/min.

RPM Revoluciones por minuto; tambien rpm o r/min.

Running design change A design change made during a current model/year production.

Cambio al diseno corriente Un cambio al diseno hecho durante la fabricacion del modelo/ano actual.

RV Recreational vehicle.

RV Vehiculo para el recreo.

Saddle valve A two-part accessory valve that may be clamped around the metal part of a system hose to provide access to the air conditioning system for service.

Valvula de silleta Valvula accesoria de dos partes que puede fijarse con una abrazadera a la parte metalica de una manguera del sistema para proveer acceso al sistema de acondicionamiento de aire para llevar a cabo servicio.

SAE Society of Automotive Engineers.

SAE Sociedad de Ingenieros Automotrices.

Safety Freedom from danger or injury; the state of being safe.

Seguridad Libre de peligro o dano, calidad o estado de seguro.

Scan tool A portable computer that may be connected to the vehicle's diagnostic connector to read data from the vehicle's onboard computer.

Schrader valve A spring-loaded valve similar to a tire valve. The Schrader valve is located inside the service valve fitting and is used on some control devices to hold refrigerant in the system. Special adapters must be used with the gauge hose to allow access to the system.

Valvula Schrader Valvula con cierre automatico parecida al vastago del neumatico. La valvula Schrader esta ubicada dntro del adjuste de la valvula de servicio y se utiliza en algunos dispositivos de regulacion para guardar refrigerante dentro del sistema. Deben utilizarse adaptadores especiales con una manguera calibrador para permitir acceso al sistema.

Seal Generally refers to a compressor shaft oil seal; matching shaft-mounted seal face and front head-mounted seal seat to prevent refrigerant and/or oil from escaping. May also refer to any gasket or O-ring used between two mating surfaces for the same purpose.

Junta hermetica Generalmente se refiere a la junta hermetica del arbol del compresor; la frente de junta hermetica montada en el arbol y el asiento de junta hermetica montado en el cabezal delantero empareja-dos para evitar la fuga de refrigerante y/o de aceite. Puede referirse tambien a cualquier guarnicion o junta torica utilizada ente dos superficies emparejadas para el mismo proposito.

Seal seat The part of a compressor shaft seal assembly that is stationary and matches the rotating part, known as the seal face or shaft seal.

Asiento de la junta hermetica Parte del conjunto de la junta hermetica del arbol del compresor que es inmovil y que se empareja a la parte rotativa; conocido como la frente de junta hermetica o la junta hermetica del arbol.

Serpentine belt A flat or V-groove belt that winds through all of the engine accessories to drive them off the crankshaft pulley.

Correa serpentina Correa plana o con ranuras en V que attraviesa todos los accesorios del motor para forzarlos fuera de la polea del ciguenal.

Service port A fitting found on the service valves and some control devices; the manifold set hoses are connected to this fitting.

Orificio de servicio Ajuste ubicado en las valvulas de servicio y en algunos dispositivos de regulacion; las mangueras del conjunto del colector se conectan a este ajuste.

Service procedure A suggested routine for the step-by-step act of troubleshooting, diagnosing, and/or repairs.

Procedimiento de servico Rutina sugerida para la accion a seguir paso a paso para detectar fallas, diagnosticar, y/o reparar.

Service valve See High-side (Low-side) service valve.

Valvula de servicio Ver High-side (Low-side) service valve [Valvula de servicio del lado de alta presion (baja presion)].

Servo motor An electrical motor that is used to control a mechanical device, such as a heater coolant flow control valve.

Shaft A long cylindrical-shaped rod that rotates to transmit power, such as a compressor crankshaft.

Shaft key A soft metal key that secures a member on a shaft to prevent it from slipping.

Chaveta del arbol Chaveta de metal blando que fija una pieza a un arbol para evitar su deslizamiento.

Shaft seal See Compressor shaft seal.

Junta hermetica del arbol Ver Compressor shaft seal [Junta hermetica del arbol del compresor].

Short Of brief duration, e.g., short cycling. Also refers to an intentional or unintentional grounding of an electrical circuit.

Breve/corto De una duracion breve; p.e., funcionamiento ciclico breve. Se refiere tambien a un puesto a tierra previsto o imprevisto de un circuito electrico.

Shut-off valve A valve that provides positive shut-off of a fluid or vapor passage.

Valvula de cierre Valvula que provee el cierre positivo del pasaje de un fluido o un vapor.

Sliding resistor A resistor having the provision of varying its resistance depending on the position of a sliding member. Also may be referred to as a "rheostat" or "pot."

SNAP An acronym for "Significant New Alternatives Policy."

Snap ring A metal ring used to secure and retain a component to another component.

Anillo de muelle Anillo metalico utilizado para fijar y sujetar un componente a otro.

Snapshot A feature of OBD II that shows, on various scanners, the conditions that the vehicle was operating under when a particular trouble code was set. For example, the vehicle was at 125°F, ambient temperature was 55°F, throttle position was part throttle at 1.45 volts, rpm was 1,450, brake was off, transmission was in third gear with torque converter unlocked, air conditioning system was off, and so on.

Society of Automotive Engineers A professional organization of the automotive industry. Founded in 1905 as the Society of Automobile Engineers, the SAE is dedicated to providing technical information and standards to the automotive industry. Present goals are to ensure a skilled engineering and technical work force for the year 2000 and beyond. The goal, known as VISION 2000, encompasses all of SAE's educational programs, including student competitions/scholarships, teacher recognition, and more.

Sociedad de Ingenieros automotrices Organizacion profesional de la industria automotriz. Establecido en 1905 coma la Sociedad de Ingenieros de Automoviles (SAE por sus siglas en ingles), dicha sociedad se dedica a proveerle informacion technia y normas a la industria automotriz. Sus metas actuales son asegurar una fuerza laboral capicitada en la ingenieria y en el campo tecnico para el ano 2000 y despues. La meta, conocida como VISION 2000, abarca todas los programas educativos de la SAE, e incluye concursos entre estudiantes, el otorgar becas, el reconocer al profesor, y mas.

Socket The concave part of a joint that receives a concave member. A term generally used for "socket wrench," referring to a female 6 , 8 , or 12-point wrench so designed to fit over a nut or bolt head.

Solenoid See Solenoid valve.

Solenolde Ver Solenoid valve [Valvula de solenoide].

Solenoid valve An electromagnetic valve controlled remotely by electrically energizing and deenergizing a coil.

Valvula de solenoide Valvula electromagnetica regulada a distancia por una bobina electronicamente.

Solid state Referring to electronics consisting of semiconductor devices and other related nonmechanical components.

Estado solido Se refiere a componentes electronicos que consisten en dispositivos semiconductores otros componentes relacionados no mecanicos.

Spade-type connector A single or multiple electrical connector that has flat spade-like mating provisions.

Conectador de tipo azadon Conectador unico o multiple provisto de dispositivos planos de tipo azadon para emparejarse.

Specifications Design characteristics of a component or assembly noted by the manufacturer. Specifications for a vehicle include fluid capacities, weights, and other pertinent maintenance information.

Especificaciones Caracteristicas de diseno de un componente o conjunto indicadas por el fabricante. Las especificaciones para un vehiculo incluyen capacidades del fluido, peso y otra informacion pertinente para mantenimiento del vehiculo.

Spike In our application, an electrical spike. An unwanted momentary high-energy electrical surge.

Impulso afilado En nuestro campo, un impulso afilado electrico. Una elevacion repentina electrica de alta energia no deseada.

Spring lock Part of a spring lock fitting. A special fitting used to form a leak-proof joint.

Spring lock fitting A special fitting using a spring to lock the mating parts together forming a leak-proof joint.

Ajuste de clerre automatico Ajuste especial utilizando un resorte para cerrar piezas emparejadas para formar asi una junta hermetica contra fugas.

Squirrel-cage blower A blower wheel designed to provide a large volume of air with a minimum of noise. The blower is more compact than the fan and air can be directed more efficiently.

Soplador con jaula de ardilla Rueda de soplador disenada para proveer un gran caudal de aire con un minimo de ruido. El soplador es mas compacto que el ventilador y el aire puede dirigirse con un mayor rendimiento.

Stabilize To make steady.
Estabilizar Quedarse detenida una cosa.

Standing vacuum test A leak test performed on an air conditioning system by pulling a vacuum and then determining, by observation, if the vacuum holds for a predetermined period of time to ensure that there are no leaks.

Stratify Arrange or form into layers. To fully blend.
Estratificar Arreglar o formar en capas. Mezclar completamente.

Subsystem A system within a system.
Subsistema Sistema dentro de un sistema.

Sun load Heat intensity and/or light intensity produced by the sun.
Carga del sol Intensidad calorifica y/o de la luz generada por el sol.

Sun-load sensor A device that senses heat or light intensity that is place on the dashboard to determine the amount of sun entering the vehicle.

Superheat switch An electrical switch activated by an abnormal temperature-pressure condition (a superheated vapor); used for system protection.
Interruptor de vapor sobrecalentado Interruptor electrico accionado por una condicion anormal de presion y temperatura (vapor sobrecalentado); utilizado para la proteccion del sistema.

System All of the components and lines that make up an air conditioning system.
Sistema Todos los componentes y lineas que componen un sistema de acondicionamiento de aire.

Tank See Header tanks and Expansion tank.
Tanque Ver Header tanks [Tanques para alimentacion por gravedad] y Expansion tank [Tanque de expansion].

Tare weight The weight of the packaging material. See Net weight and Gross weight.
Taraje Peso del material de encajonamiento. Ver Net weight [Peso neto] y Gross weight [Peso bruto].

Technical service bulletin (TSB) Periodic information provided by the vehicle manufacturer regarding any problems and offering solutions to problems encountered in their vehicles.

Technician One concerned and involved in the design, service, or repair in a specific area, such as an automotive service technician or, more specifically, automotive air conditioning service technician.

Temperature door A door within the case/duct stem to direct air through the heater and/or evaporator core.
Puerta de temperatura Puenta ubicada dentro del vastago de caja/conducto para conducir el aire a traves del nucleo del calentador y/o del evaporador.

Temperature switch A switch actuated by a change in temperature at a predetermined point.
Interruptor de temperatura Interruptor accionado por un cambio de temperatura a un punto predeterminado.

Tensioner A device used to impart tension, such as an automatic belt tensioner.

Tension gauge A tool for measuring the tension of a belt.
Manometro para tension Herramienta para medir la tension de una correa.

Thermistor A temperature-sensing resistor that has the ability to change values with changing temperature.
Termistor Resistor sensible a temperatura que tiene la capacidad de cambiar valores al ocurrir un cambio de temperatura.

Torque A turning force, e.g., the force required to seal a connection; measured in (English) foot-pounds (ft.-lb.) or inch-pounds (in.-lb.); (metric) Newton-meters (n·m).

Par de torsion Fuerza de torcimiento; por ejemplo, la fuerza requerida para sellar una conexion; medido en libras-pies (ft.-lb.) (inglesas) o en libras pulgadas (in.-lb.); metros-Newton (N·m (metricos).

Triple evacuation A process of evacuation that involves three pump downs and two system purges with an inert gas such as dry nitrogen (N).
Evacuacion triple Proceso de evacuacion que involucra tres envios con bomba y dos purgas del sistema con un gas inerte, como por ejemplo el nitrogeno seco (N).

Troubleshoot The act or act of diagnosing the cause of various system malfunctions.
Deteccion de fallas Procedimiento o arte de diagnosticar la causa de varias fallas del sistema.

TXV Thermostatic expansion valve.
TXV Valvula de expansion termosatica.

Ultraviolet (UV) The part of the electromagnetic spectrum emitted by the sun that lies between visible violet light and X-rays.
Ultravioleta Parte del espectro electromagnetico generado por el sol que se encuentra entre la luz violeta visible y los rayos X.

Ultraviolet dye A fluid that may be injected into the air conditioning system for leak testing purposes. An ultraviolet (UV) lamp is used to locate the leak.

Vacuum gauge A gauge used to measure below atmospheric pressure.
Vacuometro Calibrador utilizado para medir a una presion inferior a la de la atmosfera.

Vacuum motor A device designed to provide mechanical control by the use of a vacuum.
Motor de vacio Dispositivo disenado para proveer regulacion mecanica mediante un vacio.

Vacuum pump A mechanical device used to evacuate the refrigeration system to rid it of excess moisture and air.
Bomba de vacio Dispositivo mecanico utilizado para evacuar el sistema de refrigeracion para purgarlo de un exceso de humedad y aire.

Vacuum signal The presence of a vacuum.
Senal de vacio Presencia de un vacio.

V-belt A rubber-like continuous loop placed between the engine crankshaft pulley and accessories to transfer rotary motion of the crankshaft to the accessories.
Correa en V Bucle continuo parecido a caucho ubicado cntre la polea del ciguenal del motor y los accesorios para transferir el moviemiento giratorio del aquel a estos.

Ventilation The act of supplying fresh air to an enclosed space such as the inside of an automobile.
Ventilacion Proceso de suministrar el aire fresco a un espacio cerrado, como por ejemplo al interior de un automovil.

V-groove belt See V-belt.
Correa con ranuras en V Ver V-belt [Correa en V].

VIN An acronym for "vehicle identification number."

Voltmeter A device used to measure volt(s).
Voltimetro Dispositivo utilizado para la medicion de voltios.

Wiring harness A group of wires wrapped in a shroud for the distribution of power from one point to another point.
Cableado preformado Grupo de alambres envuelto por una gualdera para distribuir potencia de un punto a otro.

INDEX

Note: Page numbers in bold print reference non-text material.

C